AF445477

The Role of Metal Ions in Biology, Biochemistry and Medicine

The Role of Metal Ions in Biology, Biochemistry and Medicine

Editor

Michael Moustakas

MDPI • Basel • Beijing • Wuhan • Barcelona • Belgrade • Manchester • Tokyo • Cluj • Tianjin

Editor
Michael Moustakas
Aristotle University of
Thessaloniki
Greece

Editorial Office
MDPI
St. Alban-Anlage 66
4052 Basel, Switzerland

This is a reprint of articles from the Special Issue published online in the open access journal *Materials* (ISSN 1996-1944) (available at: https://www.mdpi.com/journal/materials/special_issues/role_metal_ions_biology_biochemistry_medicine).

For citation purposes, cite each article independently as indicated on the article page online and as indicated below:

LastName, A.A.; LastName, B.B.; LastName, C.C. Article Title. *Journal Name* **Year**, *Volume Number*, Page Range.

ISBN 978-3-0365-1054-5 (Hbk)
ISBN 978-3-0365-1055-2 (PDF)

Cover image courtesy of Anetta Hanć.

Contents

About the Editor

Michael Moustakas holds a Ph.D. in Botany (1989) and from 2013 a full professor position at the Department of Botany in the School of Biology at Aristotle University of Thessaloniki (AUTH). He has taught a wide variety of courses related to plant morphology, plant physiology and plant ecophysiology for graduate and master's students at the Aristotle University of Thessaloniki. He served at the administrative council of the Hellenic Botanical Society (1993–1999) and as a national representative at the Federation of European Societies for Plant Biologists (FESPB) (2000–2011). He was director of the laboratory of botany at the School of Biology at AUTH from 2013–2019 and has been the project leader on a wide range of research projects on the responses of plants to abiotic stress factors. He has authored or co-authored more than 65 scientific articles indexed in the database Web of Science with >2200 citations in Scopus, and has >60 communications to scientific meetings. He specializes in photosynthesis and his current research is focused on the photoprotective and antioxidative mechanisms of plant responses to biotic and abiotic stress.

Editorial

The Role of Metal Ions in Biology, Biochemistry and Medicine

Michael Moustakas

Department of Botany, Aristotle University of Thessaloniki, GR-54124 Thessaloniki, Greece; moustak@bio.auth.gr

Citation: Moustakas, M. The Role of Metal Ions in Biology, Biochemistry and Medicine. *Materials* **2021**, *14*, 549. https://doi.org/10.3390/ma14030549

Received: 14 January 2021
Accepted: 20 January 2021
Published: 24 January 2021

Publisher's Note: MDPI stays neutral with regard to jurisdictional claims in published maps and institutional affiliations.

Metal ions are fundamental elements for the maintenance of the lifespan of plants, animals and humans. Their substantial role in biological systems was recognized a long time ago. They are essential for the maintenance of life and their absence can cause growth disorders, severe malfunction, carcinogenesis or death. They are protagonists as macro or microelements in several structural and functional roles, participating in many biochemical reactions, and arise in several forms. They participate in intra and inter cellular communications, in maintaining electrical charges and osmotic pressure, in photosynthesis and electron transfer processes, in the maintenance of pairing, stacking and the stability of nucleotide bases, and also in the regulation of DNA transcription. They contribute to the proper functioning of nerve cells, muscle cells, the brain and the heart, the transport of oxygen and in many other biological processes up to the point that we cannot even imagine a life without metals.

In the present work, the 10 papers published in this Special Issue are summarized, providing a picture of metal ions uses in Biology, Biochemistry and Medicine, but also pointing out the toxicity impacts on plants, animals, humans, and the environment.

Metals (Cobalt (Co), Nickel (Ni) and Chromium (Cr)) are used in prosthesis manufacturing because of their high mechanic stability and good biological compatibility, but their corrosion products have been shown to elicit biological responses [1]. Monocytes and macrophages are the first barrier of the innate immune system, which interact with abrasion and corrosion products, leading to the release of proinflammatory mediators and reactive oxygen species (ROS) [1]. The inflammation-relevant changes in monocytes and macrophages after exposure to several concentrations of metal salts ($CoCl_2$, $NiCl_2$, $CrCl_3 \times 6H_2O$) were studied by analyzing viability, gene expression, protein release and ROS production [1]. It was proved that monocytes and macrophages react very sensitively to corrosion products and that high concentrations of bivalent ions lead to cell death, while lower concentrations trigger the release of inflammatory mediators, mainly in macrophages [1].

Metal particles, as well as their corrosion products, have been shown to elicit a biological response in the aseptic loosening of endoprosthetic implants [2]. Owing to different metal alloy components, the response may vary depending on the nature of the released corrosion product [2]. Human osteoblasts were incubated with different metal salts ($CoCl_2$, $NiCl_2$ and $CrCl_3 \times 6H_2O$) in order to compare the biological effects of different ions released from metal alloys [2]. The results demonstrated the pro-osteolytic capacity of metal ions in osteoblasts, while the metal ions examined intervene much earlier in inflammatory processes compared to $CoCr_{28}Mo_6$ particles [2].

Titanium (Ti), being one of the most abundant elements in the earth's crust with many examples of bioactive properties, is widely used in the cosmetic industry. Titanium dioxide (TiO_2) nanoparticles (NPs) are often incorporated in sunscreens as physical sun blockers absorbing ultraviolet (UV) radiation [3]. By examining the reactivity of TiO_2 NPs in the presence and absence of UV, Sharma et al. [3] reassessed in a review article what their seepage into bodies of water can cause to the environment and aquatic life, and examined TiO_2 NPs' effects on human skin and health in general, and especially on the human body and the bloodstream.

Phenolic compounds play a significant role in plant tolerance to toxic metals, together with the prevention and reduction of biotic and abiotic oxidative stress [4]. Chlorogenic acid

(5-caffeoylquinic acid, 5-CQA) is a phenolic compound considered to play an essential role in defense against biotic and abiotic stresses and is accumulated in plant tissues exposed to different stress factors that result in increased ROS production [4]. Zinc (Zn) cations and Zn complexes have anti-/pro-oxidant and antimicrobial activities [4]. The antimicrobial activity of the Zn(II) complex of 5-CQA and 5-CQA against *Escherichia coli*, *Pseudomonas aeruginosa*, *Bacillus subtilis*, *Staphylococcus aureus*, *Salmonella enteritidis* and *Candida albicans* was tested by Kalinowska et al. [4]. The pro-oxidant activity of Zn(II) 5-CQA was higher than the ligand and increased with the rise in the compound concentration [4]. The type of Zn(II) coordination by the chlorogenate ligand strongly affected the antioxidant activity of the complex [4].

The impact of Zn oxide nanoparticles (ZnO NPs) on photosystem II (PSII) photochemistry and ROS production was evaluated in the seagrass *Cymodocea nodosa* exposed to 5 and 10 mg L^{-1} ZnO NPs [5]. A disturbance of PSII functionality was observed 4 h after exposure to 10 mg L^{-1} ZnO NPs, but later (24 h) a stimulatory effect on PSII photochemistry was noticed and described as a hormetic response [5]. A hormetic response suggests that a basal stress level is needed for adaptive responses, which in the case of the seagrass *Cymodocea nodosa* exposed to ZnO NPs was found to be 10 mg L^{-1} ZnO NPs, with 24 h exposure time to be required for the induction of this adaptive response mechanism [5].

Both Zn and copper (Cu) are essential elements for plant growth, and an adequate supply of both is suggested to improve crop productivity. Sperdouli et al. [6] exposed young and mature *Arabidopsis thaliana* leaves to CuZn nanoparticles (NPs) in order to evaluate their effect on PSII function. PSII function in young *A. thaliana* leaves was detected to be negatively influenced by the foliar spray of CuZn NPs, while a beneficial effect on PSII function in mature leaves was observed. The explanation of the differential response of young and mature *Arabidopsis* leaves was suggested to be due to the nutrient remobilization that occurs in mature-senescing leaves, which results in nutrient deficiencies [6]. The use of CuZn NPs as foliar spray fertilizers to improve nutrient crop deficiencies is proposed, and it is concluded that leaf age must be taken into consideration in order to compare leaves of the same developmental stage [6].

Aluminium (Al), the most abundant metal in the earth's crust, is not considered to be essential for plant life, but it can be toxic to plant growth in acid soils (pH < 5.5) at the ionic form of Al^{3+} species, limiting crop production through negatively affecting root growth, nutrient uptake and other metabolic processes, especially photosynthesis [7]. Moustaka et al. [7] evaluated the responses of photosystem II (PSII) and photosystem I (PSI) to Al^{3+} phytotoxicity in the durum wheat (*Triticum turgidum* L. cv. 'Appulo E') and the triticale (X Triticosecale Witmark cv. 'Dada') and concluded that PSII was more affected than PSI by Al^{3+} phytotoxicity. Thus, PSII is documented for another time to be the most sensitive component of the photosynthetic apparatus, playing key roles in photosynthetic responses to environmental perturbations.

Cadmium (Cd), a non-essential heavy metal to plants, can occur in soil at high concentrations, thus becoming toxic to all organisms. However, some plant species that can take up heavy metals and translocate them into the above-ground parts, in a manner that is not detrimental, are called hyperaccumulators, and they can be used for phytoremediation. *Noccaea caerulescens* can accumulate Zn–Cd–Ni at enormously high concentrations in its aboveground tissues, and has been suggested as a model species for examining metal tolerance and hyperaccumulation [8]. Bayçu et al. [8] provided new data on the mechanism of *N. caerulescens* acclimation to Cd exposure by elucidating the process of PSII acclimation. The authors concluded that acclimation to 120 μM Cd exposure, possibly achieved through a reduced plastoquinone (PQ) pool signaling, mediated stomatal closure, probably by the generation of mesophyll chloroplastic hydrogen peroxide (H$_2$O$_2$). Stomatal closure decreased Cd supply at the spatially affected leaf area, while it was suggested that H$_2$O$_2$ signaling activated the Cd detoxification mechanism through its vacuolar sequestration. As a result, a lower ROS production as singlet oxygen (^{1}O$_2$) was detected [8]. It was concluded that the response of *N. caerulescens* to Cd exposure fits the 'Threshold for Tolerance

Model', with a lag time of 4 days and a threshold concentration of 40 µM Cd to be required for the induction of the acclimation mechanism [8].

Chlorophyll fluorescence analysis has been commonly used as an extremely sensitive marker of photosynthetic efficiency [9]. Nevertheless, photosynthesis is not uniform at the leaf area, denoting conventional chlorophyll fluorescence measurements as non-characteristic of the physiological status of the entire leaf. This disadvantage overcomes chlorophyll fluorescence imaging analysis, which permits the detection of spatiotemporal heterogeneity at the total leaf surface [9]. By combining the chlorophyll fluorescence imaging analysis (CF-IA) and laser ablation inductively coupled plasma mass spectrometry (LA-ICP-MS) methods, the impact of Cd accumulation on the photosynthetic efficiency of *Salvia sclarea* was examined [9]. The spatial heterogeneity of a decreased effective quantum yield of electron transport (Φ_{PSII}), which was observed after exposure to Cd, was linked to the spatial pattern of high Cd accumulation [9]. However, the increased photoprotective heat dissipation (NPQ) in the whole leaf under Cd exposure was sufficient enough to retain the same fraction of open reaction centers (q_P) as control leaves, also resulting in a decreased quantum yield of non-regulated energy loss (Φ_{NO}), even more than that of control leaves, demonstrating the tolerance of *S. sclarea* to Cd exposure [9]. The results revealed the advantages of combining CF-IA and LA-ICP-MS to monitor heavy metal effects on plants and to elucidate plant tolerance mechanisms [9].

Mercury (Hg) is a toxic metal frequently used in the illegal extraction of gold and silver, consequently resulting in environmental poisoning [10]. *Lysinibacillus sphaericus* strains characterized by scanning electron microscopy (SEM) coupled with energy dispersive spectroscopy (EDS-SEM) were assessed for Hg removal ability [10]. Sorption was evaluated in live and dead bacterial biomass by free and immobilized cells assays. EDS-SEM analysis showed that the bacteria strains used could adsorb Hg as particles of nanometric scale, removing over 95% of Hg, suggesting that *L. sphaericus* could be used as a novel biological method of mercury removal from polluted wastewater [10].

Funding: This research received no external funding.

Acknowledgments: The administrative help of Beth Bai for this Special Issue is gratefully acknowledged.

Conflicts of Interest: The author declares no conflict of interest.

References

1. Loeffler, H.; Jonitz-Heincke, A.; Peters, K.; Mueller-Hilke, B.; Fiedler, T.; Bader, R.; Klinder, A. Comparison of Inflammatory Effects in THP-1 Monocytes and Macrophages after Exposure to Metal Ions. *Materials* **2020**, *13*, 1150. [CrossRef] [PubMed]
2. Jonitz-Heincke, A.; Sellin, M.-L.; Seyfarth, A.; Peters, K.; Mueller-Hilke, B.; Fiedler, T.; Bader, R.; Klinder, A. Analysis of Cellular Activity and Induction of Inflammation in Response to Short-Term Exposure to Cobalt and Chromium Ions in Mature Human Osteoblasts. *Materials* **2019**, *12*, 2771. [CrossRef] [PubMed]
3. Sharma, S.; Sharma, R.K.; Gaur, K.; Cátala Torres, J.F.; Loza-Rosas, S.A.; Torres, A.; Saxena, M.; Julin, M.; Tinoco, A.D. Fueling a Hot Debate on the Application of TiO$_2$ Nanoparticles in Sunscreen. *Materials* **2019**, *12*, 2317. [CrossRef] [PubMed]
4. Kalinowska, M.; Sienkiewicz-Gromiuk, J.; Świderski, G.; Pietryczuk, A.; Cudowski, A.; Lewandowski, W. Zn(II) Complex of Plant Phenolic Chlorogenic Acid: Antioxidant, Antimicrobial and Structural Studies. *Materials* **2020**, *13*, 3745. [CrossRef] [PubMed]
5. Malea, P.; Charitonidou, K.; Sperdouli, I.; Mylona, Z.; Moustakas, M. Zinc Uptake, Photosynthetic Efficiency and Oxidative Stress in the Seagrass *Cymodocea nodosa* Exposed to ZnO Nanoparticles. *Materials* **2019**, *12*, 2101. [CrossRef] [PubMed]
6. Sperdouli, I.; Moustaka, J.; Antonoglou, O.; Adamakis, I.-D.S.; Dendrinou-Samara, C.; Moustakas, M. Leaf Age-Dependent Effects of Foliar-Sprayed CuZn Nanoparticles on Photosynthetic Efficiency and ROS Generation in *Arabidopsis thaliana*. *Materials* **2019**, *12*, 2498. [CrossRef] [PubMed]
7. Moustaka, J.; Ouzounidou, G.; Sperdouli, I.; Moustakas, M. Photosystem II Is More Sensitive than Photosystem I to Al^{3+} Induced Phytotoxicity. *Materials* **2018**, *11*, 1772. [CrossRef] [PubMed]
8. Bayçu, G.; Moustaka, J.; Gevrek, N.; Moustakas, M. Chlorophyll Fluorescence Imaging Analysis for Elucidating the Mechanism of Photosystem II Acclimation to Cadmium Exposure in the Hyperaccumulating Plant *Noccaea caerulescens*. *Materials* **2018**, *11*, 2580. [CrossRef] [PubMed]

9. Moustakas, M.; Hanć, A.; Dobrikova, A.; Sperdouli, I.; Adamakis, I.-D.S.; Apostolova, E. Spatial Heterogeneity of Cadmium Effects on *Salvia sclarea* Leaves Revealed by Chlorophyll Fluorescence Imaging Analysis and Laser Ablation Inductively Coupled Plasma Mass Spectrometry. *Materials* **2019**, *12*, 2953. [CrossRef] [PubMed]
10. Vega-Páez, J.D.; Rivas, R.E.; Dussán-Garzón, J. High Efficiency Mercury Sorption by Dead Biomass of *Lysinibacillus sphaericus*—New Insights into the Treatment of Contaminated Water. *Materials* **2019**, *12*, 1296. [CrossRef] [PubMed]

Article

Zn(II) Complex of Plant Phenolic Chlorogenic Acid: Antioxidant, Antimicrobial and Structural Studies

Monika Kalinowska [1,*], Justyna Sienkiewicz-Gromiuk [2], Grzegorz Świderski [1], Anna Pietryczuk [3], Adam Cudowski [3] and Włodzimierz Lewandowski [1]

[1] Department of Chemistry, Biology and Biotechnology, Institute of Civil Engineering and Energetics, Faculty of Civil Engineering and Environmental Sciences, Bialystok University of Technology, Wiejska 45E Street, 15-351 Bialystok, Poland; g.swiderski@pb.edu.pl (G.Ś.); w-lewando@wp.pl (W.L.)

[2] Department of General and Coordination Chemistry and Crystallography, Institute of Chemical Sciences, Faculty of Chemistry, Maria Curie-Sklodowska University, Maria Curie-Sklodowska Sq. 2, 20-031 Lublin, Poland; j.sienkiewicz-gromiuk@poczta.umcs.lublin.pl

[3] Department of Water Ecology, Faculty of Biology, University of Bialystok, Ciolkowskiego 1J Street, 15-245 Bialystok, Poland; annapiet@uwb.edu.pl (A.P.); cudad@uwb.edu.pl (A.C.)

* Correspondence: m.kalinowska@pb.edu.pl

Received: 16 July 2020; Accepted: 20 August 2020; Published: 24 August 2020

Abstract: The structure of the Zn(II) complex of 5-caffeoylquinic acid (chlorogenic acid, 5-CQA) and the type of interaction between the Zn(II) cation and the ligand were studied by means of various experimental and theoretical methods, i.e., electronic absorption spectroscopy UV/Vis, infrared spectroscopy FT-IR, elemental, thermogravimetric and density functional theory (DFT) calculations at B3LYP/6-31G(d) level. DPPH (2,2-diphenyl-1-picrylhydrazyl), ABTS (2,2′-azino-bis(3-ethylbenzothiazoline-6-sulfonic acid), FRAP (ferric reducing antioxidant power), CUPRAC (cupric reducing antioxidant power) and trolox oxidation assays were applied in study of the anti-/pro-oxidant properties of Zn(II) 5-CQA and 5-CQA. The antimicrobial activity of these compounds against *Escherichia coli*, *Pseudomonas aeruginosa*, *Bacillus subtilis*, *Staphylococcus aureus*, *Salmonella enteritidis* and *Candida albicans* was tested. An effect of Zn(II) chelation by chlorogenic acid on the anti-/pro-oxidant and antimicrobial activities of the ligand was discussed. Moreover, the mechanism of the antioxidant properties of Zn(II) 5-CQA and 5-CQA were studied on the basis of the theoretical energy descriptors and thermochemical parameters. Zn(II) chlorogenate showed better antioxidant activity than chlorogenic acid and commonly applied natural (L-ascorbic acid) and synthetic antioxidants (butylated hydroxyanisol (BHA) and butylated hydroxytoluene (BHT)). The pro-oxidant activity of Zn(II) 5-CQA was higher than the ligand and increased with the rise of the compound concentration The type of Zn(II) coordination by the chlorogenate ligand strongly affected the antioxidant activity of the complex.

Keywords: 5-caffeoylquinic acid; chlorogenic acid; zinc; plant phenolic compounds; oxidative stress

---

1. Introduction

Phenolic compounds play an important role in plant tolerance to toxic metals as well as in prevention and reduction of the biotic and abiotic oxidative stress. The chelation of metal ions by phenolic compounds is widely discussed in the literature as a possible hypothesis for the importance of phenolics in toxic metal tolerance in plants [1–3]. This mechanism relies on: (i) the secretion of chelating agents by e.g., roots to prevent metal uptake or (b) production of chelating agents to bind metals in the cell wall, symplast or vacuole [1]. For example, an increase in the synthesis of phenolic compounds and activity of shikimate dehydrogenase (SKDH), cinnamyl alcohol dehydrogenase (CAD) and polyphenol oxidase (PPO) was observed in roots and leaves of *Kandelia obovata* under Zn and

Cd stress [4]. According to authors the high level of Zn and Cd activated the phenolic metabolism pathways which participated in heavy metal tolerance process. Moreover the presence of Zn decreased the oxidative stress caused by Cd. Chlorogenic acid, among other phenolics, was found in the roots in woody plants as potential aluminum-detoxifying agents [5]. The study of Mongkhonsin et al. revealed that not only the increase in the level of phenolic compounds but also the lignification process played crucial roles in protecting *G. pseudochina* against an excess of Zn [6]. Among the phenolic compounds that were engaged in the esterification of the cell wall were derivatives of caffeic acid, including chlorogenic acid. The application of bulk X-ray absorption near edge structure (XANES) spectra indicated that Zn cations were bonded by O-ligands which could be phenolic compounds. Studies of other authors also confirmed that the exposure of the plant to toxic metals caused an increase in the level of phenolics which intermediated in lignin biosynthesis [7].

Chlorogenic acid (5-caffeoylquinic acid, 5-CQA) is a phenolic compound found in all part of many plants (seeds, roots, tubers, fruits, leaves, flowers or bark) [8–13]. 5-CQA plays an important role in defense against biotic and abiotic stress in plants, and is considered to be an intermediate molecule in lignin biosynthesis pathway [14]. Many studies reported accumulation of chlorogenic acid in tissue exposed to different stress factors, e.g., an increase in salinity [15], bacterial infection [16], ultraviolet (UV-B) radiation [17], toxic metals [18,19] and other related with an increase in production of reactive oxygen species (ROS) which damage cellular membrane and react with plant cell components causing disruption of metabolic pathways.

The metal chelating ability, participation in lignification process and scavenging of free radical are the main important mechanisms of antioxidant activity of 5-CQA involved in the decrease in the ROS level in plants under stress condition. However, the complexation with metal ions may change the antioxidant potential of chlorogenic acid. The alkali metal salts of chlorogenic acid possessed higher antioxidant activity than the ligand alone [10]. The complexation of chlorogenic acid with oxidovanadium(IV) also improved the antioxidant and anti-cancer properties against the human breast cancer cell line SKBR3 [20]. The higher antioxidant activity of metal complexes compared to free phenolic compounds was observed in other cases as well, e.g., Fe(II), Cu(II), Ce(IV), La(III) and oxovanadium(IV) complexes of chrysin; complexes of apigenin with La(III), Mg(II) and luteolin with oxovanadium [21] or calcium complex of gentisic acid [22]. It should be kept in mind that the antioxidants may act as pro-oxidants depending on their concentration or the presence of some metal cations. This may be due to the stabilization of the phenoxyl radicals by metal cations and prolonging their lifetime [23]. The phenoxyl radicals may accelerate lipid peroxidation, disrupt mitochondrial membrane potential or induce DNA damage. In part, this mechanism may also explain the toxic effect of some metals on plants. Therefore the question arises: do metal ions affect the anti-/pro-oxidant properties of chlorogenic acid in plant tissue? What is the stability and structure of formed metal complexes? Do the type of metal-ligand bonding and the structure of metal complexes affect the antioxidant activity of the molecules? The answers will help us to understand the features that make the phenolic compounds' effective antioxidants and detoxifying agents. It will help to design plants which are resistant to stressful conditions and oxidative stress.

Zinc is as a plant nutrient but at higher concentrations, it shows high toxicity, causes growth inhibition, and affects plant metabolism [24]. Therefore, chelation of zinc by chlorogenic acid plays an important role in protection against oxidative stress and enhances the tolerance to toxic metal stress in plants. Moreover, plant phenolic compounds and their metal complexes that possess strong biological activities are intensively studied as potential new effective antioxidants or antimicrobial agents with low toxicity. There are many studies aiming at searching for plants rich in these compounds and new non-destructive methods for chlorogenic acid extraction [25]. 5-CQA isolated from plant product possesses not only strong antioxidant activity, but among others antimicrobial, anti-inflammatory and anti-cancer [25]. Although the complexing abilities of chlorogenic acid toward different metal cations in aqueous solution and solid state were studied [10,20,26–30], the biological properties of metal chlorogenates (including Zn chlorogenate) are scarcely described. Therefore, the studies of the structure and anti-/pro-oxidant activity of zinc complex with chlorogenic acid are of great importance. The aim of the paper was to synthesize the Zn(II) chlorogenate in the solid state and study the structure and type of interactions between the Zn(II) cation and chlorogenate ligand by the use of UV/Vis, FT-IR, elemental, thermogravimetric analysis and DFT (density functional theory) calculations in Gaussian program. Anti-/pro-oxidant and microbiological activity of Zn(II) chlorogenate was studied and compared with ligand properties. Moreover, the mechanism of antioxidant activity of the complex and the effect of the coordination mode on the antioxidant activity of the complex were studied on the basis of the calculated energy descriptors and thermochemical parameters.

2. Materials and Methods

2.1. Materials

All chemicals were purchased from Sigma-Aldrich Co. (St. Louis, MO, USA) and used without purification. Only CH_3OH was bought from Merck (Darmstadt, Germany). *Escherichia coli* (PCM 2857), *Pseudomonas aeruginosa* (PCM 2720), *Bacillus subtilis* (PCM 2850), *Staphylococcus aureus* (PCM 2267), *Salmonella enteritidis* (NCTC 4776) and *Candida albicans* (PCM 2566-FY) were bought from the Polish Collection of Microorganisms (Wroclaw, Poland) or American Type Culture Collection.

2.2. Sample Preparation

The Zn (II) complex of chlorogenic acid was obtained as follows. First, sodium salt of chlorogenic acid was synthesized in a solid state: the aqueous solutions of NaOH (0.05 M) and chlorogenic acid (0.05 M) were mixed in the stoichiometric molar ratio 1:1. The solution was allowed to evaporate on a water bath [10]. The composition of the dry residue was examined by elemental and thermogravimetric analysis. The formula of sodium chlorogenate was $C_{16}H_{17}O_9Na \cdot 1.5H_2O$ (results of elemental analysis: exp.%C = 47.84, calc.%C = 47.65; exp.%H = 4.94, calc.%H = 5.00). Then, aqueous solution of the sodium salt at the concentration of 0.05 M was prepared and mixed with aqueous solution of $ZnCl_2$ (0.05 M) in the stoichiometric molar ratio 2:1. The nude precipitate occurred immediately. It was filtered and washed with the distilled water. The solid residue was dried in room temperature for 5 days. The results of the elemental analysis showed the formula of the complex to be: $Zn(C_{16}H_{17}O_9)_2 \cdot 3H_2O$ ($\%C_{exp.}$ = 46.13; $\%C_{calc.}$ = 46.49; $\%H_{exp.}$ = 4.78, $\%H_{calc.}$ = 4.84, $\%Zn_{exp.}$ = 7.59; $\%Zn_{calc.}$ = 7.90).

2.3. Anti-/Pro-Oxidant Study

The antioxidant activity of the tested compounds was measured by the use of DPPH, ABTS, FRAP and CUPRAC tests. The DPPH assay was described by [31] and our previous work [10]. The solutions of DPPH (C = 60 µM) and tested substances (C = 50 and 500 µM) were prepared in methanol. To the glass tubes the appropriate volumes of tested substances were added, then diluted with methanol and 2 mL of DPPH. The final concentrations of antioxidants were in the range 0.1–70 µM. Each tube was vortexed and incubated in dark place for 1 h at 23 °C. Then the absorbance was measured at 516 nm against methanol (the blank). The control sample was the mixture of 2 mL of DPPH and 1 mL of methanol. The antiradical activity of tested substances was calculated from the equation:

$$\% I = \frac{A_{control} - A_{sample}}{A_{control}} \times 100\% \tag{1}$$

where: % I—the percent of inhibition of DPPH· radicals, $A_{control}$—the absorbance of the control, A_{sample}—the absorbance of the sample. Then, the concentration of the tested substances was plotted versus the % inhibition and the EC_{50} values were read from the curves. The EC_{50} means the concentration of antioxidant that inhibits 50% of the DPPH· radicals.

The aqueous solutions of 2,2'-azino-bis(3-ethylbenzothiazoline-6-sulfonic acid) diammonium salt (ABTS; C = 7 mM) and $K_2S_2O_8$ (C = 2.45 mM) were mixed in a volumetric ratio 1:1 and left for 16h at 23 °C. The solution of cation radical of ABTS was prepared by mixing 1 mL of the mixture and 60 mL of methanol. 1 mL of methanolic solution of tested substance was added to 1 mL of solution of ABTS·$^+$, incubated for 7 min at 23 °C. The final concentration of tested substances was 25 µM. The absorbance was read at λ = 734 nm against methanol. The antiradical activity against ABTS·$^+$ was expressed as the percent of ABTS·$^+$ inhibition and calculated according to the aforementioned formula.

The FRAP assay determines the ferric-reducing antioxidant activity of antioxidants [10,31]. The FRAP reagents: 0.3 M acetate buffer (pH 3.6), 10 mM 2,4,6-tripyridyl-s-triazine (in 40 mM HCl) and 20 mM FeCl$_3$ (in water) were mixed in a volumetric ratio 10:1:1. Then, 3 mL of the mixture was added to 0.4 mL of tested substance (final concentration 50 µM) and incubated for 7 min at 23 °C. The absorbance was measured at 595 nm against blank (i.e., 3 mL of FRAP mixture and 0.4 mL of methanol). The antioxidant activity was expressed as Fe^{2+} equivalents (µM) using the calibration curve prepared for FeSO$_4$ (y = 2.0172x − 0.0708; R^2 = 0.9992).

The CUPRAC assay determines the cupric reducing antioxidant activity of antioxidants. The solutions of CuCl$_2$ (C = 10 mM; in water), ammonium acetate (pH = 7; in water) and neocuproine (C = 75 mM; in ethanol) were mixed in a volumetric ration 1:1:1. 3 mL of the mixture was added to 0.5 mL of tested substance and 0.6 mL of distilled water. The final concentration of tested substances was 50 µM. The samples were incubated for 1h at 23 °C. Then the absorbance was read at 450 nm against blank (i.e., 3 mL of CUPRAC mixture, 0.6 mL of water and 0.5 mL of methanol). The antioxidant activity was expressed as trolox equivalents (µM) using the calibration curve prepared for trolox (y = 4.5758x − 0.0271; R^2 = 0.9919).

The pro-oxidant activity of the tested substances were measured as the rate of oxidation of trolox according to the procedure of Zeraik et al. [32] and described previously [10]. The chemicals were added in the following order: 100 µM trolox, 50 µM H$_2$O$_2$, 0.01 µM horseradish peroxide in phosphate buffer (pH = 7) and tested substance with a final concentration from 0.025 to 0.15 µM. The mixture was vortexed and incubated at 25 °C. The absorbance measurements (at 272 nm) were made every 5 min through 15 min.

All measurements were taken in five repetition in three independent experiments. The absorbance was measured using an Agilent Carry 5000 spectrophotometer (Agilent, Santa Clara, CA, USA).

2.4. Antimicrobial Study

Escherichia coli, Pseudomonas aeruginosa, Bacillus subtilis, Staphylococcus aureus, Salmonella enteritidis and *Candida albicans* were grown overnight and then were resuspended in physiological saline to an optical density OD = 0.60 at 600 nm. It corresponds to 5.0×10^8 colony-forming units (CFU)/mL. Microorganisms (0.1 mL of the reconstituted suspension) were seeded onto sterile Mueller-Hinton agar plates. The suitable amounts of the tested compounds (dissolved in an agar medium) were added to give their desired concentration. The range of studied concentration was 0.05–10 mM. The negative controls were agar plates without tested substances. The positive control was gentamycin (in the case of bacteria) or flucanozole (in the case of fungi). The plates were incubated at 37 °C for 24 h. The antimicrobial activity of tested compounds was expressed as MIC (minimum inhibitory concentration).

2.5. Statistical Analysis

For parametric data one-way analysis of variance (ANOVA) followed Tukey's test was applied using Statistica 13.1 program. Results from three independent experiments were expressed as mean ± standard deviation (SD) of mean for parametric data. Significance was considered when $p \leq 0.05$.

2.6. Structural Studies

The FT-IR spectra for the solid samples were recorded in KBr matrix pellets with an Alfa Bruker spectrometer (Bremen, Germany) within the range of 400–4000 cm^{-1} with the resolution of 2 cm^{-1}. UV/Vis spectra of chlorogenic acid and its Zn complex were recorded with UV/VIS/NIR Carry 5000 spectrophotometer. To determine the molar ratio of zinc(II) to 5-CQA the Jobs' method was used.

The spectra were recorded for a solution with different molar ratios of zinc ion and chlorogenic acid, but constant total amount of zinc(II) plus 5-CQA moles. The concentration of 5-CQA was 0.1 mM, the concentration of $ZnCl_2$ changed from 0 to 0.09 mM. All solutions were prepared in Tris-HCl buffer (pH = 7.4; C = 50 mM). Thermogravimetric analysis (TG) along with differential scanning calorimetry (DSC) was performed by using Setsys 16/18 thermal analyzer of Setaram Instrumentation brand (SETARAM, Caluire, France). The sample of zinc(II) complex (4.27 mg) was heated in the range of 30–750 °C in ceramic crucible at a heating rate of 10 °C min^{-1} in a flowing air atmosphere ($v = 1$ dm^3 h^{-1}).

2.7. DFT Calculations

The quantum-chemical calculations for Zn chlorogenate were done in B3LYP/6-31G(d) using the GAUSSIAN 09W and GaussView 6 software (Gaussian Inc., Wallingford, CT, USA) package running on a Dell PC computer (Manufacture, Round Rock, TX, USA) [33]. The optimized structure for chlorogenic acid was published before [12]. The geometry, IR, the highest occupied molecular orbital (HOMO) and lowest unoccupied molecular orbital (LUMO), selected chemical reactivity parameters [34] and thermodynamic parameters [35,36] were calculated.

3. Results and Discussion

3.1. Anti-/Pro-Oxidant Study

The results of reactions of studied compounds with DPPH• radicals were given as the EC_{50} parameter which signifies the concentration of compound that inhibits 50% of the radicals (Table 1). The assay showed that Zn 5-CQA possessed better antioxidant properties ($EC_{50} = 5.45 \pm 0.37$ µM) than the ligand alone ($EC_{50} = 7.23 \pm 0.76$ µM) or natural (L-ascorbic acid $EC_{50} = 10.32 \pm 0.98$ µM) and synthetic antioxidants (i.e., BHA $EC_{50} = 13.54 \pm 1.61$ µM, BHT $EC_{50} = 53.14 \pm 1.05$ µM). Zn 5-CQA was a better scavenger of cation radicals $ABTS^{•+}$ than other studied compounds, including 5-CQA. At the same concentration of studied antioxidants (i.e., 25 µM), Zn 5-CQA inhibited 97.65% of the initial concentration of $ABTS^+$, whereas 5-CQA inhibited 89.53%. Two more tests for study the antioxidant activity of the compounds were used, i.e., FRAP and CUPRAC assays, which respectively allowed to determine the ferric and cupric reducing antioxidant activity of the compounds at the concentration of 50 µM. In both tests, Zn 5-CQA showed better reducing properties than 5-CQA and commonly applied antioxidants. The FRAP value for Zn 5-CQA was 385.56 µM Fe^{2+}, whereas 216.09 µM Fe^{2+} for 5-CQA and even lower for other antioxidants. The last two antioxidant tests measured the ability of an antioxidant to transfer an electron from the antioxidant to any compound in order to reduce it. The mechanism is called SET (single electron transfer). While the first two assays (DPPH and ABTS) based on mixed mechanism which is commonly described as mechanism involving both SET and HAT (hydrogen atom transfer). The studies of Litwinienko [37,38] and Foti [39,40] revealed that the mechanism of action of antioxidants depended on their chemical structure and the environment and therefore several particular mechanisms were described: (a) SPLET (sequential proton-loss electron-transfer), (b) PCET (proton-coupled electron-transfer), (c) ET-PT (electron-transfer proton-loss) as well as (d) HAT [37,38]. As the polarity of solvent increases, the contribution of the HAT mechanism decreases in favor of the other mechanisms [37,38]. In methanol (DPPH assay) and methanol-aqueous solution (ABTS assay) the studied antioxidants were partially ionized, the phenolate anions were formed, which can rapidly react with radicals, through an electron transfer. In ionizing solvents, slow HAT mechanism was rather marginal [37,38]. Therefore, it could be stated that the studied antioxidants act mainly through electron transfer and the ionization potential (IP) could be one of the important parameters defining the antioxidant potential of these compounds. The ionization potential describes the susceptibility of molecule to ionization, and the lower is the IP, the easier is to detach an electron from the antioxidant. The higher antioxidant property of zinc complex of chlorogenic acid compared with chlorogenic acid may be caused by easier electron abstraction from covalently bound

5-CQA to zinc ion than 5-CQA alone. This was discussed later in the work on the basis of spectroscopic study and quantum-chemical calculations.

Table 1. Antioxidant properties of the Zn 5-CQA, 5-CQA, L-ascorbic acid, butylated hydroxyanisol (BHA) and butylated hydroxytoluene (BHT) expressed as the ability to scavenge DPPH· radical (EC_{50}) and $ABTS^{·+}$ cation radical as well as FRAP and CUPRAC values (the concentrations of tested substances in the samples were * 25 µM, ** 50 µM).

Compound	DPPH/EC_{50} (µM)	ABTS * I%	FRAP ** C_{Fe2+} (µM)	CUPRAC ** C_{trolox} (µM)
Zn 5-CQA	5.45 ± 0.37	97.65 ± 1.25	385.56 ± 5.35	198.36 ± 6.28
5-CQA	7.23 ± 0.76	89.53 ± 3.61	216.09 ± 4.15	129.58 ± 9.70
L-ascorbic acid	10.32 ± 0.98	86.76 ± 1.58	156.86 ± 2.02	25.57 ± 3.50
BHA	13.54 ± 1.61	88.38 ± 2.98	141.82 ± 3.58	83.55 ± 14.97
BHT	53.14 ± 1.05	62.66 ± 5.65	134.40 ± 2.69	74.81 ± 7.57

Zinc chlorogenate was tested for pro-oxidative effect on trolox oxidation in the general procedure for study the pro-oxidant activity of phenolic compounds [32]. The radicals of chlorogenate and chlorogenic acid were produced in their direct reaction with H_2O_2 catalysed by the enzyme horse radish peroxide. The formed phenoxyl radicals reacted with trolox which was oxidizing to trolox radicals and then trolox quinones. Whereas the phenoxyl radicals were transformed to phenolic compounds. The results of this study are shown in Figure 1. The rate of trolox oxidation increased with the increase in the concentration of Zn 5-CQA from 0.025 to 0.15 µM. Zn chlorogenate showed higher pro-oxidant activity than chlorogenic acid, especially in the concentration range 0.025–0.15 µM (Figure 1). The maximum of the pro-oxidant activity was reached at higher concentration (in our experiment: 0.10 and 0.15 µM) after maximum 10 min. of the measurement. The mean absorbance of the sample measured after 10 min. reached 1.718 for Zn 5-CQA and 1.321 for 5-CQA, and after 15 min. it was equal 1.936 (Zn 5-CQA) and 1.504 (5-CQA).

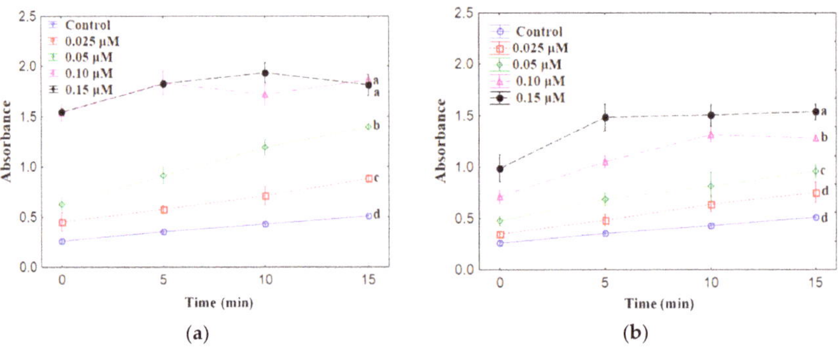

Figure 1. The effect of different concentrations (0.025–0.15 µM) of (**a**) zinc complex of chlorogenic acid and (**b**) chlorogenic acid on the oxidation of trolox. Mean values from three independent experiments ± standard deviation (SD) are shown. The same letter near the means indicates no significant difference (Tukey test, $p < 0.05$).

3.2. Antimicrobial Study

The pro-oxidant activity of chemicals could be an explanation of their antimicrobial property [41–44]. Therefore, the pro-oxidant activity of phenolic compounds may explain their antimicrobial action in plants as well. The studies revealed that Zn 5-CQA and 5-CQA did not show antimicrobial properties against *E. coli*, *P. aeruginosa*, *C. albicans*, *B. subtilis*, and *S. enteritidis* at the concentration < 10 mM (Table 2). Surprisingly, Zn 5-CQA showed higher activity than 5-CQA against *S. aureus* with the MIC = 1 mM. In the case of the rest of microorganism the same value of MIC was obtained for both 5-CQA and Zn 5-CQA. The antimicrobial properties of chlorogenic acid were widely described. The literature MIC values for 5-CQA differ significantly between each other, i.e., MIC = 14–28 mM against *K. pneumoniae, P. vulgaris, P. aeruginosa, E. faecium, C. albicans,* and *S. cerevisiae* [10], MIC = 23–45 µM for clinically isolated *Stenotrophomonas maltophilia* [45], MIC = 2.2–35.3 mM against *Aeromonas* species isolated from fish [46], MIC = 5.6 mM against *Alicyclobacillus acidoterrestris* [47], MIC = 56 µM in the cases of *S. pneumoniae* and *Shigella dysenteriae.* According to Lou et al. the mechanism of antimicrobial action of chlorogenic acid relies on binding to the outer membrane, disrupting the membrane what causes intracellular potential disorder and release cytoplasm macromolecules, which leads to cell death [25,48]. Generally, 5-CQA shows antimicrobial and anti-biofilm activity through change of the permeability of cell walls of microorganisms [28]. Wang et al. showed that 5-CQA is active toward multi-drug resistant S. aureus (IC_{50} = 96 µM) via inhibition of the activity of sortase A [49]. The formation of metal complexes with ligands of proven antimicrobial activity may enhance their antimicrobial activity. It can be explained by inter alia an increase in the lipophilicity of the complex compared with the ligand. The change of the biological activity of alkali metal salts of chlorogenic acid vs. their lipophilicity was discussed previously [10]. The lipophilicity is an important parameter of antimicrobial agents. Because most of the molecules penetrate the cell membranes by passive diffusion, so they should be lipophilic enough to cross through the biological membranes, but also hydrophilic enough to penetrate the cytoplasm. The increase in the antimicrobial properties of metal complexes of plant phenolic acids compared to ligands itself was described in many papers [21]. E.g. Zn(II) complex of p-coumaric acid possessed higher antimicrobial properties than ligand alone, especially against *S. aureus* (at the concentration of 0.1% in broth culture it caused 77 ± 3% and 75.2% of growth inhibition after 24 and 48 h of incubation, respectively) [50]. In another study, all tested Cu(II), Zn(II), Na(I) complexes of ferulic acid revealed higher antimicrobial activity than ferulic acid [51]. Zn ferulates possessed the strongest antimicrobial properties toward *E. coli, B. subtilis, S. aureus, P. vulgaris* and *C. albicans* (at the concentration of 0.1% in broth culture it caused 90.8–98.9% of growth inhibition).

Table 2. The minimum inhibitory concentration (MIC) values (mM) for 5-CQA and Zn 5-CQA against selected microorganisms.

Microorganisms	5-CQA	Zn 5-CQA
Escherichia coli	>10	>10
Pseudomonas aeruginosa	>10	>10
Staphylococcus aureus	>10	1
Candida albicans	>10	>10
Bacillus subtilis	>10	>10
Salmonella enteritidis	>10	>10

3.3. Structural Studies

3.3.1. FT-IR Spectra

To characterize the type of interaction between the Zn(II) cation and chlorogenate ligand the FT-IR method was applied. The FT-IR spectra of Zn 5-CQA and 5-CQA were shown in Figure 2, and the assignment of the selected bands were gathered in Table 3. The assignment was based on our previous publications [10,12] concerning chlorogenic acid and alkali metal chlorogenates. The comparison of the FT-IR spectra of synthesized compound and initial ligand gave information about the correctness of the synthesis and the type of metal ion coordination. In the spectrum of 5-CQA there was a strong band at 1725 cm^{-1} (derived from the stretches of the C=O from the carboxylic group) which disappeared in the spectra of Zn complex. Moreover in the spectra of zinc chlorogenate several bands assigned to the vibrations of carboxylate anion appeared, i.e., stretching asymmetric 1616 cm^{-1} and symmetric 1384 cm^{-1}, deforming in plane 814 cm^{-1}, and out of plane bending at 617 cm^{-1} (Table 3). It confirmed the metal ion coordination through the carboxylate anion. The C=O stretching bands assigned to the ester group were located at the same wavenumbers in the spectra of ligand and zinc complex, what suggested that this group did not participate in metal coordination. The band assigned to the stretching vibrations of the catechol C-O group were significantly shifted to the 1272 cm^{-1} in the spectra of Zn complex, whereas in the spectra of 5-CQA it was located at the 1289 cm^{-1}. It may suggest additional metal coordination through the catechol moiety and weakness of the C–O bond strength. The simultaneous metal ion coordination to both carboxylate and catechol moieties and the formation of oligomeric complexes is typical for ligands possessing two coordinating sites. Some other bands present in the spectra of Zn 5-CQA were slightly shifted or disappeared compared with the spectra of 5-CQA. It means that the coordinated metal affects the structure of the quinic and caffeic acid moieties.

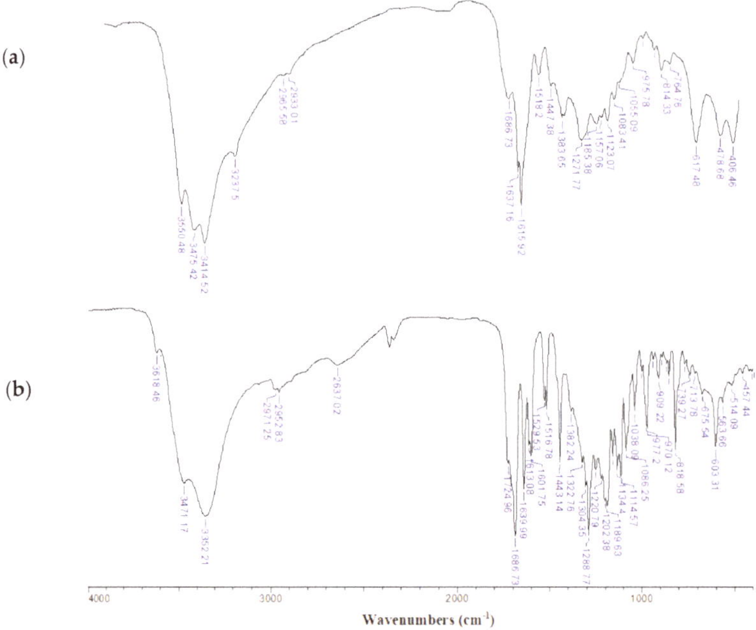

Figure 2. The FT-IR spectra of: (**a**) Zn(II) chlorogenate and (**b**) chlorogenic acid registered in the range of 400–4000 cm^{-1} for solid samples in the KBr matrix pellet.

Table 3. The wavenumbers, intesities and assignment of selected bands from the FT-IR spectra of zinc chlorogenate, sodium chlorogenate [10] and chlorogenic acid [12]; the symbols denote: ν-stretching vibrations, δ-deforming in plane and oop-out of plane bending vibrations; s-strong, m-medium, w-week, v-very, sh-on the slope.

5-CQA [12]		Na 5-CQA [10]		Zn(II) 5-CQA		Assignment
IR	**int.**	**IR**	**int.**	**IR**	**int.**	
1725	s	-	-	-	-	ν(C=O)$_{COOH\ group}$
1687	vs	1692	s	1687	s	ν(C=O)$_{ester\ group}$
1640	s	1634	s	1637	sh	ν(C=C)
-	-	1598	vs	1616	vs	ν(COO$^-$)$_{asym}$
1530	m	1528	s	1518	w	ν(CC)$_{ar}$
1517	m	-	-	-	-	ν(CC)$_{ar}$
1443	s	1450	m	1447	m	δ(COH)$_{quinic\ ring}$
-	-	1390	sh	1384	m	ν(COO$^-$)$_{sym}$
1322	m	-	-	-	-	ν(CC) + δ(CCH) + δ(COH)$_{quinic\ ring}$

Table 3. *Cont.*

5-CQA [12]		Na 5-CQA [10]		Zn(II) 5-CQA		Assignment
IR	int.	IR	int.	IR	int.	
1304	s	1323	s	-	-	$v(CO) + v(CC) + \delta(CCH)$
1289	vs	1282	vs	1272	vs	$v(C\text{-}O)_{\text{catechol group}} + \delta(CH)_{ar}$
1251	s	-	-	-	-	$v(CC) + v(CH) + \delta(CCH) + \delta(COH)$
1202	s	-	-	-	-	$\delta(COH)_{\text{quinic ring}}$
1190	vs	1178	vs	1185	m	$v(CC) + v(CH) + \delta(CCH) + \delta(COH)$
1159	s	1163	s	1157	m	$v(CC) + v(CH) + \delta(CCH) + \delta(COH)$
1134	s	-	-	-	-	$v(C\text{-}O)_{\text{COOH group}}$
1115	s	1119	s	1123	m	$v(C\text{-}O)_{\text{ester group}}$
1086	s	1081	s	1083	m	$v(CC) + v(C\text{-}O)_{\text{quinic ring}}$
1059	w	1059	w	1055	m	$v(CC)_{\text{quinic ring}} + \delta(CH_{\text{quinic ring}}$
1038	m	1037	m	-	-	$v(C_{\text{quinic ring}}\text{-}O_{\text{in ester group}}) + \delta(CH)_{\text{quinic ring}}$
1000	w	997	m	-	-	$\text{oop}(HC\text{-}C=C) + \text{oop}(HC=CH)$
970	m	969	w	976	w	$\delta(CH)_{\text{quinic ring}} + \delta(CH)_{\text{quinic ring}} + v(C\text{-}O)_{\text{quinic ring}}$
909	m	926	w	914	w	$v(CC) + v(C\text{-}O)_{\text{quinic ring}} + \delta(CC) + \delta(CH)$
853	m	854	w	851	w	$\delta(HC\text{-}CO) + \text{oop}(CH)_{\text{ester group}}$
819	s	-	-	-	-	$\delta(CC)_{\text{arom. ring}}$
-	-	808	m	814	m	$\delta(CC)_{\text{arom. ring}} + \delta(COO^-)$
-	-	615	m	617	m	$\text{oop}(COO^-)$

3.3.2. Thermogravimetric (TG)/Differential Scanning Calorimetry (DSC) Analysis

The conducted studies revealed high anti-/pro-oxidant properties of Zn 5-CQA what creates the possibility to applicate Zn 5-CQA as antioxidant or pro-oxidant in e.g., plant protection against oxidative stress or in food industry or pharmacy as a preservative or diet supplement of natural origin. Because Zn 5-CQA was synthesized in the solid state therefore its structure in solid state should be carefully described. To this aim the elemental, thermogravimetric and FT-IR analysis were applied. The general formula of the complex was $Zn(C_{16}H_{17}O_9)_2\cdot 3H_2O$. The information about thermal stability and thermal decomposition pathway of compound in question was acquired through TG/DSC analysis.

The analyzed material was stable at room temperature. The thermal decomposition of the title compound proceeded in two main stages related to the dehydration process and gradual degradation of anhydrous material into the metal oxide, which was reflected in the registered thermal profile (Figure 3, Table 4). The release of all water molecules from $(Zn(C_{16}H_{17}O_9)_2\cdot 3H_2O)$ occurred in one evident stage between 30–145 °C which was compliant with the only one observed endothermic effect on the DSC curve with peak top at 83 °C. A mass loss of 6.53% expected for the process of removal of three water molecules was in agreement with the experimentally found value of 6.83%. The anhydrous form $(Zn(C_{16}H_{17}O_9)_2)$ of the studied compound exhibited resistance to thermal decomposition up to about 235 °C. The plateaux of both TG and DSC curves seen between 145 and 235 °C meant fair thermal stability of the anhydrous derivative. Heating above 235 °C lead to the continuous weight loss up to 510 °C. The course of the DSC curve in the temperature range 235–510 °C indicated four energetic effects with the peak maxima located at 269, 413, 446 and 460 °C. First of them was probably related with the endothermic melting process preceding the intense three-step exothermic combustion of the organic part of the complex. The total mass loss of 90.58% (calculated 90.16%) harmonized to the formation of white ZnO as a final product of thermal decomposition. Because Zn(II) chlorogenate was synthesized from Na chlorogenate, the thermal studies for Na salt of chlorogenic acid was performed as well, and the results were shown in Figure 3 and Table 4.

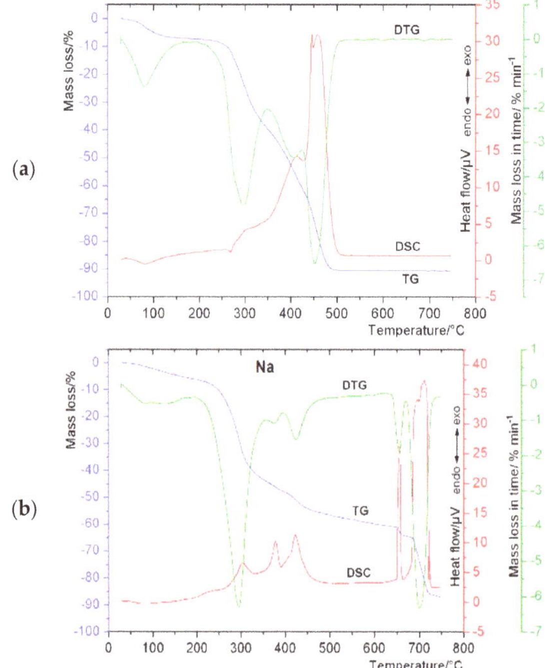

Figure 3. TG-DTG and DSC thermal curves obtained for: (**a**) zinc(II) and (**b**) sodium chlorogenates in air atmosphere.

Table 4. Results of thermal decomposition of Zn 5-CQA and Na 5-CQA in air atmosphere. *-endo effect conerns only Zn complex.

Complex	T_1/°C	T_{endo}	Mass Loss/%		Anhydrous Form	T_2/°C	T_{exo}/T_{endo} *	Residue/%		Residue
			Found	Calculated				Found	Calculated	-
$Zn(C_{16}H_{17}O_9)_2\cdot$ $3H_2O$	30–145	83	6.83	6.53	$Zn(C_{16}H_{17}O_9)_2$	235–510	269 * 413 446 460	9.42	9.84	ZnO
$Na(C_{16}H_{17}O_9)\cdot$ $1.5H_2O$	30–195	79	5.91	6.69	$Na(C_{16}H_{17}O_9)$	195–745	304 379 424 656 713 723	13.35	13.14	Na_2CO_3

3.3.3. Ultraviolet (UV) Spectra

To study of the composition of Zn 5-CQA in solution the spectrophotometric Job's method was applied. In the UV spectra of chlorogenic acid four bands at 217, 231, 299 and 325 nm appeared which derived from the $\pi \rightarrow \pi^*$ transitions in the frame of aromatic ring and the double bond (Figures 4 and 5) [12]. As a consequence of zinc complex formation the band at 299 nm disappeared, the intensity of the band at 325 nm decreased, and two more bands appeared at 272 and 374 nm. The appearance of the band around 374 nm is characteristic for metal-phenolate interaction and suggests that the catecholate mode was involved in metal ion coordination [30]. With the increase in the concentration of the Zn(II) ions the absorbance of the bands at 231 and 374 increased. The stoichiometry of the complex formed was established by the Job's method by plotting the absorbance at λ = 374 nm vs. $(Zn^{2+})/((Zn^{2+}) + (5\text{-CQA}))$. The maximum of the Job's curve corresponded to the mole fraction 0.33, confirming the molar ratio Zn:5-CQA 1:2.

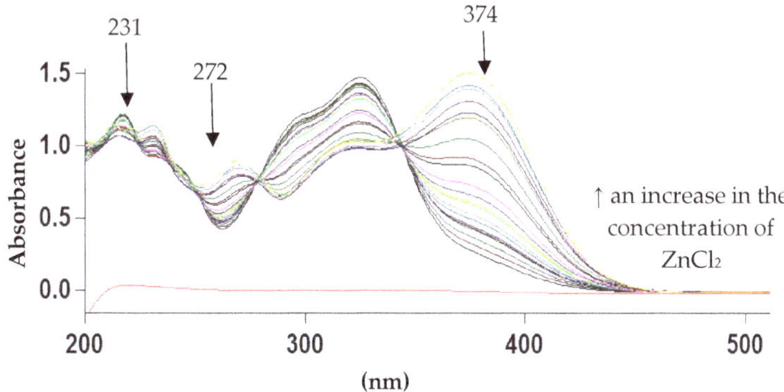

Figure 4. The ultraviolet (UV) spectra of the series of solutions prepared according to the Job's method (0.1 mM 5-CQA and 0–0.09 mM $ZnCl_2$ in Tris-HCl, pH = 7.4). The lowest line showed the spectra of 0.1 mM $ZnCl_2$.

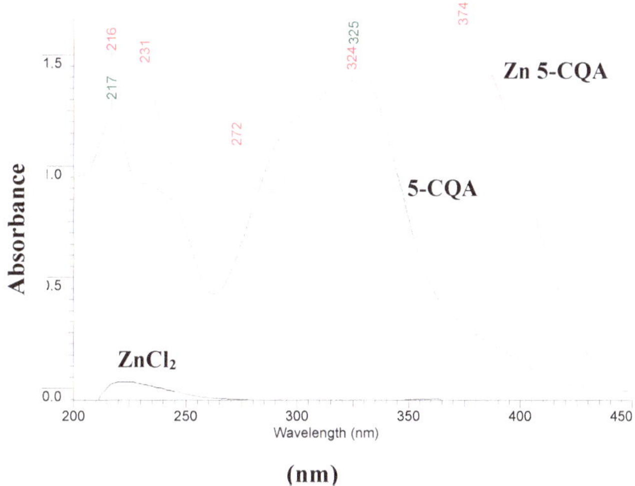

Figure 5. The UV spectra of 0.1 mM $ZnCl_2$, 0.1 mM chlorogenic acid (5-CQA), 0.05 mM $ZnCl_2$ plus 0.1 mM 5-CQA (molar ratio Zn:5-CQA 1:2) in Tris-HCl at pH = 7.4 registered in the range 200–450 nm.

In aqueous solution 5-CQA may form various oligomeric structures because of the possibility to bind metal cations through catechol and carboxylate moieties [30]. The other studies showed that Cu(II), Mn(II), Zn(II) and Fe(III) cations formed complexes with chlorogenic acid with the general formula ML_n (where L-chlorogenic acid, n = 1, 2 or 3) [29]. The stoichiometry of Cu(II), Fe(II) and Mn(II) chlorogenates at nearly neutral pH was found to be 1:1 [28]. Other studies confirmed that at pH 7.5 the stoichiometry of Cu(II) 5-CQA was 1:1 and 1:2, and for Pb(II) 5-CQA 1:1 [27]. Different coordination of Pb(II) cations by the chlorogenate ligand was shown, i.e., coordination through the carboxylate group in the complex $(PbL(H_2O)_3)^+$ with stoichiometry 1:1 or with additional binding by the catechol in the complex with stoichiometry 2:1.

3.4. Quantum-Chemical Calculations

The quantum-chemical calculations are complementary to the experimental methods and allow us to predict the structure, atomic charge distribution in molecule, spectral, electronic and thermodynamic parameters which correlate with the biological activity of molecules. The structures of chlorogenic acid and zinc chlorogenate were optimized in the B3LYP/6-31G(d) level (Figure 6). The number of water molecules in a hydrated complex agreed with the results of elemental and thermogravimetric experiments. The two possible ways of metal ion coordination were taken into consideration, i.e., (1) one of the chlorogenate ligand coordinated Zn^{2+} cation via carboxylate group and the second one through the catechol moiety (structure I) or (2) both chlorogenate ligands coordinated the Zn^{2+} cation through the COO^- groups (structure II). The selected geometrical parameters (bond lengths and angles) of the studied molecules were gathered in Table S1. The most distinct differences between the structural parameters of ligand and complex concerned the geometry of the $-C1-C7O5O4^-$ group of quinic acid moiety engaged in the metal ion coordination. In the structure II of Zn 5-CQA two COO^- groups coordinated the Zn^{2+} ion in slightly different manner. They were distinct differences in the C–O lengths and O–C–O angles of particular COO^- groups. This should be seen in the experimental FT-IR spectra of complex as a two bands derived from the asymmetric stretching vibrations of the $\nu_{as}(COO^-)$ (Figure 2). In the FT-IR spectra of Zn 5-CQA, only one band was assigned to the $\nu_{as}(COO^-)$, what may suggest metal ion coordination through only one of the carboxylate anion. In the structure I of Zn 5-CQA, the metal ion was coordinated through the COO^- of one ligand and catechol moiety of the second ligand. In such a case the metal ion affected not only the geometry of carboxylate group but also the bond lengths and angles of the skeleton $-O1'-C3'-C4'-O2'-$ in the aromatic ring. These should be seen in the FT-IR spectra of the complex by both (a) the appearance of the bands derived from the $\nu(COO^-)$ vibrations and (b) the shift of the bands assigned to the stretching vibrations of the catechol C–O group (ν (C–O)$_{catechol\ group}$). In the experimental FT-IR spectra of Zn 5-CQA the band ν (C–O)$_{catechol\ group}$ was located at much lower wavenumber (1272 cm^{-1}) compared with the spectra of acid (1289 cm^{-1}). It suggested the participation of catechol group in metal bonding.

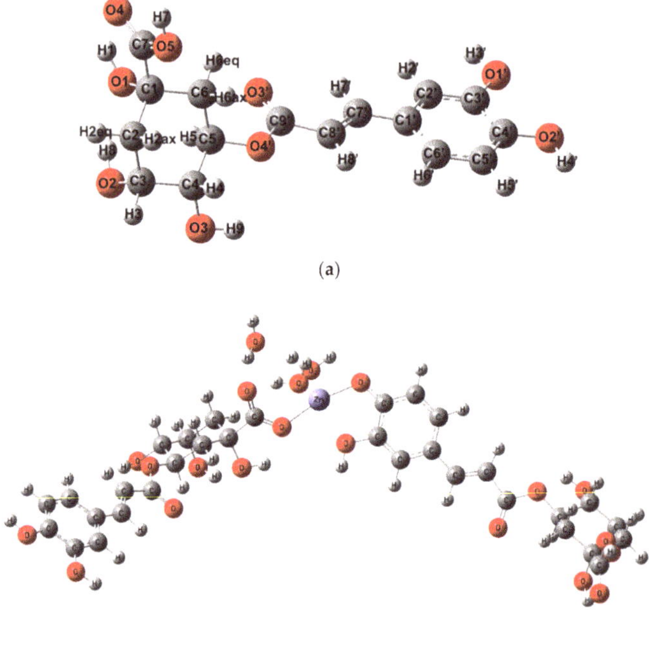

(a)

(b)

Figure 6. *Cont.*

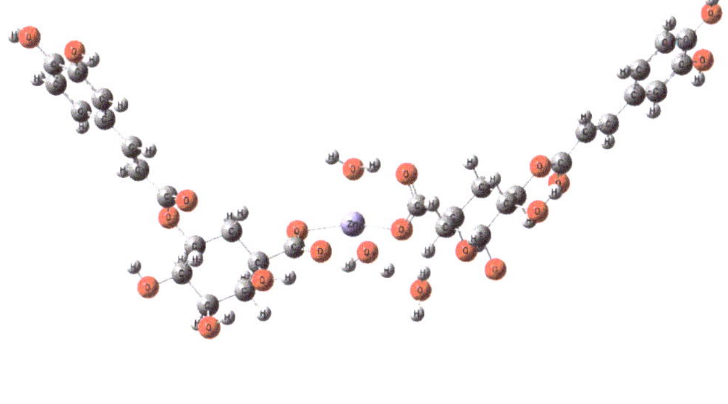

(**c**)

Figure 6. (**a**) The structure of 5-CQA with atom numbering, (**b**) Zn 5-CQA structure I and (**c**) structure II; calculated at B3LYP/6-31G(d) level. The structure I of Zn 5-CQA (**b**) can be transformed into two different phenoxyl radicals with the participation of: –OH of the catechol group engaged in the Zn^{2+} coordination; –OH from the free catechol group.

The NBO (natural bond orbital) atomic charges calculated for studied molecules were shown in Table S2. The coordination through the carboxylate group slightly increased the positive charge on the C1 and C7 atoms of the quinic acid moiety and increased the negative charge on the O4 and O5 atoms of the carboxylate group. While the coordination via the –OH group of the catechol moiety influenced on the atomic charges of all carbon atoms of the caffeic acid moiety and O3′ and O4′ of the hydroxyl substituents in the ring.

In Table 5 the electronic parameters calculated on the basis of the theoretical structures of 5-CQA and Zn(II) 5-CQA in the B3LYP/6-31G(d) level were gathered. The antioxidant activity of phenolic compounds is related to the energy of HOMO orbitals which characterizes electron-donating ability of molecule and, therefore, its free radical scavenging efficiency. With the increasing value of HOMO and lowering value of ionic potential (IP), the molecule possesses a rising tendency to donate electrons [52]. The higher HOMO energy and lower ionisation potential of Zn 5-CQA compared to 5-CQA molecule, suggested that the complex had stronger donating electron ability than the ligand. Moreover, Zn 5-CQA possessed a lower energy gap between the LUMO and HOMO orbitals than 5-CQA. This indicated that the complex was more reactive antioxidant than ligand. The type of metal coordination by chlorogenate affected the antioxidant properties of the complex. The energy of HOMO orbital and the energy gap were lower for the Zn 5-CQA where the carboxylate group and catechol moiety both coordinated Zn^{2+}, and therefore the structure I of Zn 5-CQA was stronger antioxidant than the structure II (Figure 7). In the case of structure I, the two possibilities can be discussed (Figure 6b) where: (1) * –OH of the catechol group engaged in the Zn^{2+} coordination, or (2) ** –OH from the free catechol group, may be deprotonated in the direct reaction with free radicals. The major antioxidant mechanism is HAT (hydrogen atom transfer) where as a result of a hemolytic O–H bond dissociation the hydrogen atom is transferred from the antioxidant to a free radical. BDE (bond dissociation enthalpy) is a parameter used to estimate the reactivity of the molecule in the HAT mechanism [53]. The lower value of BDE parameter the easier the O–H can be broken. The calculated BDE parameters (Table 6) showed that the abstraction of the hydrogen atom from the –OH group of catechol moiety engaged in the metal ion coordination was much easier than from any other H-atom. It may explain the higher antioxidant activity of Zn 5-CQA revealed in the experiment compared with the ligand alone.

Table 5. Calculated in B3LYP/6-31G(d) level the electronic parameters for 5-CQA and two structures of Zn(II) 5-CQA presented in Figure 6.

Electronic Parameters	5-CQA	Zn(II) 5-CQA (Structure I)	Zn(II) 5-CQA (Structure II)
HOMO (a.u.)	−0.3012	−0.2752	−0.2757
LUMO (a.u.)	−0.2037	−0.2047	−0.2039
HOMO (eV)	−8.1961	−7.4886	−7.5022
LUMO (eV)	−5.5430	−5.5710	−5.5481
GAP (eV)	2.6531	1.9176	1.9541
Ionisation potential (eV)	8.1961	7.4886	7.5022
Affinity (eV)	5.5430	5.5710	5.5481
Global hardness (eV)	1.3266	0.9588	0.9770
Chemical softness (eV)	0.6633	0.4794	0.4885
Electronegativity (eV)	6.8695	6.5298	6.5252
Chemical potential (eV)	−6.8695	−6.5298	−6.5252
Elecrophilicity index (eV)	17.7868	22.2353	21.7894
Total energy (a.u.)	−1297.531	−4602.4041	−4602.4101
Dipole moment (Debay)	5.9017	14.3747	8.8169

Table 6. The bond dissociation enthalpies (BDE) (kJ/mol), ionisation potentials (IP) (kJ/mol), proton dissociation enthalpies (PDE) (kJ/mol), proton affinities (PA) (kJ/mol), electron transfer enthalpies (ETE) (kJ/mol) calculated at in B3LYP/6-31G(d) level; Two means of –OH group deprotonation in the direct reaction with free radical are taken into account: * –OH of the catechol group engaged in the Zn^{2+} coordination, or ** –OH from the free catechol group.

Parameter *	5-CQA	Zn(II) 5-CQA (Structure I *)	Zn(II) 5-CQA (Structure I **)	Zn(II) 5-CQA (Structure II)
BDE	307.70	224.49	319.72	308.58
IP	707.46	617.62	617.62	651.71
PDE	909.93	916.58	1011.80	972.76
PA	1433.96	1266.58	1385.32	1375.23
ETE	189.63	273.81	250.30	249.24

The free radical scavenging properties can be also achieved by a donation of a single electron from the antioxidant to a radical. This mechanism is called a single-electron transfer followed by proton transfer (SET-PT) [44]. It is a two-step mechanism: (I) first the antioxidant reacts with the free radical to form a cation radical of antioxidant and an anion of the radical (ionization potential IP describes the antioxidant reactivity in this mechanism), then (II) the cation radical of the antioxidant decomposes to a radical and a proton (proton dissociation enthalpy PDE parameter describes the reaction). Structure I of the Zn 5-CQA donated the electron most easily (IP = 617.62 kJ/mol) followed by structure II of Zn 5-CQA (IP = 651.71 kJ/mol) and then chlorogenic acid (IP = 707.46 kJ/mol). The PDE parameter which describes the second step of the SET-PT mechanism was much lower for the structure I* of the Zn 5-CQA (the cation radical was formed by the –OH from the catechol moiety that coordinates Zn^{2+}) than the other structures of Zn 5-CQA. The lowest value of PDE was for chlorogenic acid what suggested that the 5-CQA contributes to the second step of SET-PT greater than the complexes of 5-CQA.

The third mechanism, a sequential proton-loss electron transfer (SPLET) is realized by a two-step mechanism [44]. First, the antioxidant dissociates into an anion and a proton (proton affinity PA is related with the mechanism). Second, the anion loses an electron and forms the corresponding free radical (electron transfer enthalpy ETE describes the reaction). The first step, i.e., the formation of the phenolate anion, was easier for Zn complexes of chlorogenic acid than for the ligand alone. The lowest PA value was for Zn 5-CQA structure I* which again proved that the antioxidant mechanism with the participation of –OH from the catechol moiety bound to Zn^{2+} is much more favorable than with the unbounded –OH. The second step of the SPLET mechanism is much easier for chlorogenic acid than the Zn(II) complexes.

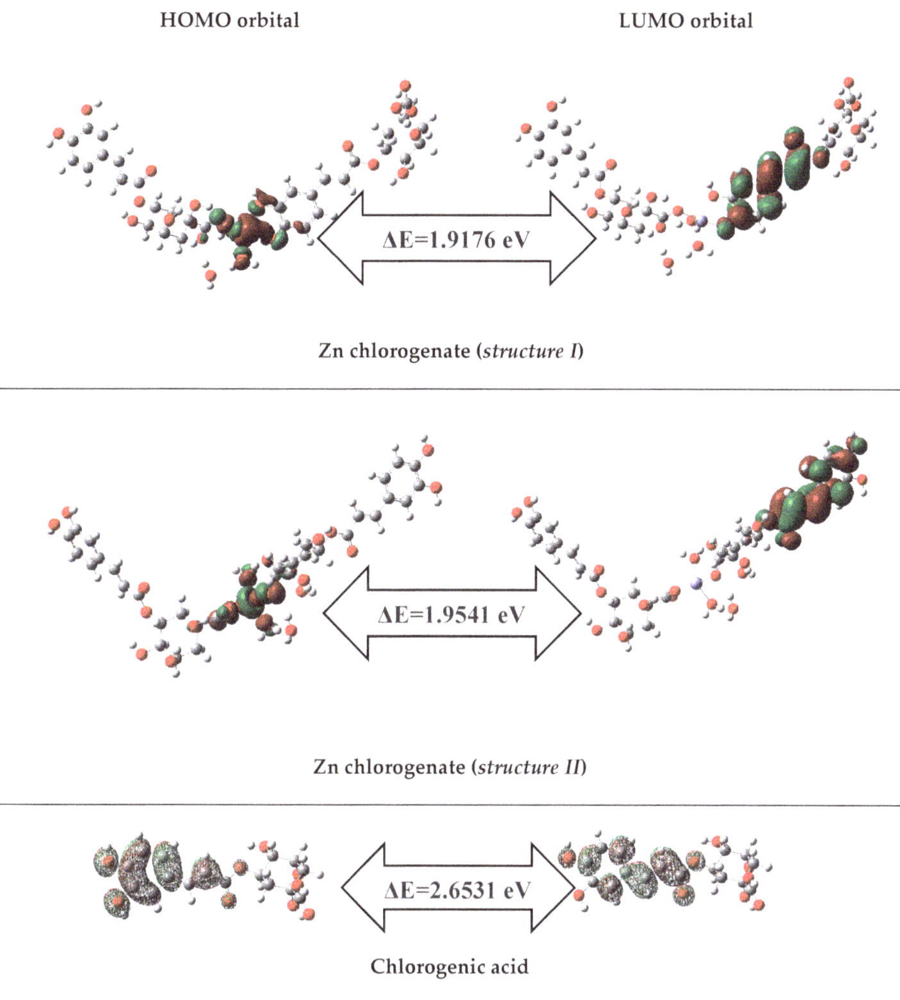

Figure 7. The shapes of frontier molecular orbitals HOMO and LUMO orbitals and energy gaps (ΔE) of chlorogenic acid and Zn chlorogenates calculated at B3LYP/6-31G(d) level.

4. Conclusions

Zn(II) cations form stable solution and solid state complexes with chlorogenic acid what cause the change in the pro-/anti-oxidant properties of the ligand. Under exposure to toxic metals, the chelation of Zn(II) by phenolic compounds may effectively increase protection against the oxidative stress in plants, but a higher concentration of formed metal complex may reveal the pro-oxidant activity of phenolic compounds. These may lead to significant damage at the cellular level. Knowledge of the structure of metal-phenolic complexes is important to understand the mechanism of action of phenolic compounds. Moreover, studies on the metal complexes of plant phenolic compounds may contribute to the development of new non-toxic antioxidant or pro-oxidant materials of natural origin which can be used in e.g., plant protection against oxidative stress, food industry or pharmacy.

Supplementary Materials: The following are available online at http://www.mdpi.com/1996-1944/13/17/3745/s1: Table S1: The geometrical parameters for 5-CQA and Zn 5-CQA complexes calculated at B3LYP/6-31G(d) level; atom numbering in Figure 6, Table S2: The NBO atomic charges calculated for 5-CQA and Zn(II) 5-CQA complexes in B3LYP/6-31(d) level; atom numbering in Figure 6.

Author Contributions: M.K. was the principal investigator who designed the research, performed the synthesis, spectroscopic, antioxidant and pro-oxidant studies and quantum–chemical calculations, discussed and analyzed the data. J.S.-G. performed and analyzed the thermal data, G.Ś. the quantum-chemical calculations, A.P. and A.C. realized and analyzed microbiological studies, W.L. provided editorial comments, discussed and analyzed the data. The first draft of the manuscript was written by M.K. and J.S.-G. and all authors commented on previous versions of the manuscript. All authors read and approved the final manuscript.

Funding: This work was supported by the National Science Centre (Poland) under Grant number DEC-2014/13/B/NZ7/02352 and Institute of Civil Engineering and Energetics (Bialystok University of Technology).

Conflicts of Interest: The authors declare that they have no conflict of interest.

References

1.   Zhang, L.; Liu, R.; Gung, B.W.; Tindall, S.; Gonzalez, J.M.; Halvorson, J.J.; Hagerman, A.E. Polyphenol–Aluminum Complex Formation: Implications for Aluminum Tolerance in Plants. *J. Agric. Food Chem.* **2016**, *64*, 3025–3033. [CrossRef] [PubMed]

2.   Cesco, S.; Mimmo, T.; Tonon, G.; Tomasi, N.; Pinton, R.; Terzano, R.; Neumann, G.; Weisskopf, L.; Renella, G.; Landi, L.; et al. Plant-borne flavonoids released into the rhizosphere: Impact on soil bio-activities related to plant nutrition. A review. *Boil. Fertil. Soils* **2012**, *48*, 123–149. [CrossRef]

3.   Chobot, V.; Hadaček, F. Iron and its complexation by phenolic cellular metabolites: From oxidative stress to chemical weapons. *Plant Signal. Behav.* **2010**, *5*, 4–8. [CrossRef] [PubMed]

4.   Chen, S.; Wang, Q.; Lu, H.; Li, J.; Yang, D.; Liu, J.; Yan, C. Phenolic metabolism and related heavy metal tolerance mechanism in *Kandelia Obovata* under Cd and Zn stress. *Ecotoxicol. Environ. Saf.* **2019**, *169*, 134–143. [CrossRef] [PubMed]

5.   Ofei-Manu, P.; Wagatsuma, T.; Ishikawa, S.; Tawaraya, K. The plasma membrane strength of the root-tip cells and root phenolic compounds are correiated with Al tolerance in several common woody plants. *Soil Sci. Plant Nutr.* **2001**, *47*, 359–375. [CrossRef]

6.   Mongkhonsin, B.; Nakbanpote, W.; Hokura, A.; Nuengchamnong, N.; Maneechai, S. Phenolic compounds responding to zinc and/or cadmium treatments in Gynura pseudochina (L.) DC. extracts and biomass. *Plant Physiol. Biochem.* **2016**, *109*, 549–560. [CrossRef]

7.   Ali, M.B.; Singh, N.; Shohael, A.M.; Hahn, E.J.; Paek, K.-Y. Phenolics metabolism and lignin synthesis in root suspension cultures of Panax ginseng in response to copper stress. *Plant Sci.* **2006**, *171*, 147–154. [CrossRef]

8.   Kalinowska, M.; Bielawska, A.; Lewandowska, H.; Priebe, W.; Lewandowski, W. Apples: Content of phenolic compounds vs. variety, part of apple and cultivation model, extraction of phenolic compounds, biological properties. *Plant Physiol. Biochem.* **2014**, *84*, 169–188. [CrossRef]

9.   Gan, R.-Y.; Zhang, D.; Wang, M.; Corke, H. Health Benefits of Bioactive Compounds from the Genus Ilex, a Source of Traditional Caffeinated Beverages. *Nutrients* **2018**, *10*, 1682. [CrossRef]

10.   Kalinowska, M.; Bajko, E.; Matejczyk, M.; Kaczyński, P.; Łozowicka, B.; Lewandowski, W. The Study of Anti-/Pro-Oxidant, Lipophilic, Microbial and Spectroscopic Properties of New Alkali Metal Salts of 5-O-Caffeoylquinic Acid. *Int. J. Mol. Sci.* **2018**, *19*, 463. [CrossRef]

11.   Tyśkiewicz, K.; Konkol, M.; Kowalski, R.; Rój, E.; Warmiński, K.; Krzyżaniak, M.; Gil, Ł.; Stolarski, M.J. Characterization of bioactive compounds in the biomass of black locust, poplar and willow. *Trees* **2019**, *33*, 1235–1263. [CrossRef]

12.   Bajko, E.; Kalinowska, M.; Borowski, P.; Siergiejczyk, L.; Lewandowski, W. 5-O-Caffeoylquinic acid: A spectroscopic study and biological screening for antimicrobial activity. *LWT* **2016**, *65*, 471–479. [CrossRef]

13.   Liang, N.; Kitts, D.D. Role of Chlorogenic Acids in Controlling Oxidative and Inflammatory Stress Conditions. *Nutrients* **2015**, *8*, 16. [CrossRef] [PubMed]

14.   Silvia, N.V.E.; Mazzafera, P.; Cesarino, I. Should I stay or should I go: Are chlorogenic acids mobilized towards lignin biosynthesis? *Phytochem.* **2019**, *166*, 112063. [CrossRef] [PubMed]

15.   Yan, K.; Cui, M.; Zhao, S.; Chen, X.; Tang, X. Salinity Stress Is Beneficial to the Accumulation of Chlorogenic Acids in Honeysuckle (*Lonicera japonica* Thunb.). *Front. Plant Sci.* **2016**, *7*, 1–16. [CrossRef]

16. Baker, C.J.; Mock, N.M.; Aver'Yanov, A.A.; Aver'Yano, A.A. A simplified technique to detect variations of leaf chlorogenic acid levels between and within plants caused by maturation or biological stress. *Physiol. Mol. Plant Pathol.* **2017**, *98*, 97–103. [CrossRef]

17. Santin, M.; Neugart, S.; Castagna, A.; Barilari, M.; Sarrocco, S.; Vannacci, G.; Schreiner, M.; Ranieri, A. UV-B pre-treatment alters phenolics response to *Monilinia fructicola* infection in a structure-dependent way in peach skin. *Front. Plant Sci.* **2018**, *9*, 1–24.

18. Kısa, D.; Elmastas, M.; Öztürk, L.; Kayır, Ö. Responses of the phenolic compounds of *Zea mays* under heavy metal stress. *Appl. Biol. Chem.* **2016**, *59*, 813–820. [CrossRef]

19. Michalak, A. Phenolic compounds and their antioxidant activity in plants growing under heavy metal stress. *Pol. J. Environ. Stud.* **2006**, *15*, 523–530.

20. Naso, L.G.; Valcárcel, M.; Roura-Ferrer, M.; Kortazar, D.; Salado, C.; Lezama, L.; Lu, J.; González-Baró, A.C.; Williams, P.A.; Ferrer, E.G. Promising antioxidant and anticancer (human breast cancer) oxidovanadium(IV) complex of chlorogenic acid. Synthesis, characterization and spectroscopic examination on the transport mechanism with bovine serum albumin. *J. Inorg. Biochem.* **2014**, *135*, 86–99.

21. Samsonowicz, M.; Regulska, E.; Kalinowska, M. Hydroxyflavone metal complexes—Molecular structure, antioxidant activity and biological effects. *Chem. Interact.* **2017**, *273*, 245–256. [CrossRef] [PubMed]

22. Kalinowska, M.; Mazur, L.; Jabłońska-Trypuć, A.; Lewandowski, W. A new calcium 2,5-dihydroxybenzoate: Synthesis, characterization and antioxidant studies and stress mediated cytotoxity in MCF-7 cells. *J. Saudi Chem. Soc.* **2018**, *22*, 742–756. [CrossRef]

23. Yamasaki, H.; Grace, S.C. EPR detection of phenoxyl radicals stabilized by zinc ions: Evidence for the redox coupling of plant phenolics with ascorbate in the H_2O_2-peroxide system. *FEBS Lett.* **1998**, *422*, 377–380. [CrossRef]

24. Rout, G.R.; Das, P. Effect of Metal Toxicity on Plant Growth and Metabolism: I. Zinc. *Agronomy* **2003**, *23*, 3–11. [CrossRef]

25. Wianowska, D.; Gil, M. Recent advances in extraction and analysis procedures of natural chlorogenic acids. *Phytochem. Rev.* **2018**, *18*, 273–302. [CrossRef]

26. Naveed, M.; Hejazi, V.; Abbas, M.; Kamboh, A.; Khan, G.J.; Shumzaid, M.; Ahmad, F.; Babazadeh, D.; FangFang, X.; Modarresi-Ghazani, F.; et al. Chlorogenic acid (CGA): A pharmacological review and call for further research. *Biomed. Pharmacother.* **2017**, *97*, 67–74. [CrossRef]

27. Milić, S.Z.; Potkonjak, N.; Gorjanović, S.Ž.; Veljovic-Jovanovic, S.; Pastor, F.T.; Sužnjević, D.Ž. A Polarographic Study of Chlorogenic Acid and Its Interaction with Some Heavy Metal Ions. *Electroanalysis* **2011**, *23*, 2935–2940. [CrossRef]

28. Mahal, H.S.; Kapoor, S.; Satpati, A.K.; Mukherjee, T. Radical Scavenging and Catalytic Activity of Metal–Phenolic Complexes. *J. Phys. Chem. B* **2005**, *109*, 24197–24202. [CrossRef]

29. Ameziane, J.; Aplincourt, M.; Dupont, L.; Heirman, F.; Pierrard, J.-C. Thermodynamic stability of copper (II), manganese (II), zinc 9II) and Iron(III) complexes with chlorogenic acid. *Bull. Soc. Fr.* **1996**, *133*, 243–249.

30. Kiss, T.; Nagy, G.; Pesci, M.; Kozłowski, H.; Micera, G.; Erre, L.S. Cupric ion binding by dihydroxybenzoic acids. *Polyhedron* **1989**, *8*, 2345–2349. [CrossRef]

31. Rice-Evans, C.A.; Diplock, A.T. *Symons MCR. Techniques in Free Radical Research*; Elsevier: Amsterdam, The Netherlands, 1991.

32. Zeraik, M.L.; Petrônio, M.S.; Coelho, D.; Regasini, L.O.; Silva, D.H.S.; Da Fonseca, L.M.; Machado, S.A.S.; Bolzani, V.D.S.; Ximenes, V.F. Improvement of Pro-Oxidant Capacity of Protocatechuic Acid by Esterification. *PLoS ONE* **2014**, *9*, e110277. [CrossRef] [PubMed]

33. Ogliaro, F.; Bearpark, M.J.; Heyd, J.J.; Brothers, E.N.; Kudin, K.N.; Staroverov, V.N.; Keith, T.A.; Kobayashi, R.; Normand, J.; Raghavachari, K.; et al. *Gaussian 16, Revision C.01*; Gaussian, Inc.: Wallingford, CT, USA, 2016.

34. Zhan, C.-G.; Nichols, J.A.; Dixon, D.A. Ionization Potential, Electron Affinity, Electronegativity, Hardness, and Electron Excitation Energy: Molecular Properties from Density Functional Theory Orbital Energies. *J. Phys. Chem. A* **2003**, *107*, 4184–4195. [CrossRef]

35. Bizarro, M.M.; Cabral, B.J.C.; De Santos, R.M.B.; Simões, J.A.M. Substituent effects on the O-H bond dissociation enthalpies in phenolic compounds: Agreements and controversies. *Pure Appl. Chem.* **1999**, *71*, 1609–1610. [CrossRef]

36. Wright, J.S.; Johnson, E.R.; DiLabio, G.A. Predicting the activity of phenolic antioxidants: Theoretical method, analysis of substituent effects, and application to major families of antioxidants. *J. Am. Chem. Soc.* **2001**, *123*, 1173–1183. [CrossRef] [PubMed]

37. Litwinienko, G.; Ingold, K.U. Abnormal Solvent Effects on Hydrogen Atom Abstractions. 1. The Reactions of Phenols with 2,2-Diphenyl-1-picrylhydrazyl (DPPH) in Alcohols. *J. Org. Chem.* **2003**, *68*, 3433–3438. [CrossRef]

38. Litwinienko, G.; Ingold, K.U. Abnormal Solvent Effects on Hydrogen Atom Abstraction. 2. Resolution of the Curcumin Antioxidant Controversy. The Role of Sequential Proton Loss Electron Transfer. *J. Org. Chem.* **2004**, *69*, 5888–5896. [CrossRef]

39. Foti, M.C.; Daquino, C.; Geraci, C. Electron-Transfer Reaction of Cinnamic Acids and Their Methyl Esters with the DPPH Radical in Alcoholic Solutions. *J. Org. Chem.* **2004**, *69*, 2309–2314. [CrossRef]

40. Foti, M.C. Use and Abuse of the DPPH Radical. *J. Agric. Food Chem.* **2015**, *63*, 8765–8776. [CrossRef]

41. Kohanski, M.A.; Dwyer, D.J.; Hayete, B.; Lawrence, C.A.; Collins, J.J. A Common Mechanism of Cellular Death Induced by Bactericidal Antibiotics. *Cell* **2007**, *130*, 797–810. [CrossRef]

42. Dwyer, D.J.; Kohanski, M.A.; Collins, J.J. Role of reactive oxygen species in antibiotic action and resistance. *Curr. Opin. Microbiol.* **2009**, *12*, 482–489. [CrossRef]

43. Foti, J.J.; Devadoss, B.; Winkler, J.A.; Collins, J.J.; Walker, G.C. Oxidation of the Guanine Nucleotide Pool Underlies Cell Death by Bactericidal Antibiotics. *Science* **2012**, *336*, 315–319. [CrossRef] [PubMed]

44. Grant, S.S.; Kaufmann, B.B.; Chand, N.S.; Haseley, N.; Hung, D.T. Eradication of bacterial persisters with antibiotic-generated hydroxyl radicals. *Proc. Natl. Acad. Sci. USA* **2012**, *109*, 12147–12152. [CrossRef] [PubMed]

45. Karunanidhi, A.; Thomas, R.; Van Belkum, A.; Neela, V.K. In Vitro Antibacterial and Antibiofilm Activities of Chlorogenic Acid against Clinical Isolates of Stenotrophomonas maltophilia including the Trimethoprim/Sulfamethoxazole Resistant Strain. *BioMed Res. Int.* **2012**, *2013*, 1–7. [CrossRef] [PubMed]

46. Kot, B.; Kwiatek, K.; Janiuk, J.; Witeska, M.; Pękala-Safińska, A. Antibacterial Activity of Commercial Phytochemicals against Aeromonas Species Isolated from Fish. *Pathogens* **2019**, *8*, 142. [CrossRef] [PubMed]

47. Cai, R.; Miao, M.; Yue, T.; Zhang, Y.; Cui, L.; Wang, Z.; Yuan, Y. Antibacterial activity and mechanism of cinnamic acid and chlorogenic acid against *Alicyclobacillus acidoterrestris* vegetative cells in apple juice. *Int. J. Food Sci. Technol.* **2018**, *54*, 1697–1705. [CrossRef]

48. Lou, Z.; Wang, H.; Zhu, S.; Ma, C.; Wang, Z. Antibacterial Activity and Mechanism of Action of Chlorogenic Acid. *J. Food Sci.* **2011**, *76*, M398–M403. [CrossRef]

49. Wang, L.; Bi, C.; Cai, H.; Liu, B.; Zhong, X.; Deng, X.; Wang, T.; Xiang, H.; Niu, X.; Wang, D. The therapeutic effect of chlorogenic acid against Staphylococcus aureus infection through sortase A inhibition. *Front. Microbiol.* **2015**, *6*, 1–12. [CrossRef]

50. Kalinowska, M.; Mazur, L.; Piekut, J.; Rzaczynska, Z.; Laderiere, B.; Lewandowski, W. Synthesis, crystal structure, spectroscopic properties, and antimicrobial studies of a zinc(II) complex of *p*-coumaric acid. *J. Coord. Chem.* **2013**, *66*, 334–344. [CrossRef]

51. Kalinowska, M.; Piekut, J.; Bruss, A.; Follet, C.; Sienkiewicz-Gromiuk, J.; Świsłocka, R.; Rzączyńska, Z.; Lewandowski, W. Spectroscopic (FT-IR, FT-Raman, ^{1}H, ^{13}C NMR, UV/VIS), thermogravimetric and microbiological studies of Ca(II), Mn(II), Cu(II), Zn(II) and Cd(II) complexes of ferulic acid. *Spectrochim. Acta A Mol. Biomol. Spectrosc.* **2014**, *122*, 631–638. [CrossRef]

52. Ur-Rahman, A. *Frontiers in Clinical Drug Research*; Bentham Science Publishers: Sharjah, UAE, 2016; Volume 2.

53. Litwinienko, G.; Ingold, K.U. Solvent effects on the rates and mechanisms of reaction of phenols with free radicals. *Acc. Chem. Res.* **2007**, *40*, 222–230. [CrossRef]

 *materials*

Article

Comparison of Inflammatory Effects in THP-1 Monocytes and Macrophages after Exposure to Metal Ions

Henrike Loeffler [1], Anika Jonitz-Heincke [1], Kirsten Peters [2], Brigitte Mueller-Hilke [3], Tomas Fiedler [4], Rainer Bader [1] and Annett Klinder [1,*]

[1] Biomechanics and Implant Technology Research Laboratory, Department of Orthopedics, Rostock University Medical Centre, Doberaner Strasse 142, 18057 Rostock, Germany; henrike.loeffler@uni-rostock.de (H.L.); anika.jonitz-heincke@med.uni-rostock.de (A.J.-H.); rainer.bader@med.uni-rostock.de (R.B.)

[2] Department of Cell Biology, Rostock University Medical Center, Schillingallee 69, 18057 Rostock, Germany; kirsten.peters@med.uni-rostock.de

[3] Institute for Immunology, Rostock University Medical Center, Schillingallee 70, 18057 Rostock, Germany; brigitte.mueller-hilke@med.uni-rostock.de

[4] Institute for Medical Microbiology, Virology and Hygiene, Rostock University Medical Center, Schillingallee 70, 18057 Rostock, Germany; tomas.fiedler@med.uni-rostock.de

* Correspondence: annett.klinder@med.uni-rostock.de; Tel.: +49-381-494-9386

Received: 20 December 2019; Accepted: 3 March 2020; Published: 5 March 2020

Abstract: Monocytes and macrophages are the first barrier of the innate immune system, which interact with abrasion and corrosion products, leading to the release of proinflammatory mediators and free reactive molecules. The aim of this study was to understand inflammation-relevant changes in monocytes and macrophages after exposure to corrosion products. To do this, the THP-1 cell line was used to analyze the effects of metal ions simultaneously in monocytes and differentiated macrophages. Cells were stimulated with several concentrations of metal salts ($CoCl_2$, $NiCl_2$, $CrCl_3 \times 6H_2O$) to analyze viability, gene expression, protein release and ROS production. Untreated cells served as negative controls. While exposure to Cr(3+) did not influence cell viability in both cell types, the highest concentration (500 µM) of Co(2+) and Ni(2+) showed cytotoxic effects mirrored by significantly reduced metabolism, cell number and a concomitant increase of ROS. The release of IL-1β, IL-8, MCP-1 and M-CSF proteins was mainly affected in macrophages after metal ion exposure (100 µM), indicating a higher impact on pro-inflammatory activity. Our results prove that monocytes and macrophages react very sensitively to corrosion products. High concentrations of bivalent ions lead to cell death, while lower concentrations trigger the release of inflammatory mediators, mainly in macrophages.

Keywords: aseptic loosening; corrosion; metal ions; monocytes; macrophages; inflammation

1. Introduction

Degenerative joint diseases contribute to the decrease in quality of life during aging. Due to increased aging in the general population, therapeutic measures, including total joint replacement, progressively gain importance in tackling these diseases [1]. One major problem associated with total joint replacement is the necessity of revisions caused by the septic and aseptic loosening of the implant. According to Herberts et al., aseptic loosening accounts for more than 70% of knee prosthesis failures and 44% of hip prosthesis failures [2]. In order to improve the long-term outcome after total joint replacement, orthopedic research targets elevating the implant success rates and understanding the reasons for revision.

One approach is to improve implant materials by understanding the mechanisms at the implant surface responsible for inflammation and loosening. Commonly used materials in prosthesis manufacturing are metal alloys, because of their high mechanic stability and good biological compatibility. Main materials are stainless steel (consisting of iron, chromium, nickel), cobalt-chromium–molybdenum and titanium [3]. Due to abrasion and corrosion processes, wear particles and metal ions occur in periprosthetic tissue [4].

These wear products interact with the defense barrier of innate immunity that is mainly driven by monocytes and macrophages. Phagocytosis of wear particles by macrophages is considered the beginning of these reactions, finally resulting in endoprosthesis failure [4,5]. Secretion of proinflammatory cytokines, osteomodulating mediators as well as reactive oxygen and nitrogen species initiates and maintains the inflammation processes around the implant [4].

Initial immune response to particle exposition is mainly accompanied by the production of proinflammatory TNF-α, chemokine IL-8, and cytokines IL-1β and IL-6 [6]. The release of monocyte chemotactic protein 1 (MCP-1) triggers the recruitment of more and more monocytes, thus further promoting inflammation processes [4,7,8]. Besides the invading monocytes, the release of various differentiation factors like receptor activator of NF-κB ligand (RANKL) and monocyte colony-stimulating factor (M-CSF) also triggers the maturation of local macrophages and their differentiation into osteoclasts [9]. Thus, in vivo, there is always a combination of mature macrophages and freshly invading monocytes that drive the inflammation. Often, only one cell type was investigated, or reactivity was attributed equally to both of them. However, considering that after phagocytosis of metallic wear particles by mature macrophages, these macrophages release metal ions from their lysosomes into their microenvironment [10], they might be better equipped to withstand the effects of those ions. Therefore, we hypothesized that macrophages might react differently to metal ions compared to monocytes.

Therefore, our study aimed to understand inflammation-relevant changes in monocytes and macrophages after exposure to corrosion products. The use of the THP-1 cell line, which can be cultured as monocytes but also differentiated into macrophages, allowed us to analyze the effects of metal ions in parallel with monocytes and macrophages from the same origin. In the in vitro investigation, viability assays as well as gene and protein expression analyses and the quantification of reactive oxygen species (ROS) after metal salt exposure were carried out.

2. Materials and Methods

2.1. Preparation of Metal Salt Solutions

The following metal salts were purchased from Sigma-Aldrich (Sigma-Aldrich Chemie GmbH, Munich, Germany): Cobalt(II) chloride (purum p.a., anhydrous, purity $\geq$ 98.0% (KT)), Nickel(II) chloride (anhydrous, powder, purity 99.99% trace metals basis) and Chromium(III) chloride hexahydrate (purum p.a., purity $\geq$ 98.0% (RT)) and stock solutions of a concentration of 100 mM were produced as described previously [11]. For cell culture experiments, the stock solutions were diluted with cell culture media to various concentrations.

2.2. Cell Culture

THP-1 monocytes were cultivated in Roswell Park Memorial Institute (RPMI) 1640 medium supplemented with 20% fetal calf serum (FCS; Pan Biotech GmbH, Aidenbach, Germany), 2% L-Alanyl-L-Glutamin (Biochrom GmbH, Berlin, Germany), 1% amphotericin b and 1% penicillin/streptomycin (both: Sigma-Aldrich, Munich, Germany) at 37 °C and 5% CO_2.

For each experiment, cells were seeded into two cell culture plates. While cells of one plate remained in suspension, cells of another plate were differentiated for 24 h using 100 ng/mL Phorbol-12-myristat-13-acetat (PMA; Sigma-Aldrich, Munich, Germany), so monocytes and macrophages could be examined simultaneously. Cells were stimulated with several concentrations of metal salts and viability, and gene expression and protein biosynthesis analyses as well as ROS assay were carried out. Untreated cells served as negative controls.

The effects of metal salts were tested after 48 h incubation, since longer incubation periods were not feasible. As found by us (data not shown) and also reported by Lund et al. [12], THP-1 derived macrophages de-differentiate, detach from the surface of the cell culture dish and show a round monocyte-like morphology after more than 48 h without PMA. However, the simultaneous presence of PMA in THP-1 macrophage culture was shown to interfere with the effects of nickel ions [13]. We therefore decided to differentiate for 24 h with PMA, to then remove PMA from the cell culture and to limit the incubation time with metal ions to 48 h.

2.3. Cellular Activity

The viability of THP-1 monocytes and macrophages after exposure to metal ions was determined by the metabolic activity assay WST-1 (Roche, Penzberg, Germany) and CyQUANT NF Cell Proliferation Assay (Invitrogen (Thermo Fisher Scientific), Waltham, MA, USA).

A total of 10,000 cells per well were seeded into black 96-well cell culture plates (Thermo Fisher Scientific Inc., Waltham, MA, USA). Cells were treated with 10, 50, 100 and 500 µM of metal ions for 48 h. For the determination of cell activity, ion solution was removed and cells were incubated with a defined volume of WST-1/medium reagent (ratio 1:10) at 37 °C and 5% CO_2 for 30 min. Afterwards, supernatants of the respective culture medium were transferred into 96-well cell culture plates (ThermoFisher Scientific, Waltham, MA, USA) to measure the absorption at 450 nm (reference wave length: 630 nm) in a microplate reader (Tecan Reader Infinite® 200 Pro, Tecan Trading AG, Maennedorf, Switzerland)

CyQUANT cell proliferation assay was performed to determine the absolute cell number according to the manufacturer's recommendations. Cells were covered with 100 µL 1× Dye Binding Solution (consisting of 1:500 Dye Reagent and 1× HBSS) and incubated for 60 min, protected from light. Fluorescence intensity was measured at 530 nm (excitation wavelength: 485 nm) using the Tecan-Reader Infinite® 200 Pro. In order to relate the fluorescence signal to an actual cell number, a cell number calibration curve was prepared with previously defined cell numbers in duplicate.

Cellular activity was calculated by dividing WST-1 results by the respective cell number.

2.4. Analysis of Gene Expression

The following experimental setup was used to determine gene as well as protein expression and formation of reactive oxygen species: 60,000 cells per well of a 24-well cell culture plate were treated with 100 µM of metal salts for 48 h. Untreated cells served as negative control for metal salt exposure. Supernatants were collected and stored at −20 °C.

RNA was isolated using the peqGOLD Total line RNA Kit and the related manufacturer's protocol (VWR International GmbH, Hanover, Germany). RNA was eluted into a fresh sterile tube using RNase-free water and RNA concentration was measured using the Tecan Reader Infinite® 200 Pro microplate reader and NanoQuantTM Plate (Tecan Trading AG, Maennedorf, Switzerland) with RNase-free water as blank. Afterwards, RNA was transcribed into amplifiable cDNA using the High Capacity cDNA Reverse Transcription Kit (Applied Biosytems, Foster City, CA, USA) according to manufacturer's recommendations. A total of 50 ng RNA was attained by transferring appropriate amounts of RNA-containing sample into PCR tubes and adding RNase free water to reach a volume of 10 µL. A total of 10 µL of master mix was added and PCR was carried out using following RT-PCR protocol: 10 min at 25 °C, 120 min at 37 °C, 15 s at 85 °C in a thermocycler (Analytik Jena, Jena, Germany). Afterwards, samples were diluted in additional 20 µL RNase free water and stored at −20 °C.

Relative quantification of target cDNA levels was done by semi-quantitative realtime PCR (qTower 2.0, Analytik Jena AG, Jena/Germany) using innuMIX qPCR MasterMix SyGreen (Analytik Jena AG, Jena, Germany) and the primers (Sigma-Aldrich, Darmstadt, Germany), as listed in Table 1.

A master mix was prepared for each gene, containing 0.5 µL of forward and reverse primer, 3 µL Aqua dest. and 5 µL of SyGreen qPCR MasterMix. A total of 1 µL of template cDNA of each sample was pipetted onto the bottom of a 96-well PCR plate in duplicates and filled up with 9 µL of master mix. RNase-free water served as a negative control. The plate was sealed with adhesive foil and placed in the qTower 2.0. qPCR was performed under the following conditions: 2 min at 95 °C and 40 cycles of 95 °C (5 s) and 65 °C (25 s). A cycle threshold (Ct) of 30 was set as the limit of interpretation. The relative expression of each mRNA compared with the housekeeping gene HPRT was calculated by the equation $\Delta Ct = Ct_{target} - Ct_{HPRT}$. The relative amount of target mRNA in the unstimulated cells and treated cells was expressed as $2^{(-\Delta\Delta Ct)}$, where $\Delta\Delta Ct = \Delta Ct_{treated} - \Delta Ct_{control}$.

Table 1. Primer sequences for qPCR.

Primer	Sequences (5′-3′)
Hypoxanthine-guanine phosphoribosyl transferase (HPRT)	*for:* CCCTGGCGTCGTGATTAGTG *rev:* TCGAGCAAGACGTTCAGTCC
Interleukin 1β (IL-1β)	*for:* TACTCACTTAAAGCCCGCCT *rev:* ATGTGGGAGCGAATGACAGA
Interleukin 8 (IL-8)	*for:* TCTGTGTGAAGGTGCAGTTTTG *rev:* ATTTCTGTGTTGGCGCAGTG
Monocyte chemotactic protein 1 (MCP-1)	*for:* CCGAGAGGCTGAGACTAACC *rev:* GGCATTGATTGCATCTGGCTG
Macrophage colony-stimulating factor (M-CSF)	*for:* TCCAGCCAAGATGTGGTGAC *rev:* AGTTCCCTCAGAGTCCTCCC

2.5. Quantification of Cytokine Release in Cell Culture Supernatants

The protein contents of interleukin 1β (IL-1β), unterleukin 8 (IL-8), monocyte chemotactic protein 1 (MCP-1) and macrophage colony-stimulating factor (M-CSF) were quantified in cell culture supernatants using corresponding ELISA Ready-SET-Go! Kits (ThermoFisher Scientific, Waltham, MA, USA) according to the manufacturer's recommendations. Absorbance was measured at 450 nm (reference wave length: 570 nm). Sample concentrations were calculated using a standard curve and set in ratio to total protein concentrations, quantified by the Qubit Protein Assay Kit and Qubit 1.0 (both: Invitrogen, Waltham, MA, USA).

2.6. ROS Assay

To detect the presence of total free-reactive oxygen species, the OxiSelect™ In vitro ROS/RNS Assay (Cell Biolabs, Inc., San Diego, CA, USA) was used. Firstly, medium supernatants of exposed monocytes and macrophages were centrifuged. Afterwards, 50 µL of standard or sample were transferred into black 96-well cell culture plates (Thermo Fisher Scientific Inc., Waltham, MA, USA) and incubated with a catalyst that sped up the oxidative reaction. Next dichlorodihydrofluorescin was added to the samples. Finally, samples were fluorometrically measured against the standard curve to determine the content of ROS.

2.7. Statistical Analyses

Statistical and graphic data interpretation was performed using GraphPad Prism 7.02 (GraphPad Software Inc., San Diego/USA).

Cellular viability assay results are shown as box plots. Boxes depict interquartile ranges, horizontal lines within boxes depict medians and whiskers depict maximum and minimum values. Viability assay data were interpreted using repeated measures two-way ANOVA followed by Bonferroni multiple comparison tests.

ROS assay results are shown as mean ± SD, including the single datapoints related to the untreated control values set as 100%. Statistical analyses were performed using the reactive oxygen species amounts divided by control values. Statistical analysis was performed using two-way ANOVA and Bonferroni's multiple comparison test as the post hoc test.

Gene expression results are shown as mean ± SD, including the single datapoints as percentage of $2^{(-\Delta\Delta Ct)}$ with untreated controls set as 100%. Statistical analysis was performed using two-way ANOVA with the ΔCt values and Bonferroni's multiple comparison test as the post hoc test.

Protein expression values are shown as mean ± SD, including the single datapoints normalized to total protein. For these data, there is no depiction related to the untreated controls as, for some of the determined proteins, no protein release was detected in the untreated controls. Statistical analyses were performed using the protein concentration values normalized to total protein. Two-way ANOVA and Bonferroni's multiple comparisons test were performed.

3. Results

3.1. Effects of Metal Ions on Cell Number and Cellular Activity in THP-1 Monocytes and Macrophages

After exposure to the highest concentration of 500 µM, the metabolic cell activity of monocytes was reduced by cobalt and nickel ions as measured by WST-1 conversion assay (Figure 1a) or metabolic activity normalized to the cell number (Figure 1e).

The total cell number was less affected by the treatment with cobalt and nickel ions (Figure 1c). However, it has to be taken into consideration that, due to the nature of the CyQuant assay, which quantifies cell-number-based fluorescence labelling of DNA content, dead or damaged cells were also counted. Exposure to chromium ions at the used concentrations had no effect on cell activity or cell number in monocytes (Figure 1a–c). Similar results were observed in the THP-1 macrophages (Figure 1b,d,f) with Ni(2+) exposure, showing a clear concentration-dependent reduction (Figure 1f). When comparing the effects between monocytes and macrophages, there were no significant differences for the higher concentrations. However, lower concentrations of cobalt ions led to an increase in WST-1 activity and cell number in macrophages (monocytes vs. macrophages: p = 0.0089 and p = 0.0049 for WST-1 activity at 10 and 50 µM cobalt ions, respectively, as well as p < 0.0001 for cell number at 10 µM cobalt ions).

The result that concentrations of 500 µM of cobalt and nickel ions were quite toxic for monocytes as well as macrophages was confirmed by the morphological changes observed by light microscopy. Monocytes as well as macrophages showed a more irregular shape after 500 µM cobalt and nickel ion exposure (Figures 2 and 3).

Macrophage attachment seemed less strong after cobalt and nickel treatment with a higher concentration and, instead of the spread morphology that was observed in untreated controls and for lower concentrations, here, macrophages showed a rounded cell shape similar to that of monocytes (Figure 3e,g). Chromium ion exposure had no visible effect compared to untreated controls.

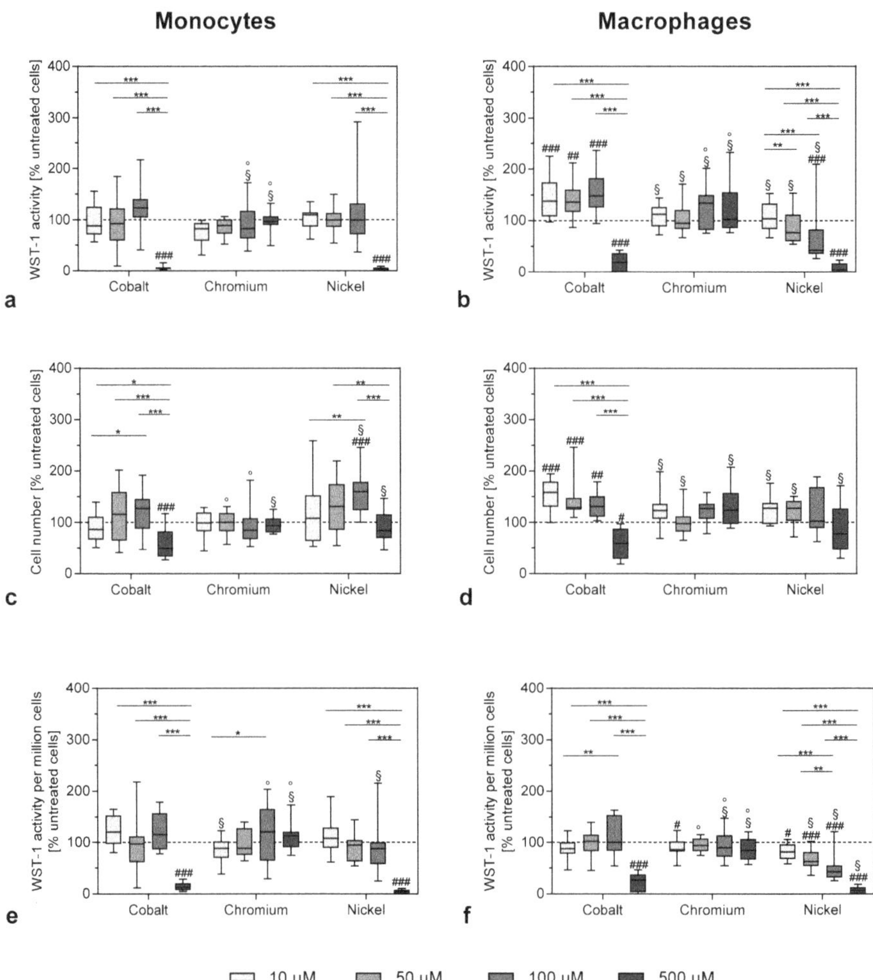

Figure 1. Cell viability of THP-1 monocytes (**a–c**) and macrophages (**d–f**) after exposure to metal salts. WST-1 assay (**a,b**) and CyQUANT NF Cell Proliferation Assay (**c,d**) were carried out after 48 h of stimulation with different concentrations of cobalt, chromium and nickel ions. Figure 1e,f show the WST-1 activity per 1,000,000 cells. All values were normalized to untreated cells and shown as a percentage, with negative control set as 100% (dotted line). Data (n = 12) are shown as box plots with minimum, 25th percentile, median, 75th percentile and maximum values. Significant differences were calculated by repeated measures two-way ANOVA and Bonferroni's multiple comparisons test as post hoc test from the original values (non-normalized). Significantly different between different concentrations: * $p < 0.05$, ** $p < 0.01$, *** $p < 0.001$; significantly different from untreated control: # $p < 0.05$, ## $p < 0.01$, ### $p < 0.001$; significantly different from cobalt at the same concentration: §; significantly different from nickel at the same concentration: °.

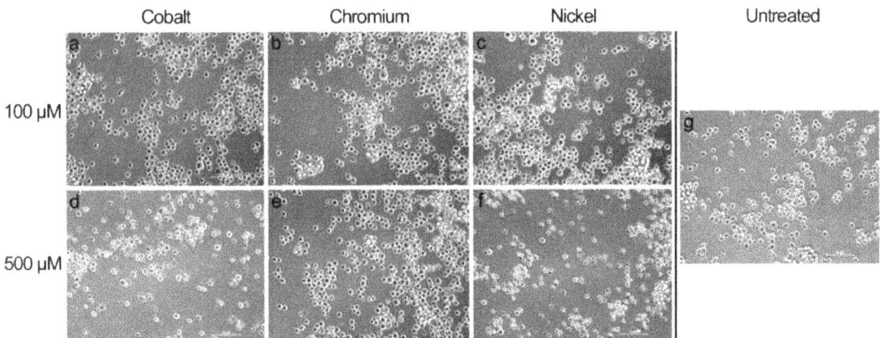

Figure 2. Light microscopic pictures of THP-1 monocytes exposed to metal salts and untreated cells. Monocytes were treated with 100 µM (**a–c**) and 500 µM (**d–f**) of cobalt, chromium and nickel ions. Untreated cells are depicted in Figure 2g. Pictures were taken after 48 h of exposure in 20× magnification. Scale bar = 50 µm.

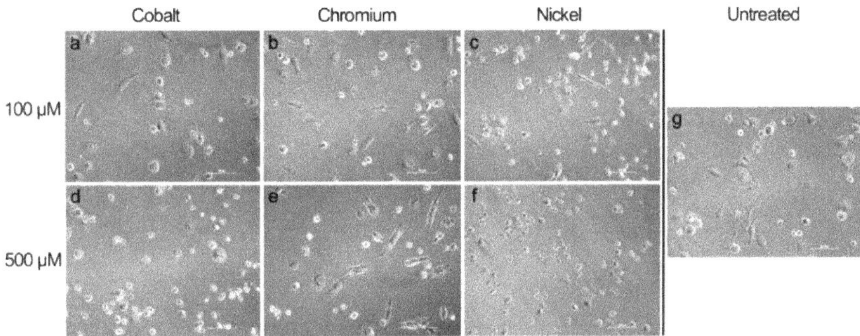

Figure 3. Light microscopic pictures of THP-1 macrophages exposed to metal salts and untreated cells. Macrophages were treated with 100 µM (**a–c**) and 500 µM (**e–g**) of cobalt, chromium and nickel ions). Untreated cells are depicted in Figure 3g. Pictures were taken after 48 h of exposure in 20× magnification. Scale bar = 50 µm.

3.2. Effects of Metal ions on ROS Production in THP-1 Monocytes and Macrophages

The determination of reactive oxygen species in 100 and 500 µM cobalt-, chromium- and nickel ion-treated THP-1 monocytes and macrophages revealed a massive increase in ROS formation in reaction to exposure to the higher cobalt ion concentration for monocytes and macrophages in comparison to control cells (both: $p < 0.0001$) as well as to 500 µM chromium- and nickel-stimulated cells ($p < 0.0001$), respectively (Figure 4). While reaction to nickel ion exposure in THP-1 monocytes only showed an association in the 500 µM stimulated sample ($p = 0.0384$), macrophages showed a stronger reaction to nickel ion treatment. Exposure to 100 and 500 µM nickel increased in ROS release into the supernatant compared to untreated control ($p = 0.0231$ and $p < 0.0001$, respectively) in THP-1 macrophages, and the effect was more pronounced for 500 compared to 100 µM nickel ion exposure ($p = 0.0203$). Incubation with 500 µM nickel ions also led to a higher release than observed in 500 µM chromium-stimulated macrophages ($p = 0.0004$).

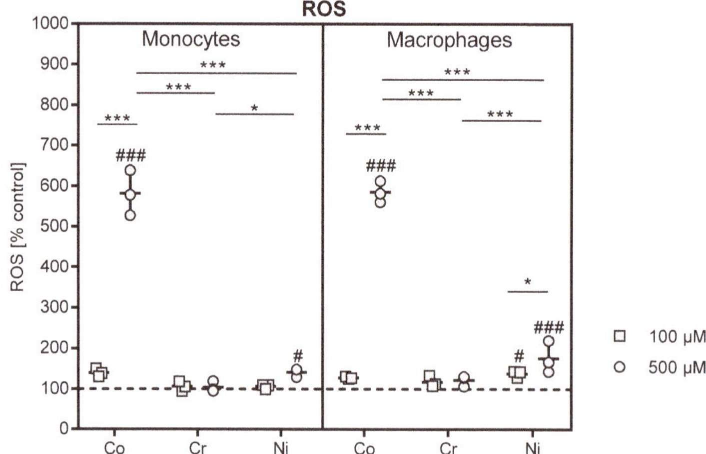

Figure 4. Reactive oxygen species (ROS) quantification in cell culture supernatants. THP-1 monocytes and macrophages were treated with different concentrations of cobalt, chromium and nickel ions over 48 h. Untreated cells served as negative control. ROS formation was determined in cell culture supernatants using OxiSelect™ In vitro ROS/RNS Assay. Data are shown as percentage in relation to control cells (100%, dotted line) as single datapoints with mean ± SD (n = 3). Significances between groups were calculated with Two-way ANOVA and Bonferroni's multiple comparisons test as a post hoc test. Significantly different between concentrations or different metal salt treatments at the same concentration: * p < 0.05, ** p < 0.01, *** p < 0.001; significantly different from untreated control: # p < 0.05, ## p < 0.01, ### p < 0.001.

3.3. Influence of Metal Ions on Gene Expression and Protein Release of Inflammatory Mediators in THP-1 Monocytes and Macrophages

As exposure to cobalt and nickel at the concentration of 500 µM was shown to be toxic for monocytes and macrophages, the experiments regarding the effects of metal ion on gene and protein were only carried out with the highest subtoxic concentration of 100 µM.

3.3.1. Gene Expression and Protein Release of the Proinflammatory Cytokine IL-1β

While the gene expression of *IL-1β* was upregulated after exposure to 100 µM cobalt ions (Figure 5a), this effect only reached significance in THP-1 monocytes (p = 0.0304 compared to untreated control, Bonferroni post hoc test). There were no significant differences regarding *IL-1β* gene expression between monocytes and macrophages. The increase in *IL-1β* gene expression in THP-1 monocytes was mirrored by increased protein release into the cell culture medium (Figure 5b). Indeed, only after exposure to metal ions IL-1β protein release was detected, while untreated monocytes did not release IL-1β (p = 0.0429 for variable "treatment" in two-way ANOVA).

Interestingly, THP-1 macrophages showed IL-1β protein release already at baseline, i.e., in the untreated macrophages (Figure 5b), and the protein production was not influenced by the treatment with metal ions. The release of IL-1β from THP-1 macrophages was considerably higher than from monocytes (monocytes vs. macrophages: p = 0.0415, p = 0.0159 and p = 0.0109 for untreated cells, as well as treatment with chromium and nickel ions, respectively).

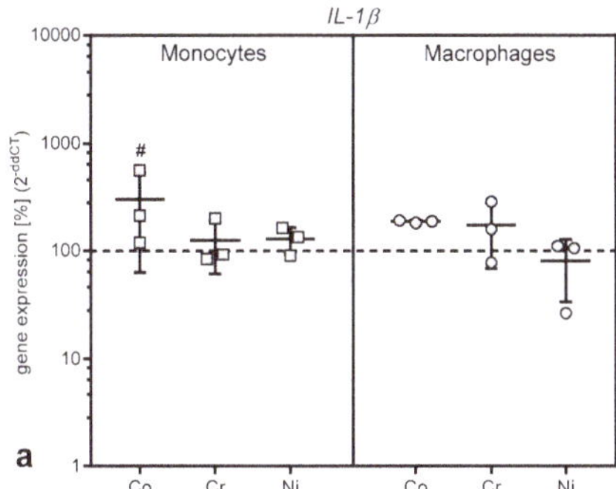
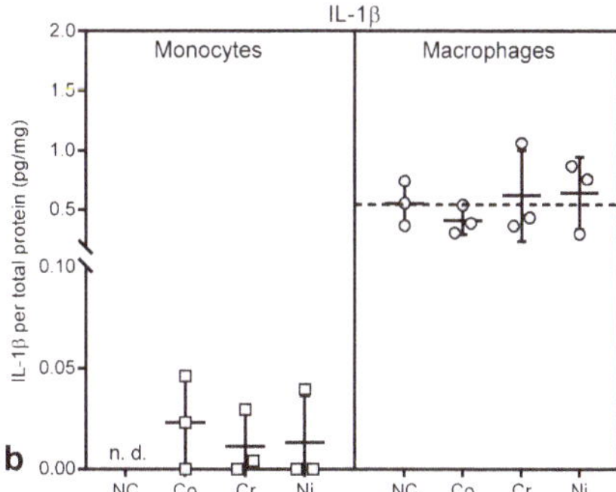

Figure 5. Gene expression and protein release of interleukin (IL-) 1β following exposure to metal salts. Gene expression (**a**) and protein release into cell culture supernatants (**b**) was determined for THP-1 monocytes and macrophages after 48 h of treatment with the concentration of 100 μM of cobalt, chromium and nickel salts. Untreated cells served as negative controls. Data (n = 3) are depicted as single datapoints with mean ± SD. Gene expression data are shown as percentage of untreated cells ($2^{(-\Delta\Delta Ct)}$, 100%, dotted line), while protein release data represent values of the specific protein amount normalized to total protein content in cell culture supernatants. Significances between groups were calculated with two-way ANOVA with Bonferroni post hoc test using ΔCT values for gene expression and specific protein amount normalized to total protein content for protein release in cell culture supernatants. Significantly different from untreated control: # $p < 0.05$, ## $p < 0.01$, ### $p < 0.001$. Abbreviations: n.d. = not detectable; NC = negative control (untreated cells).

3.3.2. Gene Expression and Protein Release of Chemokines

MCP-1 is a chemokine which is responsible for the recruitment of monocytes, but also of natural killer cells and T-lymphocytes, to the location of inflammation [4,8,14,15]. While gene expression of *MCP-1* was slightly, but non-significantly, increased in THP-1 monocytes after exposure to cobalt ions (Figure 6a), no protein release into the cell culture supernatant for MCP-1 was detected in monocytes (Figure 6b). Contrary to the monocytes, treatment with cobalt and nickel ions led to a significant downregulation of *MCP-1* gene expression in THP-1 macrophages (p = 0.0005 and p = 0.0142 for Co and Ni vs. untreated control, Figure 6a). This decrease was also observed regarding the release of MCP-1 protein from macrophages (p = 0.0049 and p = 0.0036 for Co and Ni vs. untreated control, Figure 6b). Exposure to chromium ions did not show any effects.

IL-8 is another chemokine initially involved in the recruitment of neutrophilic granulocytes and macrophages [4,7,8]. Exposure to cobalt and nickel ions resulted in an upregulation of *IL-8* gene expression in THP-1 monocytes and macrophages compared to untreated cells (p = 0.0001 and p = 0.0238 for Co and Ni in monocytes; p = 0.0179 for Co in macrophages). Chromium ions did not affect *IL-8* gene expression (Figure 6c). While the increased gene expression resulted in an increased protein release for IL-8 in macrophages, protein release was significantly reduced in THP-1 monocytes after exposure to all three metal ions compared to untreated controls, despite increased levels of gene expression (p = 0.0004, p = 0.0002 and p = 0.0002 for Co, Cr and Ni vs. untreated control in THP-1 monocytes, Figure 6d).

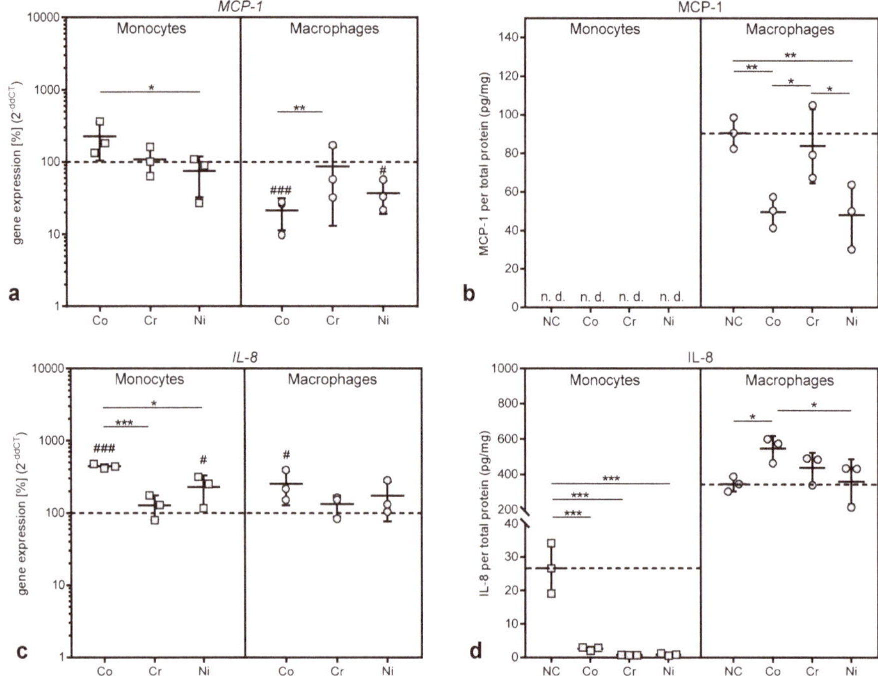

Figure 6. Gene expression and protein release of monocyte chemotactic protein (MCP-) 1 (**a,b**) and Interleukin (IL-) 8 (**c,d**) following exposure to metal salts. Gene expression (**a,c**) and protein release into cell culture supernatants (**b,d**) were determined for THP-1 monocytes and macrophages after 48 h of treatment with the concentration of 100 μM of cobalt, chromium and nickel salts. Untreated cells served as negative controls. Data (n = 3) are depicted as single datapoints with mean ± SD. Gene expression data are shown as percentage of untreated cells ($2^{(-\Delta\Delta Ct)}$, 100%, dotted line) while protein release data represent values of the specific protein amount normalized to total protein content in cell culture supernatants. Significances between groups were calculated with two-way ANOVA and Bonferroni's multiple comparison test as post hoc test using ΔCT values for gene expression and specific protein amount normalized to total protein content for protein release in cell culture supernatants. Significantly different between stimulation groups: * $p < 0.05$, ** $p < 0.01$, *** $p < 0.001$; significantly different from untreated control: # $p < 0.05$, ## $p < 0.01$, ### $p < 0.001$. Abbreviations: n.d. = not detectable; NC = negative control (untreated cells).

3.3.3. Gene Expression and Protein Release of Mediators of Macrophage Differentiation

M-CSF and RANK are both proteins involved in the differentiation and maturation of macrophages, the latter especially in the differentiation into osteoclasts [5,14]. Gene expression of both mediators was only detectable in THP-1 macrophages but not in monocytes (Figure 7a,c). Exposure to cobalt and chromium ions increased gene expression of *M-CSF* and *RANK*, with the association reaching significance for *RANK* (p = 0.0030 and p = 0.0024 for Co and Cr vs. untreated control, respectively). The upregulation of *M-CSF* in THP-1 macrophages was mirrored by an increased release of the protein after cobalt and chromium treatment (Figure 7b). It is notable that chromium ions, which did not show any effects in the other assays presented here, affected both mediators.

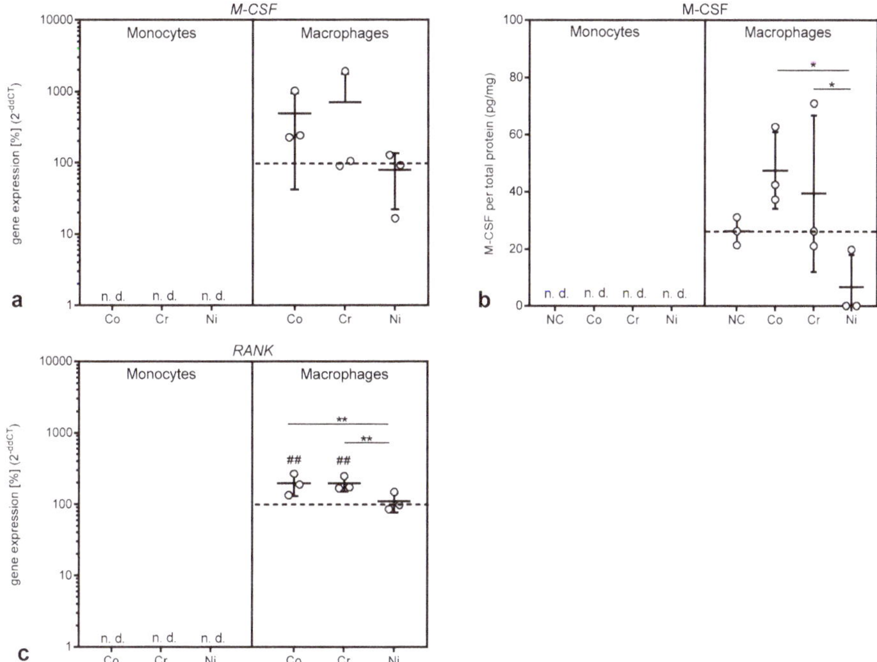

Figure 7. Gene expression and protein release of macrophage colony-stimulating factor (M-CSF) (**a,b**) and receptor activator of NF-κB (RANK) (**c**) following exposure to metal salts. Gene expression (**a,c**) and protein release into cell culture supernatants (**b**) was determined for THP-1 monocytes and macrophages after 48 h of treatment with the concentration of 100 µM of cobalt, chromium and nickel salts. Untreated cells served as negative controls. Data (n = 3) are depicted as single datapoints with mean ± SD. Gene expression data are shown as percentage of untreated cells ($2^{(-\Delta\Delta Ct)}$, 100%, dotted line) while protein release data represent values of the specific protein amount normalized to total protein content in cell culture supernatants. Significances between groups were calculated with two-way ANOVA and Bonferroni's multiple comparison test as post hoc test using ΔCT values for gene expression and specific protein amount normalized to total protein content for protein release in cell culture supernatants. Significantly different between stimulation groups: * $p < 0.05$, ** $p < 0.01$, *** $p < 0.001$; significantly different from untreated control: # $p < 0.05$, ## $p < 0.01$, ### $p < 0.001$. Abbreviations: n.d. = not detectable; NC = negative control (untreated cells).

4. Discussion

In our study, the effects on THP-1 monocytes and macrophages from metal ions at concentrations ranging from 10 to 500 µM were investigated. The evidence that these concentrations are indeed similar to concentrations found in periprosthetic tissue was discussed in detail in a previous publication [11]. The analysis of cell number and cellular activity revealed the impact of metal ions on metabolic activity by showing a significant decrease in activity that was mainly prominent in 500 µM cobalt- and nickel ion-treated cells. Similar results for nickel ions were reported by Chana et al. [13]. The loss of metabolic activity was accompanied by a massive release of reactive oxygen species. The mechanisms of concentration-dependent cytotoxicity for cobalt and nickel ions are based on their ability to interfere with DNA replication and DNA repair after cell membrane penetration, and these changes finally initiate necrosis [16]. These events are partially mediated by reactive oxygen species. It was suggested that metal ions potentiate electron exchange reactions and induce radical formation [17]. Reactive oxygen species are signal molecules produced and broken down in the cells under physiological conditions. Increased production of reactive oxygen species is originally applied by cells of the innate immune system to fight pathogens and initiate pathogen destruction [18]. However, if the production of reactive oxygen species exceeds the elimination capacity of cells, ROS accumulate and induce oxidative cell stress. Cobalt and nickel ions were able to induce ROS production and might therefore cause oxidation of cellular proteins, lipids and DNA, finally leading to cell damage and cell death [19]. Thus, the phagocytosis of particles or elevated cell stress induced by exposure to corrosion product, i.e., high concentrations of cobalt and nickel ions, results in overshooting ROS production that can lead to the observed decrease in cell activity, or even to cell destruction, as indicated by the light microscopy images [20,21].

In contrast, chromium ions did not cause a significant decrease in cell activity. Investigations by Ferko et al. [22] and Kwon et al. [16] further support the assumption of cobalt and nickel ions having a stronger impact on monocytes and macrophages in the applied concentrations. Cell death studies by Huk et al. [23] proved that chromium ion toxicity needed much higher concentrations than in cobalt stimulation to manifest. Bivalent cobalt ions, for example, can penetrate cell membranes passively using different ion channels [24]. Trivalent chromium ions, as used in this study, are not known to possess mechanisms like this. That chromium(III) ions have no known processes of uptake via the cell membrane might also impact on ROS production, as in the chromium-stimulated samples no increase in ROS was observed. In solution, they form aggregates in the cell culture medium. However, very high concentrations of chromium ions that exceed the investigated range may be able to induce hypoxia in the cell and cause a decrease in viability [22,25]. However, since chromium (VI) ions exhibit up to 1000-fold higher toxicity than chromium(III) ions, our results may not adequately reflect the in vivo effects of chromium [26]. In the body, chromium ions may also be present in other forms, such as chromate $[CrO_4]_2-$ or dichromate, during aseptic loosening [27]. Chromate is the predominant form of chromium 6+ in solutions and is able to cross cell membranes through nonspecific anion carriers [17,26]. This may be the reason for the observed in vivo toxicity of chromium components. It is also possible that Cr(3+) in the investigated concentrations might be involved in aseptic loosening via other mechanisms, as, indeed, chromium ions were able to increase the gene expression of mediators of macrophage differentiation *M-CSF* and *RANK* in our study. This finding was rather surprising, as Cr(3+) at the investigated concentration showed no cellular or metabolic effects in this nor in a previous study in osteoblasts [11]. While M-CSF is known to drive differentiation of macrophages into a M2 phenotype [28], and might thus rather ameliorate the inflammation, RANK as the membrane bound receptor for the RANK ligand is directly involved in the initiation and persistence of osteolysis [4,5,7,29,30]. The induction of RANK might thus link chromium ions to the osteolysis processes observed in aseptic loosening.

Apart from chromium ions, cobalt ions were also able to upregulate *RANK* expression in THP-1 macrophages. Indeed, cobalt ions had the most pronounced effects in all the performed experiments. While the highest concentration of 500 µM resulted in a loss of cell number and WST-1 activity, in both THP-1 monocytes and macrophages, the lower concentrations of 10, 50 and 100 µM cobalt ions specifically increased proliferation and cellular activity in THP-1 macrophages. This effect might also be due to the low-level oxidative stress and the production of ROS as, for example, the stimulation of macrophage proliferation by ceramide 1-phosphate was mediated through the generation of ROS [31]. It is possible that the stimulation of proliferation and metabolic activity is a means by which macrophages adapt to the changed microenvironment and fulfill their function as "cleaners" in the body [32,33]. However, this seems to be a finely balanced process, as the higher ROS concentrations observed after nickel treatment of THP-1 macrophages were already cytotoxic.

The induction of oxidative stress in macrophages is furthermore considered a main cause of cytokine release. A recent review by Hallab and Jacobs [7] summarized the danger signal pathway that finally leads to the release of mature IL-1β, IL-18, IL-33, and other cytokines and chemokines as follows. The "inflammasome" pathway senses "danger-associated molecular patterns" and induces danger signaling through mechanisms such as lysosomal destabilization. The cascade of nicotinamide adenine dinucleotide phosphate (NADPH) oxidase caused by the lysosomal destabilization and an associated increase in reactive oxygen species in turn activate the intracellular multi-protein "inflammasome" complex composed of NALP3 (NACHT-, LRR-, and pyrin domain-containing protein 3) in association with apoptosis-associated speck-like protein containing a CARD domain (ASC). This further activates Caspase-1, which, in this case, does not act as an apoptosis stimulus but rather converts cytokines such as IL-1β and IL-18 from their inactive into their active form. This mechanism could account for the observation that IL-1β release was only detected in monocytes after treatment with metal salts as the activation of IL-1β secretion requires a "second signal" [34]. However, THP-1 macrophages already showed a high basic IL-1β release in the untreated cells, which was not influenced by metal ion treatment. It cannot be ruled out that the artificial situation in cell culture, e.g., the attachment to the cell culture dish or the differentiation with Phorbol-12-myristat-13-acetat, provided enough stimulus for IL-1β secretion. Danger signaling might not be the only mechanisms involved in IL-1β secretion, as it was shown that HIFα, which is stabilized by cobalt ions [24], is involved in IL-1β biosynthesis [35]

Interestingly, MCP-1, which was upregulated in debris-induced inflammation [7], was shown to be decreased in gene expression as well as protein biosynthesis after treatment with cobalt and nickel ions in THP-1 macrophages, and was not released from THP-1 monocytes at all despite increased gene expression. A similar effect was observed in THP-1 monocytes for IL-8, which showed a significant elevation in gene expression level, while the protein release was significantly decreased compared to untreated controls. In THP-1 macrophages, however, we found the expected increase in *IL-8* mRNA as well as IL-8 protein after incubation with cobalt ions [4,8,11,36]. To our knowledge, a mechanism for the suppression of IL-8 release in monocytes by metal ions has not been described and we can only speculate that it represents a self-limiting mechanism in inflammation by reducing the further recruitment of neutrophilic granulocytes and monocytes once metallic debris is broken down into ions by macrophages. We assume that the reduction in MCP-1 and IL-8 protein in some of the supernatants, which contradicted the increased gene expression results, was due to reduced protein synthesis and release, e.g., by influencing the mRNA stability, which has been described as one mechanism to fine-tune chemokine availability [37]. However, other fine-tuning mechanisms may also influence the availability of chemokines in the supernatants. It was reported that both chemokines can bind to atypical chemokine receptors [38]. For example, the binding of a chemokine to D6, which is an atypical receptor for MCP-1, leads to the internalization of the receptor-ligand complex followed by the rapid degradation of the ligand [38,39].

We can conclude from this study that metal ions induced different effects in monocytes and macrophages, which were especially apparent for the concentration of inflammatory mediators in the supernatants. We cannot rule out that the use of a cell line influenced our results and further experiments in primary human monocytes and macrophages have to be carried out to confirm the results.

Author Contributions: All authors were fully engaged in the study and in producing the manuscript. A.J.-H. and A.K. projected the study. H.L. and A.J.-H. performed the experiments and evaluated the data with support of A.K. A.K. made the statistical analysis. H.L., A.J.-H. and A.K. wrote the primary manuscript with support of K.P., B.M.-H., T.F. and R.B. R.B. provided all laboratory equipment. A.J.-H., K.P., T.F. and B.M.-H. organized the research funding. All authors ensured the accuracy of the data and the analyses and reviewed the manuscript in its current state.

Funding: This work was financially supported by Rostock University Medical Center (Joint Project: KOBE).

Acknowledgments: We gratefully acknowledge Katharina Ekat and Doris Hansmann for their technical support.

References

1. Kurtz, S.; Ong, K.; Lau, E.; Mowat, F.; Halpern, M. Projections of primary and revision hip and knee arthroplasty in the United States from 2005 to 2030. *J. Bone Jt. Surg. Am.* **2007**, *89*, 780–785. [CrossRef]

2. Herberts, P.; Malchau, H. Long-term registration has improved the quality of hip replacement: A review of the Swedish THR Register comparing 160,000 cases. *Acta Arthop. Scand.* **2000**, *79*, 111–121. [CrossRef] [PubMed]

3. Patel, N.R.; Gohil, P.P. A Review on Biomaterials: Scope, Applications & Human Anatomy Significance. *IJETAE* **2012**, *2*, 91–101.

4. Bitar, D.; Parvizi, J. Biological response to prosthetic debris. *World J. Orthop.* **2015**, *6*, 172–189. [CrossRef] [PubMed]

5. Purdue, P.E. Alternative macrophage activation in periprosthetic osteolysis. *Autoimmunity* **2008**, *41*, 212–217. [CrossRef]

6. Gallo, J.; Konttinen, Y.T.; Goodman, S.B.; Thyssen, J.P.; Gibon, E.; Pajarinen, J.; Takakubo, Y.; Schalock, P.; Mackiewicz, Z.; Jämsen, E.; et al. Aseptic Loosening of Total Hip Arthroplasty as a Result of Local Failure of Tissue Homeostasis. In *Recent Advances in Arthroplasty*; Fokter, S.K., Ed.; IntechOpen: London, UK, 2012; pp. 319–362. [CrossRef]

7. Hallab, N.J.; Jacobs, J.J. Chemokines Associated with Pathologic Responses to Orthopedic Implant Debris. *Front. Endocrinol.* **2017**, *8*. [CrossRef]

8. Gu, Q.; Shi, Q.; Yang, H. The Role of TLR and Chemokine in Wear Particle-Induced Aseptic Loosening. *J. Biomed. Biotechnol.* **2012**, 596870. [CrossRef]

9. Sundfeld, M.; Carlsson, L.V. Aseptic loosening, not only a question of wear: A review of different theories. *Acta Orthop.* **2006**, *77*, 177–197. [CrossRef]

10. Sabella, S.; Carney, R.P. A general mechanism for intracellular toxicity of metal-containing nanoparticles. *Nanoscale* **2014**, *6*, 7052–7061. [CrossRef]

11. Jonitz-Heincke, A.; Sellin, M.L.; Seyfarth, A.; Peters, K.; Mueller-Hilke, B.; Fiedler, T.; Bader, R.; Klinder, A. Analysis of Cellular Activity Short-Term Exposure to Cobalt and Chromium Ions in Mature Human Osteoblasts. *Materials* **2019**, *12*, 2771. [CrossRef]

12. Lund, M.E.; To, J.; O'Brien, B.A.; Donnelly, S. The choice of phorbol 12-myristate 13-acetate differentiation protocol influences the response of THP-1 macrophages to a pro-inflammatory stimulus. *J. Immunol. Methods* **2016**, *430*, 64–70. [CrossRef] [PubMed]

13. Chana, M.; Lewis, J.B.; Davis, R.; Elam, Y.; Hobbs, D.; Lockwood, P.E.; Wataha, J.C.; Messer, R.L. Biological effects of Ni(II) on monocytes and macrophages in normal and hyperglycemic environments. *J. Biomed. Mater. Res. A* **2018**, *106*, 2433–2439. [CrossRef] [PubMed]

14. Italiani, P.; Boraschi, D. From Monocytes to M1/M2 Macrophages: Phenotypical vs. Functional Differentiation. *Front. Immunol.* **2014**, *5*. [CrossRef]

15. Shi, C.; Pamer, E.G. Monocyte recruitment during infection and inflammation. *Nat. Rev. Immunol.* **2011**, *11*, 762–774. [CrossRef] [PubMed]

16. Kwon, Y.M.; Xia, Z.; Glyn-Jones, S.; Beard, D.; Gill, H.S.; Murray, D.W. Dose-dependent cytotoxicity of clinically relevant cobalt nanoparticles and ions on macrophages in vitro. *Biomed. Mater.* **2009**, *4*, 25018. [CrossRef]

17. Stohs, S.J.; Bagchi, D. Oxidative mechanisms in the toxicity of metal ions. *Free Radic. Biol. Med.* **1995**, *18*, 321–336. [CrossRef]

18. Warnatsch, A.; Tsourouktsoglou, T.D.; Branzk, N.; Wang, Q.; Reincke, S.; Herbst, S.; Gutierrez, M.; Papayannopoulos, V. Reactive Oxygen Species Localization Programs Inflammation to Clear Microbes of Different Size. *Immunity* **2017**, *46*, 421–432. [CrossRef]

19. Petit, A.; Mwale, F.; Tkaczyk, C.; Antoniou, J.; Zukor, D.J.; Huk, O.L. Induction of protein oxidation by cobalt and chromium ions in human U937 macrophages. *Biomaterials* **2005**, *26*, 4416–4422. [CrossRef]

20. Scharf, B.; Clement, C.C.; Zolla, V.; Perino, G.; Yan, B.; Elci, S.G.; Purdue, E.; Goldring, S.; Macaluso, F.; Cobelli, N.; et al. Molecular analysis of chromium and cobalt-related toxicity. *Sci. Rep.* **2014**, *4*, 5729. [CrossRef]

21. Sansone, V.; Pagani, D.; Melato, M. The effects on bone cells of metal ions released from orthopaedic implants. *A review. Clin. Cases Miner. Bone Metab.* **2013**, *10*, 34–40. [CrossRef]

22. Ferko, M.A.; Catelas, I. Effects of metal ions on caspase-1 activation and interleukin-1β release in murine bone marrow-derived macrophages. *PLoS ONE* **2018**, *13*, e0199936. [CrossRef]

23. Huk, O.L.; Catelas, I.; Mwale, F.; Antoniou, J.; Zukor, D.J.; Petit, A. Induction of apoptosis and necrosis by metal ions in vitro. *J. Arthoplast.* **2004**, *19*, 84–87. [CrossRef]

24. Simonsen, L.O.; Harbak, H.; Bennekou, P. Cobalt metabolism and toxicology—A brief update. *Sci. Total Environ.* **2012**, *432*, 210–215. [CrossRef]

25. Shrivastava, R.; Upreti, R.K.; Seth, P.K.; Chaturvedi, U.C. Effects of chromium on the immune system. *FEMS Immunol. Med. Microbiol.* **2002**, *34*, 1–7. [CrossRef]

26. Bagchi, D.; Stohs, S.J.; Downs, B.W.; Bagchi, M.; Preuss, H.G. Cytotoxicity and oxidative mechanisms of different forms of chromium. *Toxicology* **2002**, *180*, 5–22. [CrossRef]

27. Fleury, C.; Petit, A.; Mwale, F.; Antoniou, J.; Zukor, D.J.; Tabrizian, M.; Huk, O.L. Effect of cobalt and chromium ions on human MG-63 osteoblasts in vitro: Morphology, cytotoxicity, and oxidative stress. *Biomaterials* **2006**, *27*, 3351–3360. [CrossRef]

28. Boyette, L.B.; Macedo, C.; Hadi, K.; Elinoff, B.D.; Walters, J.T.; Ramaswami, B.; Chalasani, G.; Taboas, J.M.; Lakkis, F.G.; Metes, D.M. Phenotype, function, and differentiation potential of human monocyte subsets. *PLoS ONE* **2017**, *12*, e0176460. [CrossRef]

29. Gallo, J.; Goodman, S.B.; Konttinen, Y.T.; Raska, M. Particle disease: Biologic mechanisms of periprosthetic osteolysis in total hip arthroplasty. *Innate Immun.* **2013**, *19*, 213–224. [CrossRef]

30. Boyce, B.F.; Xing, L. The RANKL/RANK/OPG pathway. *Curr. Osteoporos. Rep.* **2007**, *5*, 98–104. [CrossRef]

31. Arana, L.; Gangoiti, P.; Ouro, A.; Rivera, I.G.; Ordoñez, M.; Trueba, M.; Lankalapalli, R.S.; Bittman, R.; Gomez-Muñoz, A. Generation of reactive oxygen species (ROS) is a key factor for stimulation of macrophage proliferation by ceramide 1-phosphate. *Exp. Cell Res.* **2012**, *318*, 350–360. [CrossRef]

32. Mantovani, A.; Biswas, S.K.; Galdiero, M.R.; Sica, A.; Locati, M. Macrophage plasticity and polarization in tissue repair and remodelling. *J. Pathol.* **2013**, *229*, 176–185. [CrossRef]

33. Glanz, V.; Myasoedova, V.A.; Sukhorukov, V.; Grechko, A.; Zhang, D.; Romaneneko, E.B.; Orekhova, V.A.; Orekhov, A. Transcriptional Characteristics of Activated Macrophages. *Curr. Pharm. Des.* **2019**, *25*, 213–217. [CrossRef]

34. Sitia, R.; Rubartelli, A. The unconventional secretion of IL-1β: Handling a dangerous weapon to optimize inflammatory responses. *Semin. Cell Dev. Biol.* **2018**, *83*, 12–21. [CrossRef]

35. Talwar, H.; Bauerfeld, C.; Bouhamdan, M.; Farshi, P.; Liu, Y.; Samavati, L. MKP-1 negatively regulates LPS-mediated IL-1β production through p38 activation and HIF-1α expression. *Cell. Signal.* **2017**, *34*, 1–10. [CrossRef]

36. Hallab, N.J.; Jacobs, J.J. Biologic effects of implant debris. *Bull. NYU Hosp. Jt. Dis.* **2009**, *67*, 182–188.

37. Mortier, A.; Van Damme, J.; Proost, P. Overview of the mechanisms regulating chemokine activity and availability. *Immunol. Lett.* **2012**, *145*, 2–9. [CrossRef]

38. Graham, G.J.; Locati, M. Regulation of the immune and inflammatory responses by the 'atypical' chemokine receptor D6. *J. Pathol.* **2013**, *229*, 168–175. [CrossRef]

39. Savino, B.; Borroni, E.M.; Torres, N.M.; Proost, P.; Struyf, S.; Mortier, A.; Mantovani, A.; Locati, M.; Bonecchi, R. Recognition versus adaptive up-regulation and degradation of CC chemokines by the chemokine decoy receptor D6 are determined by their N-terminal sequence. *J. Biol. Chem.* **2009**, *284*, 26207–26215. [CrossRef]

Article

Spatial Heterogeneity of Cadmium Effects on *Salvia sclarea* Leaves Revealed by Chlorophyll Fluorescence Imaging Analysis and Laser Ablation Inductively Coupled Plasma Mass Spectrometry

Michael Moustakas [1,*], **Anetta Hanć** [2], **Anelia Dobrikova** [3], **Ilektra Sperdouli** [4], **Ioannis-Dimosthenis S. Adamakis** [5] **and Emilia Apostolova** [3]

[1] Department of Botany, Aristotle University of Thessaloniki, 54124 Thessaloniki, Greece
[2] Department of Trace Element Analysis by Spectroscopy Method, Faculty of Chemistry, Adam Mickiewicz University, 61 614 Poznań, Poland
[3] Institute of Biophysics and Biomedical Engineering, Bulgarian Academy of Science, 1113 Sofia, Bulgaria
[4] Institute of Plant Breeding and Genetic Resources, Hellenic Agricultural Organization–Demeter, Thermi, 57001 Thessaloniki, Greece
[5] Department of Botany, Faculty of Biology, National and Kapodistrian University of Athens, 157 84 Athens, Greece
* Correspondence: moustak@bio.auth.gr; Tel.: +30-2310-99-8335

Received: 28 July 2019; Accepted: 9 September 2019; Published: 12 September 2019

Abstract: In this study, for a first time (according to our knowledge), we couple the methodologies of chlorophyll fluorescence imaging analysis (CF-IA) and laser ablation inductively coupled plasma mass spectrometry (LA-ICP-MS), in order to investigate the effects of cadmium (Cd) accumulation on photosystem II (PSII) photochemistry. We used as plant material *Salvia sclarea* that grew hydroponically with or without (control) 100 µM Cd for five days. The spatial heterogeneity of a decreased effective quantum yield of electron transport (Φ_{PSII}) that was observed after exposure to Cd was linked to the spatial pattern of high Cd accumulation. However, the high increase of non-photochemical quenching (NPQ), at the leaf part with the high Cd accumulation, resulted in the decrease of the quantum yield of non-regulated energy loss (Φ_{NO}) even more than that of control leaves. Thus, *S. sclarea* leaves exposed to 100 µM Cd exhibited lower reactive oxygen species (ROS) production as singlet oxygen ($^1O_2$). In addition, the increased photoprotective heat dissipation (NPQ) in the whole leaf under Cd exposure was sufficient enough to retain the same fraction of open reaction centers (q_P) with control leaves. Our results demonstrated that CF-IA and LA-ICP-MS could be successfully combined to monitor heavy metal effects and plant tolerance mechanisms.

Keywords: bioimaging; clary sage; effective quantum yield (Φ_{PSII}); non-photochemical quenching (NPQ); photochemical quenching (q_P); photoprotective mechanism; photosynthetic heterogeneity; phytoremediation; reactive oxygen species (ROS); singlet oxygen ($^1O_2$)

---

1. Introduction

Cadmium (Cd), a non-essential element for plants, is considered to be as one of the most toxic elements for plants because it is not biodegradable in soil and it accumulates in the environment exhibiting toxic effects [1–3]. It can appear in the environment at high concentrations, due to several human activities (industrial and agricultural activities, such as mining and smelting of metalliferous ores, electroplating, wastewater irrigation, and abuse of chemical fertilizers) and subsequently becomes toxic to all living organisms [3–6]. However, some plants have established several mechanisms for Cd detoxification that result in acclimation and tolerance [2,6].

The photosynthetic process has been shown to be very sensitive to Cd action either directly or indirectly [6–13]. Decrease in the photosynthetic efficiency by Cd may result from stomatal closure, a disorder in enzymatic activities of Calvin-Benson cycle, a decrease in the pigment content, and disturbances in the photosynthetic electron transport [2,6,14–18]. The absorbed light energy is converted into chemical energy via photosystem II (PSII) and photosystem I (PSI) that work co-operatively to transfer efficiently photosynthetic electrons from H_2O to $NADP^+$ (forming NADPH) through the formation of a proton gradient that is used to drive ATP synthesis. PSII catalyzes one of the most exciting reactions in nature, the light-driven oxidation of water and liberation of molecular oxygen [19]. PSII eventually provides the electrons required for the conversion of inorganic molecules into the organic molecules and establishes itself as the "engine of life" [20]. Cd-induced inhibition of PSII photochemistry and linear electron transport may also be due to the limited use of ATP and NADPH by the Calvin-Benson cycle [21] and/or influence on the energy transfer between pigment-protein complexes of the photosynthetic apparatus [13,14]. The Cd toxicity has been reported to affect both photosystems [13,22,23], however, PSII was found to be more sensitive and more affected than PSI [10,13–16,22,24–26]. It has also been suggested that the Cd toxicity influences both donor and acceptor sites of PSII [10,16,21,25]. Nevertheless, Cd has been recently regarded to be PSII donor-side inhibitor as it affects the oxygen-evolving complex [13,14,27]. The effects of Cd toxicity differ by the applied concentrations, the period of exposure, and the plant species [11,26,28,29].

In order to understand the ways of distribution and transport of elements in plant tissues, different techniques of elemental imaging are used, starting from microscopic techniques through techniques based on mass spectrometry and ending with synchrotron X-ray-based techniques [30]. However, undoubtedly, techniques based on mass spectrometry, such as LA-ICP-MS offer a range of advantages in high detection limits and high sensitivity for many elements [31,32]. The coupling of a laser with an ICP-MS as a detector, gives the possibility of wide-range of imaging analysis with good spatial resolution, high sensitivity and capability to image/locate many elements in a single analysis of plant samples [33,34], providing unique data that renders LA-ICP-MS a powerful technique in bioimaging.

Chlorophyll fluorescence analysis has been commonly used as a highly sensitive indicator of the photosynthetic efficiency [35–42], but the photosynthetic function is not uniform at the whole leaf, particularly under abiotic stress conditions [43,44]. This renders conventional chlorophyll fluorescence measurements non-typical of the physiological status of the entire leaf [6,45]. This disadvantage overcomes chlorophyll fluorescence imaging analysis (CF-IA) that reveals spatial heterogeneity of the total leaf area [45,46].

We have previously observed in the metallophyte *Noccaea caerulescens* exposed in hydroponic culture to Cd for three and four days, a spatial leaf heterogeneity of the effective quantum yield of electron transport (Φ_{PSII}) [6] that was suggested to arise from differences in the distribution of Cd across the leaf as it was observed by LA-ICP-MS analysis [47]. A future research direction was proposed then to combine CF-IA with LA-ICP-MS to evaluate Cd stress on whole leaves in order to verify this suggestion [6].

Plants used for phytoremediation of polluted soils should be highly tolerant and produce a great quantity of biomass in contaminated conditions despite the accumulation of high amounts of heavy metals in their tissues [2,48,49]. In recent years, increasing attention is paid to the aromatic plants as an alternative for conducting environmentally safe and cost-effective phytoremediation, since these species are mainly grown for secondary products and the contamination of the food chain with heavy metals is eliminated [50].

The herbal plant *Salvia sclarea* L. (clary sage, belongs to the family Lamiaceae) that is tolerant to heavy metals has been attributed to the Zn and Cd accumulators, and has the potential for phytoremediation of soils contaminated with heavy metals [51,52]. It has also been discovered that *S. sclarea* accumulated heavy metals through the root system and the distribution of the heavy metals in organs of the clary sage decreases in the following order: Leaves > roots> stems > seeds [51,52].

However, heavy metal accumulation does not influence its development, as well as the quality and quantity of the essential oils, which can be used in the perfumery and cosmetics [50–52]. To the best of our knowledge, the effects of Cd action on the photosynthetic apparatus and PSII photochemistry of Salvia leaves, and especially Cd distribution in the leaf area, have not been studied before.

In the present work we tested the hypothesis whether exposure of *Salvia sclarea* plants to Cd will result in spatial leaf heterogeneity of the effective quantum yield of electron transport (Φ_{PSII}), and if it does, whether the decreased Φ_{PSII} values in the leaf area will correspond to the respective pattern of high Cd accumulation obtained by LA-ICP-MS analysis.

2. Materials and Methods

2.1. Plant Material and Growth Conditions

Seeds of *Salvia sclarea* L. collected from a field in the Rose Valley region of Bulgaria were kindly provided by Bio Cultures Ltd (Karlovo, Bulgaria), which is focused on growing several types of herbs for the production of essential oils.

Salvia seeds were germinated and grown on soil in a growth room for about a month. When one pair of true leaves fully expanded (height 4–5 cm), the seedlings were transferred to 1-L vessels (two seedlings per vessel) filled with a continuously aerated modified Hoagland nutrient solution composed of 1.5 mM KNO_3, 1.5 mM $Ca(NO_3)_2$, 0.5 mM NH_4NO_3, 0.5 mM $MgSO_4$, 0.25 mM KH_2PO_4, 50 µM NaFe(III)EDTA, 23 µM H_3BO_3, 4.5 µM $MnCl_2$, 5 µM $ZnSO_4$, 0.2 µM $CuSO_4$, and 0.2 µM Na_2MoO_4, adjusted to pH 6.0 and changed regularly every three days. The plants were kept under a photon flux density of about 220 µmol m^{-2} s^{-1}, 25/20 °C and 14/10 h day/night photoperiod.

2.2. Cd Treatment

About 2-month-old uniform plants were selected and subjected to treatment with 0 and 100 µM Cd (applied as $3CdSO_4\ 8H_2O$) for five days. The nutrient solution with or without Cd was renewed every three days.

2.3. Chlorophyll Fluorescence Imaging Analysis

An Imaging-PAM Fluorometer M-Series MINI-Version (Walz, Effeltrich, Germany) was used to measure in 15 min dark-adapted leaves of *S. sclarea* plants, grown with 0 (control) or 100 µM Cd for five days, the effects of Cd on PSII function. Five leaves were measured from five different plants with the actinic light intensity of 220 µmol photons m^{-2} s^{-1}. In each leaf, 14–16 areas of interest were selected from which chlorophyll fluorescence values were measured. The chlorophyll fluorescence parameters, measured as described in detail previously [53], were the minimum chlorophyll *a* fluorescence in the dark (F_o), the maximum chlorophyll *a* fluorescence in the dark (F_m), the maximum chlorophyll *a* fluorescence in the light (F_m'), and the steady-state photosynthesis in the light (F_s). The minimum chlorophyll *a* fluorescence in the light was computed by the Imaging Win V2.41a software (Heinz Walz GmbH, Effeltrich, Germany) as $F_o' = F_o/(F_v/F_m + F_o/F_m')$. By using Win software, we calculated the allocation of absorbed light energy at PSII, that is the effective quantum yield of photochemistry (Φ_{PSII}), the quantum yield of regulated non-photochemical energy loss (Φ_{NPQ}), and the quantum yield of non-regulated energy loss (Φ_{NO}). The relative PSII electron transport rate (ETR), the fraction of open PSII reaction centers, the so-called photochemical quenching (q_P) and the non-photochemical quenching that reflects heat dissipation of excitation energy (NPQ) were also calculated.

2.4. Laser Ablation Inductively Coupled Plasma Mass Spectrometry

Leaf tissues were analysed in vivo using an ICP-QMS spectrometer (Elan DRC II, Perkin-Elmer Sciex, Guelph, ON, Canada) equipped with a laser ablation system (LA; model LSX-500, CETAC Technologies, Omaha, NE, USA) operating at a wavelength of 266 nm. The instrumentation was optimized on a daily basis by ablating the standard reference glass material NIST SRM 610 and adjusting the nebulizer gas flow, RF generator power and ion lens voltage in order to obtain the maximum signal intensity for $^{24}Mg^+$, $^{115}In^+$, $^{238}U^+$. Plasma robustness was monitored via the $^{232}Th^{16}O^+/^{232}Th$, doubly charged ions and the $^{238}U/^{232}Th$ intensity ratios. ThO^+/Th^+ intensity ratios were always below 0.2%, doubly charged ions $^{42}Ca^{2+}/^{42}Ca^+ < 0.5\%$ and $^{238}U^+/^{232}Th^+$ intensity ratio was less than 1.2. For leaf sample analysis optimization of the parameters, such as energy of laser beam, spot size, shot frequency and scanning speed, was performed. The laser ablation conditions were chosen so that the ablation of the sample was completed. In the experiment, the volume of standard ablation chamber was reduced to ~10 mL, which shortened the washout time of the aerosol and improved the LA images. The instrumental and analytical conditions of LA-ICP-MS are summarized in Table 1. For bioimage generation, LA-iMageS software was used [54].

Table 1. Operating conditions for laser ablation inductively coupled plasma mass spectrometry (LA-ICP-MS) system.

Laser Ablation	
Instrument	CETAC LSX-500, Nd-YAG
Wavelength [nm]	266
Ablation frequency [Hz]	10
Spot size [µm]	100
Laser energy [mJ]	5.4
Scan rate [µm/s]	80
Distance between scan lines [µm]	20
Scan method	Mapping 2D; scanning
ICP-MS	
Instrument	PE Sciex ELAN 6100 DRC II
Nebulizer gas flow [L/min]	1.1
Auxiliary gas flow [L/min]	1.2
Plasma gas flow [L/min]	16
RF Power [W]	1350
Lens setting	Autolens calibrated
Detector mode	Dual (pulse counting and analog mode)
Measured mass to charge ratios	Cd (*m/z* 111); C (*m/z* 13)
Sweeps	1

2.5. Statistical Analysis

Chlorophyll fluorescence analysis data are presented as the mean ± SD. Statistical analysis of means from five leaves from different plants was performed using the Student's t-test. Differences were considered statistically significant at $p < 0.05$.

3. Results

3.1. Photosynthetic Heterogeneity Revealed by Chlorophyll Fluorescence Imaging Analysis in Salvia sclarea Leaves under Cd Exposure

The imaging area in the MINI-Version of the Imaging-PAM Fluorometer M-Series that was used is 24 × 32 mm. Thus, we studied such an area from the distal leaf area of *S. sclarea*. CF-IA revealed a photosynthetic heterogeneity in the studied leaf area of *S. sclarea* leaves under Cd exposure. The spatial heterogeneity was observed mainly in the effective quantum yield of photochemistry (Φ_{PSII}), the quantum yield of regulated non-photochemical energy loss (Φ_{NPQ}) and the quantum yield of non-regulated energy loss (Φ_{NO}) after five days exposure of *S. sclarea* to 100 µM Cd. We observed three clearly distinguishable leaf areas in the chlorophyll fluorescence images of Φ_{PSII}, Φ_{NPQ} and Φ_{NO}. More specifically we marked in the chlorophyll fluorescence images of Φ_{PSII}, a leaf area at the leaf edge, than a top leaf area with lower Φ_{PSII} values than those of the leaf edge, and a second leaf area with Φ_{PSII} values higher than the top leaf area (Figure 1a). The lower Φ_{PSII} values, of the top leaf area, were found near the midvein (Figure 1a). The same three areas appeared at the chlorophyll fluorescence images of Φ_{NPQ} with the top leaf area having higher Φ_{NPQ} values compared to the other two areas (leaf edge and second leaf area) (Figure 1b). The higher Φ_{NPQ} values were found near the midvein of the top leaf area (Figure 1b). In the chlorophyll fluorescence images of Φ_{NO}, the higher values were found in the second leaf area (Figure 2). Representative chlorophyll fluorescence images of Φ_{PSII}, Φ_{NPQ}, and Φ_{NO} of *Salvia sclarea* leaves from plants grown under control conditions (0 µM Cd) are shown in Figure S1. At control growth conditions, a photosynthetic homogeneity was observed in *S. sclarea* leaves.

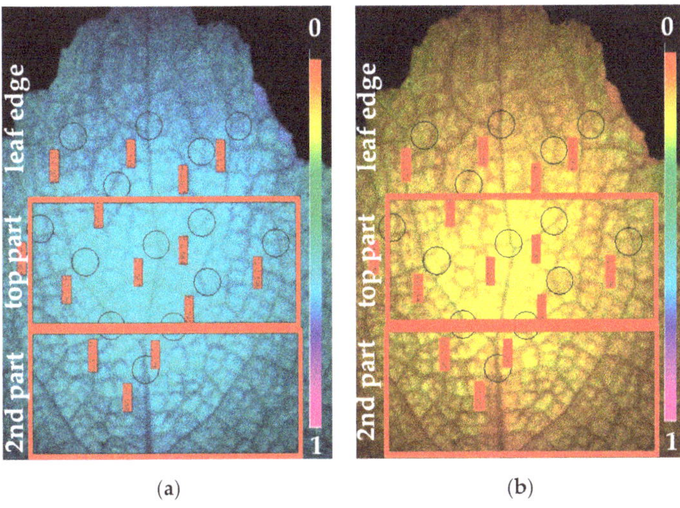

Figure 1. Representative chlorophyll fluorescence images of Φ_{PSII} (**a**) and Φ_{NPQ} (**b**) of *Salvia sclarea* leaves exposed to 100 µM Cd for five days. The different leaf areas: Leaf edge, top leaf part, and second leaf part, are marked. The color code depicted at the right of the images ranges from 0 to 1.

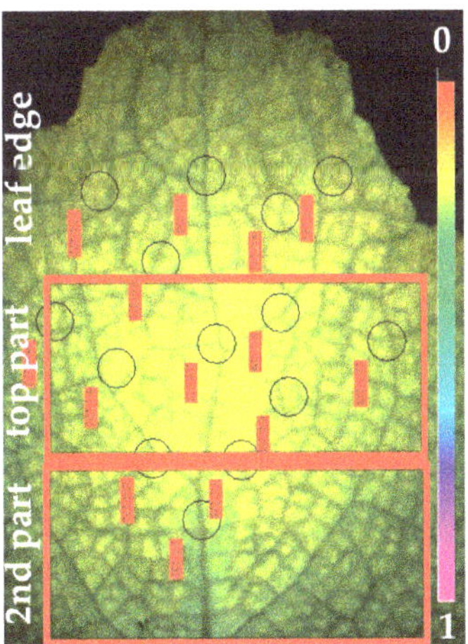

Figure 2. A representative chlorophyll fluorescence image of Φ_{NO} of *Salvia sclarea* leaves exposed to 100 µM Cd for five days. The different leaf areas: Leaf edge, top leaf part, and second leaf part, are marked. The color code depicted at the right of the image ranges from 0 to 1.

3.2. Cadmium Imaging in Salvia sclarea Leaves by Laser Ablation Inductively Coupled Plasma Mass Spectrometry

Three leaves from three different plants were studied by LA-ICP-MS. The area that was selected for analysis corresponds to the two areas that were marked in CF-IA, a top leaf part and a second leaf part. Thus, from each leaf, the corresponding area of 20 × 18 mm was cut and placed on the polyethylene terephthalate slide. In order to normalize the signal, compensating plasma variations and ablations process, two candidates for internal standards, such as ^{13}C and ^{34}S were evaluated [33]. Finally, isotope of carbon ^{13}C was selected as the internal standard as its distribution in the *S. sclarea* leaves was homogeneous. The leaves from *S. sclarea* plants grown under control conditions (0 μM Cd) were analyzed by the whole area (Figure S2), while the leaves from plants exposed to Cd were analyzed in two parts that corresponded to the two leaf parts studied by CF-IA (Figure 3).

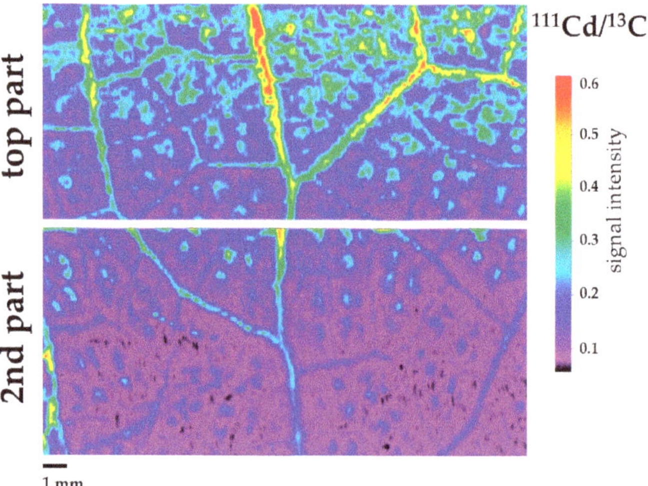

Figure 3. Laser ablation inductively coupled plasma-mass spectrometry (LA-ICP-MS) Cd distribution in a *Salvia sclarea* leaf under Cd exposure (100 μM Cd for five days). The two-leaf areas (marked in Figures 1 and 2) top leaf part and second leaf part, are shown. Cd intensity was normalized using ^{13}C.

The laser beam scanned a selected area of the sample line by line from left to right side with 3 s delay at the end of each line. Approximately 60 lines per leaf were analysed. The number and width of the ablation lines were the same for all leaf samples. Leaves were ablated from the adaxial (upper) side.

The area with high Cd signal intensity was the top leaf area, and, more specifically, the area in the midvein of the top leaf area (Figure 3). In contrast, the presence of Cd was not revealed in leaves of control plants (Figure S2).

3.3. Changes in the Light Energy Use at PSII Under Cd Exposure

We calculated for all chlorophyll fluorescence parameters whole leaf values, top leaf part area values and second leaf part area values (leaf part areas are marked in Figures 1 and 2 and Figure S1). We estimated the allocation of absorbed light energy at PSII, that is the effective quantum yield of photochemistry (Φ_{PSII}), the quantum yield of regulated non- photochemical energy loss (Φ_{NPQ}), and the quantum yield of non-regulated energy loss (Φ_{NO}) of *S. sclarea* leaves from plants exposure to 0 and 100 μM Cd. Φ_{PSII} whole leaf values, after five days exposure to Cd, decreased significantly compared to controls as did also top leaf part area values compared to their corresponding controls (Figure 4a). Φ_{PSII} values of the second leaf part area did not differ compared to the corresponding control values (Figure 4a). The second leaf part values after five days exposure to Cd were significantly higher than the top leaf part area Φ_{PSII} values (Figure 4a). Φ_{NPQ} values after five days exposure to Cd, increased significantly in the whole leaf compared to control, and also in the other two parts compared to their corresponding controls (Figure 4b). Top leaf part Φ_{NPQ} values after Cd exposure were significantly higher than second leaf part area values (Figure 4b).

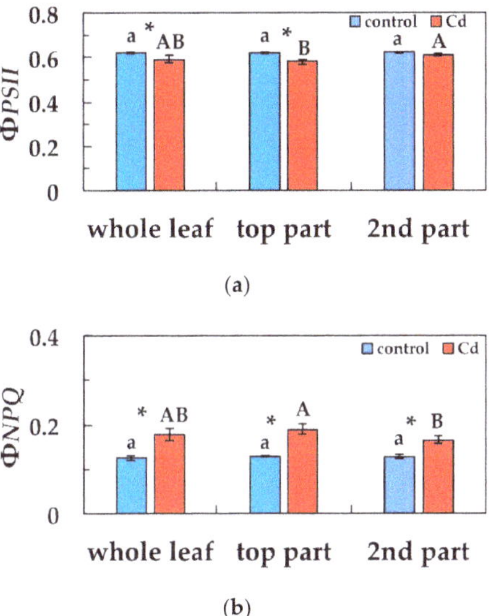

(a)

(b)

Figure 4. Changes in the quantum efficiency of photosystem II (PSII) photochemistry (Φ_{PSII}) (**a**) and the quantum yield of regulated non-photochemical energy loss (Φ_{NPQ}) (**b**). Whole leaf values, top leaf part area values and second leaf part area values are shown in *Salvia sclarea* plants grown at 0 (control), or 100 µM Cd for five days). Error bars on columns are standard deviations based on five leaves from different plants. Columns with a different letter (lower case for controls and capitals for 100 µM Cd) are statistically different between different leaf areas ($p < 0.05$). An asterisk represents a significantly different mean between controls and 100 µM Cd of the same leaf area ($p < 0.05$).

Φ_{NO} values after five days exposure to Cd, decreased significantly in the whole leaf compared to control, and also in the other two parts compared to their corresponding controls (Figure 5). Top leaf part Φ_{NO} values after five days exposure to Cd were significantly lower than second leaf part area values (Figure 5).

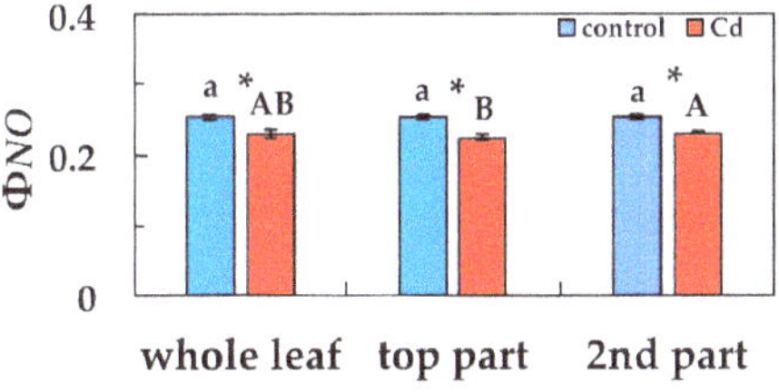

Figure 5. Changes in the quantum yield of non-regulated energy loss (Φ_{NO}) in *Salvia sclarea* plants grown at 0 (control), or 100 µM Cd for five days. Error bars on columns are standard deviations based on five leaves from different plants. Error bars on columns are standard deviations based on five leaves from different plants. Columns with a different letter (lower case for controls and capitals for 100 µM Cd) are statistically different between different leaf areas ($p < 0.05$). An asterisk represents a significantly different mean between controls and 100 µM Cd of the same leaf area ($p < 0.05$).

3.4. Changes in Non-Photochemical Quenching and the Redox State of PSII under Cd Exposure

Non-photochemical quenching (NPQ) that reflects heat dissipation of excitation energy, increased significantly after five days exposure to Cd in the whole leaf compared to control, and also in the other two parts compared to their corresponding controls (Figure 6a). Top leaf part area NPQ values after five days exposure to Cd were significantly higher than second leaf part area values (Figure 6a). The redox state of PQ pool (q_P), decreased significantly in the top leaf part area compared to the corresponding control, but remained the same with control at the whole leaf level and at the second leaf part area (Figure 6b). Top leaf part q_P values after five days exposure to Cd were significantly lower than second leaf part values (Figure 6b).

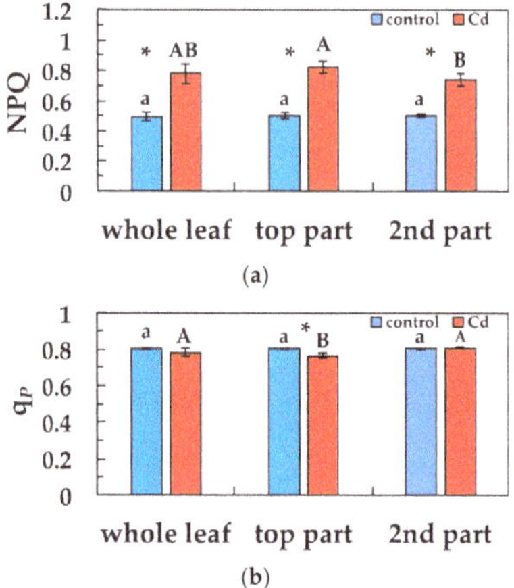

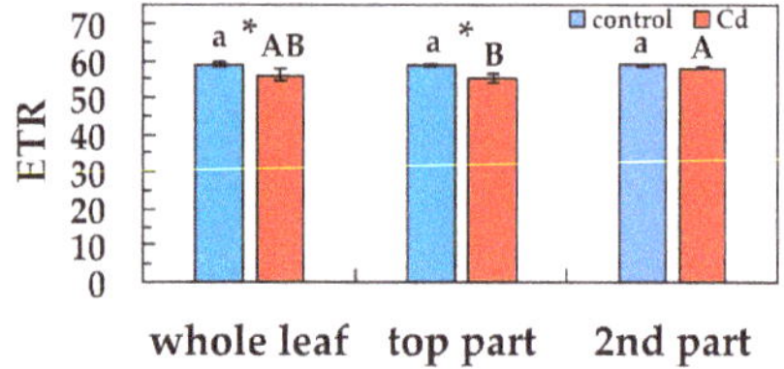

Figure 6. Changes in the non-photochemical fluorescence quenching (NPQ) (**a**) and in the relative reduction state of Q_A, reflecting the fraction of open PSII reaction centers (q_P) (**b**) in *Salvia sclarea* plants grown at 0 (control) or 100 µM Cd for five days. Error bars on columns are standard deviations based on five leaves from different plants. Columns with a different letter (lower case for controls and capitals for 100 µM Cd) are statistically different between different leaf areas ($p < 0.05$). An asterisk represents a significantly different mean between controls and 100 µM Cd of the same leaf area ($p < 0.05$).

3.5. Changes in the Electron Transport Rate in Response to Cd Exposure

The relative electron transport rate at PSII (ETR) decreased significantly at the whole leaf level and at the top leaf part, after five days exposure to Cd, compared to their corresponding controls, while retained the same ETR values with controls at the second leaf part (Figure 7). Top leaf part ETR values after five days exposure to Cd were significantly lower than second leaf part values (Figure 7).

Figure 7. Changes in the relative PSII electron transport rate (ETR) in *Salvia sclarea* plants grown at 0 (control), or 100 µM Cd for five days. Error bars on columns are standard deviations based on five leaves from different plants. Columns with a different letter (lower case for controls and capitals for 100 µM Cd) are statistically different between different leaf areas ($p < 0.05$). An asterisk represents a significantly different mean between controls and 100 µM Cd of the same leaf area ($p < 0.05$).

4. Discussion

Among the different techniques that have been developed for elemental imaging, including secondary ion mass spectrometry, X-ray fluorescence, scanning electron microscopy with energy-dispersive X-ray analysis, and LA-ICP-MS, the latter one has emerged as the prevailing, with high sensitivity and more comprehensible tool for bioimaging of mineral elements in plant tissues [55,56].

Bioimaging of mineral elements in plant tissues has revealed that the distribution of trace elements in leaves is highly heterogeneous [57–59]. The accumulation, distribution and localization of Cd in plant leaves, reported by numerous studies [33,47,60], proposed that the accumulation and distribution of Cd and also of other elements depends on the element, the plant species, the organ and the age of the organ [61–63]. In Salvia leaves information regarding the distribution of any element is lacking. In our experiment, the distribution of Cd in Salvia leaves, under 100 µM Cd, shows that high Cd signal intensity was detected in the midvein of the top leaf part (Figure 3). In contrast, no Cd could be detected in *S. sclarea* grown under control conditions (Figure S2).

Photosynthetic perturbations to heavy metal exposure do not develop homogeneously over the whole leaf area, thus, making chlorophyll fluorescence measurements at a specific point on the leaf surface non-reliable [43,44]. CF-IA detects spatial and temporal heterogeneity of photochemical efficiency under heavy metal stress and can provide further information on the particular leaf area that is most sensitive to heavy metal stress [6,45].

No significant photosynthetic heterogeneity was detected in leaves of control grown clary sage plants, but we were able to identify a spatial photosynthetic heterogeneity in the leaves of clary sage exposed to Cd. This spatial heterogeneity was observed in Φ_{PSII}, Φ_{NPQ} and Φ_{NO} after exposure to Cd. The lower Φ_{PSII} values, of the top leaf area, that were found near the midvein (Figure 1a) are linked to the high Cd signal intensity in the midvein of the top leaf area (Figure 3). These observations confirm the previous suggestion [6] that spatial leaf heterogeneity of the effective quantum yield of electron transport (Φ_{PSII}) arise from differences in the distribution of Cd across the leaf.

The high increase of Φ_{NPQ} values after five days exposure to Cd, in the whole leaf area compared to control values, and especially at the top leaf part (Figure 4b), resulted in lower Φ_{NO} values, compared to control, with the lower values to be observed at the top leaf part (Figure 5). Φ_{NO} consists of chlorophyll fluorescence internal conversions and intersystem crossing, which leads to the formation of singlet oxygen ($^1O_2$) via the triplet state of chlorophyll ($^3chl^*$) [39,64–66]. Consequently, after five days exposure to Cd, $^1O_2$ decreased in the whole leaf area, compared to control values, and especially at the top leaf part, where the higher Cd signal intensity was scored. This can be explained by the photoprotective mechanism of non-photochemical quenching, that allows for the sequestration of reactive oxygen species (ROS) below critical levels [61]. Otherwise, an increase in ROS triggers remarkable damage to the metabolic machinery, inducing photoinhibition and a generalized damage response [67–69].

Non-photochemical chlorophyll fluorescence quenching (NPQ) is a process that takes place in the photosynthetic membranes of plants, algae, and cyanobacteria in which surplus absorbed light energy is dissipated as heat [70–72]. This is a molecular adaptation process that represents the fastest response of the photosynthetic membrane to the surplus light energy [70,71]. Thus, the excess light causes rapid saturation of the photosynthetic reaction centers and their eventual closure [69,70]. The excess light energy that cannot be used for photochemistry can damage the most delicate part of the photosynthetic apparatus, the PSII reaction center, which drives the oxidation of water and liberation of molecular oxygen [18]. In order this photodamage to be avoided, the excess excitation energy has to be safely removed by the photoprotective mechanism of NPQ [71,73,74].

The presence of Cd ions increased the heat dissipation of energy as NPQ [75], but this is a photoprotective response mechanism to avoid ROS generation and damage to PSII [71,76,77]. In general, the influence of Cd on photosystems is more serious in PSII than in PSI [78,79]. The photoprotective dissipation of excess light energy (NPQ) under stress conditions can be regarded as efficient only if it is adjusted in such a way to retain the same fraction of open reaction centers as in control conditions [80]. The photoprotective mechanism of NPQ was sufficient in the leaves of clary sage exposed to Cd, since the fraction of open reaction centers at the whole leaf area remained the same to controls (Figure 6b).

Although excess Cd accumulation is harmful to plants [1,2,7], detoxification mechanisms to Cd toxicity that are involved in Cd tolerance and accumulation exist in the hyperaccumulators [6,45]. Plants have developed complicated mechanisms to control concentrations of essential nutrient elements and to diminish the injury from exposure to non-essential metals, but the mechanisms regarding the regulatory network of metal uptake, chelation, transport, sequestration and detoxification which contributes to the alleviation of heavy metal toxicity and photosynthetic tolerance remain to be further elucidated [81–84].

5. Conclusions

Exposure of *S. sclarea* plants to Cd resulted in spatial leaf heterogeneity of Φ_{PSII}, with the decreased Φ_{PSII} values in the leaf area to correspond to the respective pattern of high Cd accumulation obtained by LA-ICP-MS analysis. We propose that combining the methodologies of chlorophyll fluorescence imaging analysis (CF-IA) and laser ablation inductively coupled plasma mass spectrometry (LA-ICP-MS) can identify the effects of heavy metals on plants and provide information on tolerance mechanisms. We suggest that *S. sclarea* could be characterized as a heavy metal accumulator, as it is tolerant to Cd, and could also potentially be used for phytoremediation.

Supplementary Materials: The following are available online at http://www.mdpi.com/1996-1944/12/18/2953/s1, Figure S1: Representative chlorophyll fluorescence images of Φ_{PSII}, Φ_{NPQ} and Φ_{NO} of *Salvia sclarea* leaves from plants grown under control conditions. Figure S2: LA-ICP-MS Cd distribution of a *Salvia sclarea* leaf under control conditions.

Author Contributions: Conceptualization, M.M., A.D. and E.A.; methodology, M.M.; software, M.M. and I.S.; validation, M.M. and A.H.; formal analysis, M.M., A.H. and I.S.; investigation, A.H., I.S., I.-D.S.A.; resources, M.M., A.D. and E.A.; data curation, M.M., A.H., A.D. and E.A.; writing—original draft preparation, M.M., A.H., A.D.; writing—review and editing, M.M., A.H., A.D.; visualization, M.M. and A.H.; supervision, M.M. and A.D.; project administration, M.M., A.D. and E.A.; funding acquisition, M.M., A.H. and A.D.

Funding: This work was supported by the Agreement for scientific cooperation between the Bulgarian Academy of Sciences and Aristotle University of Thessaloniki, Greece. AH was supported by the National Science Center in Poland under the grant number 2017/01/X/ST4/00373.

Acknowledgments: Seeds of *Salvia sclarea* used for the experiments were kindly provided by Bio Cultures Ltd.

Conflicts of Interest: The authors declare no conflict of interest. The funders had no role in the design of the study; in the collection, analyses, or interpretation of data; in the writing of the manuscript, or in the decision to publish the results.

References

1. Clemens, S.; Ma, J.F. Toxic heavy metal and metalloid accumulation in crop plants and foods. *Ann. Rev. Plant Biol.* **2016**, *67*, 489–512. [CrossRef] [PubMed]
2. Dobrikova, A.G.; Apostolova, E.L. Damage and protection of the photosynthetic apparatus under cadmium stress. In *Cadmium Toxicity and Tolerance in Plants: From Physiology to Remediation*, 1st ed.; Hasanuzzaman, M., Prasad, M.N.V., Fujita, M., Eds.; Academic Press: Cambridge, MA, USA, 2019; pp. 275–298.
3. Sharma, R.K.; Agrawal, M.; Marshall, F. Heavy metal contamination of soil and vegetables in subur-ban areas of Varanasi, India. *Ecotoxicol. Environ. Saf.* **2007**, *66*, 258–266. [CrossRef] [PubMed]
4. Buchauer, M.J. Contamination of soil and vegetation near a zinc smelter by zinc, cadmium, copper, and lead. *Environ. Sci. Technol.* **1973**, *7*, 131–135. [CrossRef]
5. McBride, M.B.; Richards, B.K.; Steenhuis, T.; Russo, J.J.; Sauvé, S. Mobility and solubility of toxic metals and nutrients in soil fifteen years after sewage sludge application. *Soil Sci.* **1997**, *162*, 487–500. [CrossRef]
6. Bayçu, G.; Moustaka, J.; Gevrek-Kürüm, N.; Moustakas, M. Chlorophyll fluorescence imaging analysis for elucidating the mechanism of photosystem II acclimation to cadmium exposure in the hyperaccumulating plant *Noccaea caerulescens*. *Materials* **2018**, *11*, 2580. [CrossRef] [PubMed]

7. Greger, M.; Ögren, E. Direct and indirect effects of Cd^{2+} on photosynthesis in sugar beet (*Beta vulgaris*). *Physiol. Plant.* **1991**, *83*, 129–135. [CrossRef]
8. Ouzounidou, G.; Moustakas, M.; Eleftheriou, E.P. Physiological and ultrastructural effects of cadmium on wheat (*Triticum aestivum* L.) leaves. *Arch. Environ. Contam. Toxicol.* **1997**, *32*, 154–160. [CrossRef] [PubMed]
9. Küpper, H.; Parameswaran, A.; Leitenmaier, B.; Trtílek, M.; Šerlík, I. Cadmium induced inhibition of photosynthesis and long-term acclimation to cadmium stress in the hyperaccumulator *Thlaspi caerulescens*. *New Phytol.* **2007**, *175*, 655–674. [CrossRef] [PubMed]
10. Ekmekçi, Y.; Tanyolaç, D.; Ayhan, B. Effects of cadmium on antioxidant enzyme and photosynthetic activities in leaves of two maize cultivars. *J. Plant Physiol.* **2008**, *165*, 600–611. [CrossRef]
11. Parmar, P.; Kumari, N.; Sharma, V. Structural and functional alterations in photosynthetic apparatus of plants under cadmium stress. *Bot. Stud.* **2013**, *54*, 45. [CrossRef] [PubMed]
12. Xue, Z.C.; Gao, H.Y.; Zhang, L.T. Effects of cadmium on growth, photosynthetic rate, and chlorophyll content in leaves of soybean seedlings. *Biol. Plant.* **2013**, *57*, 587–590. [CrossRef]
13. Dobrikova, A.G.; Yotsova, E.K.; Börner, A.; Landjeva, S.P.; Apostolova, E.L. The wheat mutant DELLA-encoding gene (Rht-B1c) affects plant photosynthetic responses to cadmium stress. *Plant Physiol. Biochem.* **2017**, *114*, 10–18. [CrossRef] [PubMed]
14. Yotsova, E.K.; Dobrikova, A.G.; Stefanov, M.A.; Kouzmanova, M.; Apostolova, E.L. Improvement of the rice photosynthetic apparatus defence under cadmium stress modulated by salicylic acid supply to roots. *Theor. Exp. Plant Physiol.* **2018**, *30*, 57–70. [CrossRef]
15. Baszynski, T.; Wajda, L.; Krol, M.; Wolinska, D.; Krupa, Z.; Tukendorf, A. Photosynthetic activities of cadmium-treated tomato plants. *Physiol. Plant.* **1980**, *48*, 365–370. [CrossRef]
16. Sigfridsson, K.G.V.; Bernát, G.; Mamedov, F.; Styring, S. Molecular interference of Cd^{2+} with Photosystem II. *Biochim. Biophys. Acta* **2004**, *1659*, 19–31. [CrossRef] [PubMed]
17. Malik, D.; Sheoran, I.S.; Singh, R. Carbon metabolism in leaves of cadmium treated wheat seedlings. *Plant Physiol. Biochem.* **1992**, *30*, 223–229.
18. Shi, G.R.; Cai, Q.S. Photosynthetic and anatomic responses of peanut leaves to cadmium stress. *Photosynthetica* **2008**, *46*, 627–630. [CrossRef]
19. Derks, A.; Schaven, K.; Bruce, D. Diverse mechanisms for photoprotection in photosynthesis. Dynamic regulation of photosystem II excitation in response to rapid environmental change. *Biochim. Biophys. Acta* **2015**, *1847*, 468–485. [CrossRef] [PubMed]
20. Barber, J. Photosynthetic energy conversion: Natural and artificial. *Chem. Soc. Rev.* **2009**, *38*, 185–196. [CrossRef]
21. Krupa, Z.; Öquist, G.; Hunter, N.P.A. The effects of cadmium on photosynthesis of *Phaseolus vulgaris* L.—A fluorescence analysis. *Physiol. Plant.* **1993**, *88*, 626–630. [CrossRef]
22. Atal, N.; Saradhi, P.P.; Mohanty, P. Inhibition of the chloroplast photochemical reactions by treatment of wheat seedlings with low concentrations of cadmium: Analysis of electron transport activities and changes in fluorescence yield. *Plant Cell Physiol.* **1991**, *32*, 943–951. [CrossRef]
23. Wodala, B.; Eitel, G.; Gyula, T.N.; Ördög, A.; Horváth, F. Monitoring moderate Cu and Cd toxicity by chlorophyll fluorescence and P700 absorbance in pea leaves. *Photosynthetica* **2012**, *50*, 380–386. [CrossRef]
24. Vassilev, A.; Lidon, F.; Scotti, P.; Da Graca, M.; Yordanov, I. Cadmium-induced changes in chloroplast lipids and photosystem activities in barley plants. *Biol. Plant.* **2004**, *48*, 153–156. [CrossRef]
25. Pagliano, C.; Raviolo, M.; Dalla Vecchia, F.; Gabbrielli, R.; Gonnelli, C.; Rascio, N.; Barbato, R.; La Rocca, N. Evidence for PSII-donor-side damage and photoinhibition induced by cadmium treatment on rice (*Oryza sativa* L.). *J. Photochem. Photobiol. B Biol.* **2006**, *84*, 70–78. [CrossRef] [PubMed]
26. López-Millán, A.F.; Sagardoy, R.; Solanas, M.; Abadía, A.; Abadía, J. Cadmium toxicity in tomato (*Lycopersicon esculentum*) plants grown in hydroponics. *Environ. Exp. Bot.* **2009**, *65*, 376–385. [CrossRef]
27. Wilson, S.; Ruban, A.V. Quantitative assessment of the high-light tolerance in plants with an impaired photosystem II donor side. *Biochem. J.* **2019**, *476*, 1377–1386. [CrossRef] [PubMed]
28. Prasad, M.N. Cadmium toxicity and tolerance in vascular plants. *Environ. Exp. Bot.* **1995**, *35*, 525–545. [CrossRef]
29. Burzyński, M.; Żurec, A. Effect of copper and cadmium on photosynthesis in cucumber cotyledons. *Photosynthetica* **2007**, *45*, 239–244. [CrossRef]

30. Zhao, F.J.; Moore, K.L.; Lombi, E.; Zhu, Y.G. Imaging element distribution and speciation in plant cells. *Trends Plant Sci.* **2014**, *19*, 183–192. [CrossRef]

31. Wu, B.; Becker, J.S. Imaging techniques for elements and element species in plant science. *Metallomics* **2012**, *4*, 403–416. [CrossRef]

32. Pessoa, G.S.; Lopes Junior, C.A.; Chacon-Madrid, K.; Arruda, M.A.Z. A quantitative approach for Cd, Cu, Fe and Mn through laser ablation imaging for evaluating the translocation and accumulation of metals in sunflower seeds. *Talanta* **2017**, *167*, 317–324. [CrossRef] [PubMed]

33. Hanć, A.; Piechalak, A.; Tomaszewska, B.; Barałkiewicz, D. Laser ablation inductively coupled plasma mass spectrometry in quantitative analysis and imaging of plant's thin sections. *Int. J. Mass Spectrom.* **2014**, *363*, 16–22. [CrossRef]

34. Diniz, A.P.; Kozovits, A.R.; Lana de Carvalho, C.; de Abreu, A.T.; Leite, M.G.P. Quantitative analysis of plant leaf elements using the LA-ICP-Ms technique. *Int. J. Mass Spectrum.* **2019**, *435*, 251–258. [CrossRef]

35. Murchie, E.H.; Lawson, T. Chlorophyll fluorescence analysis: A guide to good practice and understanding some new applications. *J. Exp. Bot.* **2013**, *64*, 3983–3998. [CrossRef] [PubMed]

36. Kalaji, H.M.; Oukarroum, A.A.; Alexandrov, V.; Kouzmanova, M.; Brestic, M.; Zivcak, M.; Samborska, I.A.; Cetner, M.D.; Allakhverdiev, S.I. Identification of nutrient deficiency in maize and tomato plants by in vivo chlorophyll *a* fluorescence measurements. *Plant Physiol. Biochem.* **2014**, *81*, 16–25. [CrossRef] [PubMed]

37. Guidi, L.; Calatayud, A. Non-invasive tools to estimate stress-induced changes in photosynthetic performance in plants inhabiting Mediterranean areas. *Environ. Exp. Bot.* **2014**, *103*, 42–52. [CrossRef]

38. Sperdouli, I.; Moustakas, M. Differential blockage of photosynthetic electron flow in young and mature leaves of *Arabidopsis thaliana* by exogenous proline. *Photosynthetica* **2015**, *53*, 471–477. [CrossRef]

39. Moustaka, J.; Tanou, G.; Adamakis, I.D.; Eleftheriou, E.P.; Moustakas, M. Leaf age dependent photoprotective and antioxidative mechanisms to paraquat-induced oxidative stress in *Arabidopsis thaliana*. *Int. J. Mol. Sci.* **2015**, *16*, 13989–14006. [CrossRef]

40. Moustaka, J.; Ouzounidou, G.; Bayçu, G.; Moustakas, M. Aluminum resistance in wheat involves maintenance of leaf Ca^{2+} and Mg^{2+} content, decreased lipid peroxidation and Al accumulation, and low photosystem II excitation pressure. *BioMetals* **2016**, *29*, 611–623. [CrossRef]

41. Kalaji, H.M.; Jajoo, A.; Oukarroum, A.; Brestic, M.; Zivcak, M.; Samborska, I.A.; Cetner, M.D.; Łukasik, I.; Goltsev, V.; Ladle, R.J. Chlorophyll a fluorescence as a tool to monitor physiological status of plants under abiotic stress conditions. *Acta Physiol. Plant.* **2016**, *38*, 102. [CrossRef]

42. Moustaka, J.; Ouzounidou, G.; Sperdouli, I.; Moustakas, M. Photosystem II is more sensitive than photosystem I to Al^{3+} induced phytotoxicity. *Materials* **2018**, *11*, 1772. [CrossRef] [PubMed]

43. Sperdouli, I.; Moustakas, M. Spatio-temporal heterogeneity in *Arabidopsis thaliana* leaves under drought stress. *Plant Biol.* **2012**, *14*, 118–128. [CrossRef] [PubMed]

44. Moustaka, J.; Moustakas, M. Photoprotective mechanism of the non-target organism *Arabidopsis thaliana* to paraquat exposure. *Pestic. Biochem. Physiol.* **2014**, *111*, 1–6. [CrossRef]

45. Moustakas, M.; Bayçu, G.; Gevrek-Kürüm, N.; Moustaka, J.; Csatári, I.; Rognes, S.E. Spatiotemporal heterogeneity of photosystem II function during acclimation to zinc exposure and mineral nutrition changes in the hyperaccumulator *Noccaea caerulescens*. *Environ. Sci. Pollut. Res.* **2019**, *26*, 6613–6624. [CrossRef] [PubMed]

46. Gorbe, E.; Calatayud, A. Applications of chlorophyll fluorescence imaging technique in horticultural research: A review. *Sci. Hortic.* **2012**, *138*, 24–35. [CrossRef]

47. Callahan, D.L.; Hare, D.J.; Bishop, D.P.; Doble, P.A.; Roessner, U. Elemental imaging of leaves from the metal hyperaccumulating plant *Noccaea caerulescens* shows different spatial distribution of Ni, Zn and Cd. *RSC Adv.* **2016**, *6*, 2337–2344. [CrossRef]

48. Reeves, R.D.; Baker, A.J.M. Metal-accumulating plants. In *Phytoremediation of Toxic Metals: Using Plants to Clean Up the Environment*; Raskin, I., Ensley, B., Eds.; Wiley: New York, NY, USA, 2000; pp. 193–229.

49. Rehman, M.Z.; Rizwan, M.; Ali, S.; Ok, Y.S.; Ishaque, W.; Nawaz, M.F.; Akmal, F.; Waqar, M. Remediation of heavy metal contaminated soils by using *Solanum nigrum*: A review. *Ecotoxicol. Environ. Saf.* **2017**, *143*, 236–248. [CrossRef]

50. Pandey, J.; Verma, R.K.; Singh, S. Suitability of aromatic plants for phytoremediation of heavy metal contaminated areas: A review. *Int. J. Phytoremediation* **2019**, *21*, 405–418. [CrossRef]

51. Angelova, V.R.; Ivanova, R.V.; Todorov, G.M.; Ivanov, K.I. Potential of *Salvia sclarea* L. for phytoremediation of soils contaminated with heavy metals. *Int. J. Agric. Biosyst. Eng.* **2016**, *10*, 780–790.

52. Zheljazkov, V.D.; Nielsen, N.E. Growing clary sage (*Salvia sclarea* L.) in heavy metal-polluted areas. *Acta Hortic.* **1996**, *426*, 309–328. [CrossRef]

53. Moustaka, J.; Panteris, E.; Adamakis, I.D.S.; Tanou, G.; Giannakoula, A.; Eleftheriou, E.P.; Moustakas, M. High anthocyanin accumulation in poinsettia leaves is accompanied by thylakoid membrane unstacking, acting as a photoprotective mechanism, to prevent ROS formation. *Environ. Exp. Bot.* **2018**, *154*, 44–55. [CrossRef]

54. López-Fernandez, H.; Pessôa, G.S.; Arruda, M.A.Z.; Capelo-Martínez, J.L.; Fdez-Riverola, F.; Glez-Peña, D.; Reboiro-Jato, M. LA-iMageS: A software for elemental distribution bioimaging using LA-ICP-MS data. *J. Cheminformatics* **2016**, *8*, 1758–2946. [CrossRef] [PubMed]

55. Yamaji, N.; Ma, J.F. Bioimaging of multiple elements by high-resolution LA-ICPMS reveals altered distribution of mineral elements in the nodes of rice mutants. *Plant J.* **2019**. [CrossRef] [PubMed]

56. Persson, D.P.; Chen, A.; Aarts, M.G.M.; Salt, D.E.; Schjoerring, J.K.; Husted, S. Multi-element bioimaging of *Arabbidopsis thaliana* roots. *Plant Physiol.* **2016**, *172*, 835–847. [PubMed]

57. Kyriacou, B.; Moore, K.L.; Paterson, D.; de Jonge, M.D.; Howard, D.L.; Stangoulis, J.; Tester, M.; Lombi, E.; Johnson, A.A.T. Localization of iron in rice grain using synchrotron X-ray fluorescence microscopy and high resolution secondary ion mass spectrometry. *J. Cereal Sci.* **2014**, *59*, 173–180. [CrossRef]

58. Ozturk, L.; Yazici, M.A.; Yucel, C.; Torun, A.; Cekic, C.; Bagci, A.; Ozkan, H.; Braun, H.J.; Sayers, Z.; Cakmak, I. Concentration and localization of zinc during seed development and germination in wheat. *Physiol. Plant.* **2006**, *128*, 144–152. [CrossRef]

59. Lombi, E.; Smith, E.; Hansen, T.H.; Paterson, D.; de Jonge, M.D.; Howard, D.L.; Persson, D.P.; Husted, S.; Ryan, C.; Schjoerring, J.K. Megapixel imaging of (micro) nutrients in mature barley grains. *J. Exp. Bot.* **2011**, *62*, 273–282. [CrossRef] [PubMed]

60. Hanć, A.; Małecka, A.; Kutrowska, A.; Bagniewska-Zadworna, A.; Tomaszewska, B.; Barałkiewicz, D. Direct analysis of elemental biodistribution in pea seedlings by LA-ICP-MS, EDX and confocal microscopy: Imaging and quantification. *Microchem. J.* **2016**, *128*, 305–311. [CrossRef]

61. Galiova, M.V.; Szakova, J.; Prokes, L.; Cadkova, Z.; Coufalik, P.; Kanicky, V.; Otruba, V.; Tlustos, P. Variability of trace element distribution in *Noccaea* spp., *Arabidopsis* spp., and *Thlaspi arvense* leaves: The role of plant species and element accumulation ability. *Environ. Monit. Assess.* **2019**, *191*, 181. [CrossRef] [PubMed]

62. Kutrowska, A.; Małecka, A.; Piechalak, A.; Masiakowski, W.; Hanć, A.; Barałkiewicz, D.; Andrzejewska, B.; Tomaszewska, A. Effects of binary metal combinations on zinc, copper, cadmium and lead uptake and distribution in *Brassica juncea*. *J. Trace Elem. Med. Biol.* **2017**, *44*, 32–39. [CrossRef]

63. Pickering, I.J.; Prince, R.C.; Salt, D.E.; George, G.N. Quantitative, chemically specific imaging of selenium transformation in plants. *Proc. Natl. Acad. Sci. USA* **2000**, *97*, 10717–10722. [CrossRef] [PubMed]

64. Klughammer, C.; Schreiber, U. Complementary PS II quantum yields calculated from simple fluorescence parameters measured by PAM fluorometry and the Saturation Pulse method. *PAM Appl. Notes* **2008**, *1*, 27–35.

65. Antonoglou, O.; Moustaka, J.; Adamakis, I.D.; Sperdouli, I.; Pantazaki, A.; Moustakas, M.; Dendrinou-Samara, C. Nanobrass CuZn nanoparticles as foliar spray non phytotoxic fungicides. *ACS Appl. Mater. Interfaces* **2018**, *10*, 4450–4461. [CrossRef] [PubMed]

66. Malea, P.; Charitonidou, K.; Sperdouli, I.; Mylona, Z.; Moustakas, M. Zinc uptake, photosynthetic efficiency and oxidative stress in the seagrass *Cymodocea nodosa* exposed to ZnO nanoparticles. *Materials* **2019**, *12*, 2101. [CrossRef] [PubMed]

67. Gururani, M.A.; Venkatesh, J.; Tran, L.S.P. Regulation of photosynthesis during abiotic stress-induced photoinhibition. *Mol. Plant.* **2015**, *8*, 1304–1320. [CrossRef] [PubMed]

68. Jones, M.A. Retrograde signalling as an informant of circadian timing. *New Phytol.* **2019**, *221*, 1749–1753. [CrossRef] [PubMed]

69. Mullineaux, P.M.; Exposito-Rodriguez, M.; Laissue, P.P.; Smirnoff, N. ROS dependent signalling pathways in plants and algae exposed to high light: Comparisons with other eukaryotes. *Free Radical Biol. Med.* **2018**, *122*, 52–64. [CrossRef] [PubMed]

70. Demmig-Adams, B.; Adams, W.W., III. Photoprotection and other responses of plants to high light stress. *Annu. Rev. Plant Physiol. Plant Mol. Biol.* **1992**, *43*, 599–626. [CrossRef]

71. Ruban, A.V. Nonphotochemical chlorophyll fluorescence quenching: Mechanism and effectiveness in protecting plants from photodamage. *Plant Physiol.* **2016**, *170*, 1903–1916. [CrossRef]

72. Agathokleous, E.; Kitao, M.; Harayama, H. On the nonmonotonic, hormetic photoprotective response of plants to stress. *Dose-Response* **2019**, *17*, 1559325819838420. [CrossRef]

73. Müller, P.; Li, X.P.; Niyogi, K.K. Non-photochemical quenching. A response to excess light energy. *Plant Physiol.* **2001**, *125*, 1558–1566. [CrossRef] [PubMed]

74. Takahashi, S.; Badger, M.R. Photoprotection in plants: A new light on photosystem II damage. *Trends Plant Sci.* **2011**, *16*, 53–60. [CrossRef] [PubMed]

75. Janeczko, A.; Koscielniak, J.; Pilipowicz, M.; Szarek-Lukaszewska, G.; Skoczowski, A. Protection of winter rape photosystem 2 by 24-epibrassinolide under Cd stress. *Photosynthetica* **2005**, *43*, 293–298. [CrossRef]

76. Sperdouli, I.; Moustakas, M. Differential response of photosystem II photochemistry in young and mature leaves of *Arabidopsis thaliana* to the onset of drought stress. *Acta Physiol. Plant.* **2012**, *34*, 1267–1276. [CrossRef]

77. Sperdouli, I.; Moustakas, M. Leaf developmental stage modulates metabolite accumulation and photosynthesis contributing to acclimation of *Arabidopsis thaliana* to water deficit. *J. Plant Res.* **2014**, *127*, 481–489. [CrossRef] [PubMed]

78. Mallick, N.; Mohn, F.H. Use of chlorophyll fluorescence in metal stress research: A case study with the green microalga Scenedesmus. *Ecotoxicol. Environ. Saf.* **2003**, *55*, 64–69. [CrossRef]

79. Chu, J.; Zhu, F.; Chen, X.; Liang, H.; Wang, R.; Wang, X.; Huang, X. Effects of cadmium on photosynthesis of *Schima superba* young plant detected by chlorophyll fluorescence. *Environ. Sci. Pollut. Res.* **2018**, *25*, 10679–10687. [CrossRef] [PubMed]

80. Lambrev, P.H.; Miloslavina, Y.; Jahns, P.; Holzwarth, A.R. On the relationship between non-photochemical quenching and photoprotection of photosystem II. *Biochim. Biophys. Acta* **2012**, *1817*, 760–769. [CrossRef] [PubMed]

81. Gallego, S.M.; Pena, L.B.; Barcia, R.A.; Azpilicueta, C.E.; Iannone, M.F.; Rosales, E.P.; Zawoznik, M.S.; Groppa, M.D.; Benavides, M.P. Unravelling cadmium toxicity and tolerance in plants: Insight into regulatory mechanisms. *Environ. Exp. Bot.* **2012**, *83*, 33–46. [CrossRef]

82. Sitko, K.; Rusinowski, S.; Kalaji, H.M.; Szopiński, M.; Małkowski, E. Photosynthetic efficiency as bioindicator of environmental pressure in *A. halleri*. *Plant Physiol.* **2017**, *175*, 290–302. [CrossRef]

83. Paunov, M.; Koleva, L.; Vassilev, A.; Vangronsveld, J.; Goltsev, V. Effects of different metals on photosynthesis: Cadmium and zinc affect chlorophyll fluorescence in durum wheat. *Int. J. Mol. Sci.* **2018**, *19*, 787. [CrossRef] [PubMed]

84. Szopiński, M.; Sitko, K.; Gieroń, Ż.; Rusinowski, S.; Corso, M.; Hermans, C.; Verbruggen, N.; Małkowski, E. Toxic Effects of Cd and Zn on the photosynthetic apparatus of the *Arabidopsis halleri* and *Arabidopsis arenosa* pseudo-metallophytes. *Front. Plant Sci.* **2019**, *10*, 748. [CrossRef] [PubMed]

Article

Analysis of Cellular Activity and Induction of Inflammation in Response to Short-Term Exposure to Cobalt and Chromium Ions in Mature Human Osteoblasts

Anika Jonitz-Heincke [1,*], Marie-Luise Sellin [1], Anika Seyfarth [1], Kirsten Peters [2], Brigitte Mueller-Hilke [3], Tomas Fiedler [4], Rainer Bader [1] and Annett Klinder [1]

[1]   Biomechanics and Implant Technology Research Laboratory, Department of Orthopedics, Rostock University Medical Centre, Doberaner Strasse 142, 18057 Rostock, Germany

[2]   Department of Cell Biology, Rostock University Medical Center, Schillingallee 69, 18057 Rostock, Germany

[3]   Institute for Immunology, Rostock University Medical Center, Schillingallee 70, 18057 Rostock, Germany

[4]   Institute for Medical Microbiology, Virology and Hygiene, Rostock University Medical Center, Schillingallee 70, 18057 Rostock, Germany

*   Correspondence: anika.jonitz-heincke@med.uni-rostock.de; Tel.: +49-381-494-9306

Received: 8 July 2019; Accepted: 26 August 2019; Published: 28 August 2019

Abstract: In aseptic loosening of endoprosthetic implants, metal particles, as well as their corrosion products, have been shown to elicit a biological response. Due to different metal alloy components, the response may vary depending on the nature of the released corrosion product. Our study aimed to compare the biological effects of different ions released from metal alloys. In order to mimic the corrosion products, different metal salts ($CoCl_2$, $NiCl_2$ and $CrCl_3 \times 6H_2O$) were dissolved and allowed to equilibrate. Human osteoblasts were incubated with concentrations of 10 µM to 500 µM metal salt solutions under cell culture conditions, whereas untreated cells served as negative controls. Cells exposed to CoCr28Mo6 particles served as positive controls. The cell activity and expression of osteogenic differentiation and pro-osteolytic mediators were determined. Osteoblastic activity revealed concentration- and material-dependent influences. Collagen 1 synthesis was reduced after treatment with Co(2+) and Ni(2+). Additionally, exposure to these ions (500 µM) resulted in significantly reduced OPG protein synthesis, whereas RANKL as well as IL-6 and IL-8 secretion were increased. *TLR4* mRNA was significantly induced by Co(2+) and CoCr28Mo6 particles. The results demonstrate the pro-osteolytic capacity of metal ions in osteoblasts. Compared to CoCr28Mo6 particles, the results indicated that metal ions intervene much earlier in inflammatory processes.

Keywords: osteoblasts; corrosion; ions; particles; osteolysis; inflammation

---

1. Introduction

The wear and corrosion of metal implant materials are the main risk factors for aseptic loosening and implant failure in orthopedic surgery. Although metal surfaces are protected by a passive oxygen layer, electrochemical reactions are still present at the implant site [1]. Different components can cause galvanic effects resulting in oxidation and reduction reactions on the metal surfaces. As a consequence, a continuous exchange of electrons as well as ions between the metal and surrounding liquid is maintained [2]. Head-neck taper corrosion is the most prominent example of this in orthopedic surgery, especially for large-diameter metal-on-metal bearings [3], but also for metal-polyethylene bearings [4]. Another type of corrosion is the intracellular reduction of wear particles in cells, especially in macrophages. When metal particles are taken up via active mechanisms or internalization, the degradation of these nanostructures is promoted by the lysosomal environment [5]. Due to the acidic pH in the cellular components, the release of metal ions in different oxidation states is fostered [6]. Besides the intracellular accumulation of these toxic substances, metal ions are released into the extracellular environment where other cell types can be directly influenced. However, osteolysis is not directly induced by the release of ions but rather the reaction of ions with biomolecules, which leads to higher bioactivity and destruction [7]. A third mechanism of corrosion has been described as direct inflammatory cell-induced corrosion, which occurs on retrieved metal implants. Here, the direct cell attack and concomitant formation of a ruffled cell membrane have been assumed as a trigger for corrosive processes [8].

Apart from titanium-aluminum-vanadium, cobalt-chromium-molybdenum (CoCr28Mo6) is the most prominent alloy in orthopedic surgery. While cobalt alloys are characterized by their high strength, fracture toughness, and excellent biocompatibility, the degradation of the surface due to corrosion, or a combination of corrosion and wear, can lead to harmful local and systemic side-effects in the human body [2,6,9]. Cytotoxic effects of Co, Cr, Ti, and nickel (Ni), which present themselves mainly as apoptosis, necrosis, and inhibitory effects on the DNA repair mechanism, are caused by means of chromosomal damage and oxidative stress [2,10].

Despite these similar outcomes, different metal ions vary with regard to local and systemic effects. Cr ions are able to react with phosphate compounds to form stable chromium phosphates ($CrPo_4$) which then accumulate in the peri-implant tissue [2,6]. In contrast, Co ions can circulate within the body, affecting many organs and causing, e.g., neurological, cardiological, or endocrine symptoms [2]. The varying behavior is also reflected in the different concentrations found locally as well as in the blood of patients suffering from aseptic loosening and metallosis. As blood is easily accessible, there is a sound base of data. Serum cobalt concentrations of 2–7 µg/L (0.002–0.007 ppm) or 10 µg/L (0.010 ppm) were stated respectively in a European multidisciplinary consensus statement [11] or by the Mayo Clinic [12] as indicative of metallic wear and implant loosening. The systemic concentrations reported in patients with acute cobalt poisoning after excessive wear particle release with fatal or near fatal outcomes were approximately a hundred times higher (serum levels of around 0.40–0.64 ppm cobalt and 0.05–0.08 ppm chromium) than the suggested threshold, with even higher values excreted in urine [13–17]. While cobalt concentrations are higher than chromium concentrations in serum this is reversed in the periprosthetic tissue, partially due to the above mentioned reasons. Scharf et al. [6] measured an average of 0.17 ppm for cobalt and 1.60 ppm for chromium in tissue surrounding of total hip implants with metal on metal (MoM) bearings in nine patients with revisions due to adverse local tissue reactions [6]. In a more recent publication by Kuba et al. (2019) revision patients with long-term surviving implants displayed mean values of 6.52 ± 16.38 ppm for cobalt and 8.88 ± 26.88 ppm for chromium in periprosthetic tissue—concentrations in joint fluid were around 0.048 ppm for Co and 0.167 ppm for Cr [18]. However, in the rare cases, where wear and loosening led to acute poisoning, local concentrations of 41 ppm Co [16] up to 397.8 ppm Co and 236.0 ppm Cr [19] in periprosthetic tissue, or 76 ppm Co and 126.5 ppm Cr in locally aspirated liquid [14], were reported. These high local concentrations induce detrimental effects and an inflammatory response to metal ion intoxication can be locally found in the periprosthetic tissue. Tissue necrosis is associated with macrophage and lymphocyte infiltration [6] and the secretion of a variety of pro-inflammatory mediators (e.g., interleukins (IL), tumor necrosis factor (TNF)), especially from macrophages, is induced [7]. However, various studies have revealed that bone forming osteoblasts are also affected by metal ions, resulting in reduced proliferation and differentiation capacity as well as inducing pro-inflammatory cytokine release [2,20,21]. In this context, it is interesting that mineralized bone tissue seems to be prone to accumulate metal debris as the concentrations of 38–413 ppm Co here were considerably higher than those reported above for soft tissue [22].

In a previous in vitro study, we exposed human osteoblasts and macrophages to a mixture of Co and Cr ions (200 µg/L or 0.2 ppm) and analyzed their pro-osteolytic capacity in the cell cultures [21]. Although we observed marginal detrimental effects in single cell cultures, both a clear impact on osteoblastic function and the initiation of pro-osteolytic events were detectable under co-culture conditions of human osteoblasts and macrophages. However, a limitation of this study was the missing knowledge of the proportion of either Co or Cr ions. As a result, proven effects could not be attributed to one metal ion species or another. Thus, for further understanding of differential effects of Co and Cr ions, we now exposed human osteoblasts to different concentrations of Co and Cr metal salts. Based on previous studies by Scharf et al. (2014) and Drynda et al. (2018) we used ion concentrations between 10 µM and 500 µM [6,23]. These concentrations correspond to 0.59–29.47 ppm Co and 0.52–26.00 ppm Cr and are similar to those reported in periprosthetic tissue by Kuba et al. (2019) [18], but lower than those in bone tissue [22]. However, due to the analytical methodology it is difficult to specify whether the determined metal concentration originated from metal particles or metal ions.

Additionally to Co and Cr, nickel ions—which are known to have a higher toxicity on osteoblastic viability and proliferation than Co and Cr [2,24]—were investigated since forced corrosion of CoCr28Mo6 particles resulted in considerable levels of Ni ions in the solution that exceeded the amount of molybdenum [21]. Simultaneously, human osteoblasts were also treated with CoCr28Mo6 particles in order to clarify whether corrosion products exhibit similar or differing effects to abrasion particles in cells. Hence, the objective of this study was to assess the effects of relevant corrosion products on cell survival, bone remodeling, and the release of pro-osteolytic mediators in mature human osteoblasts. Therefore, short-term studies with different metal ion concentrations were carried out to determine the threshold at which metal salts are harmful for cell survival.

2. Materials and Methods

2.1. Preparation of Metal Salt Solutions

The following metal salts were purchased from Sigma-Aldrich (Sigma-Aldrich Chemie GmbH, Munich, Germany): Cobalt(II) chloride (purum p.a., anhydrous, purity $\geq$ 98.0% (KT)), Nickel(II) chloride (anhydrous, powder, purity 99.99% trace metals basis) and Chromium(III) chloride hexahydrate (purum p.a., purity $\geq$ 98.0% (RT)). Stock solutions of a concentration of 100 mM were produced by dissolving the appropriate amount of salt in Aqua ad iniectabilia (B. Braun Melsungen AG, Melsungen, Germany). Stock solutions were stored in Schott Duran® laboratory glass bottles (Schott AG, Mainz, Germany) at 4 °C in the dark. Before the use in the experiments the chromium salt stock solution was equilibrated according to Drynda et al. (2018) for at least two weeks until the color of the solution changed completely from emerald green to purple, indicating the presence of the stable species of $[Cr_3 + (H_2O)_6]Cl_3$ [23]. For cell experiments, the stock solutions were further diluted with cell culture media to expose cells to concentrations of 10 µM, 50 µM, 100 µM and 500 µM, respectively.

2.2. Isolation and Cultivation of Human Primary Osteoblasts

Isolation of human primary osteoblasts (n = 11; eight female donors: mean age 70 years ± 18 years; three male donors: mean age 70 years ± 13 years) was performed according to the protocol of Lochner et al. (2011) [25]. The bone marrow was extracted under sterile conditions from femoral heads of patients undergoing primary hip replacements who had given written consent (Local Ethical Committee AZ: 2010-10). Extracted spongiosa was washed three times in phosphate-buffered saline (PBS, Biochrom AG, Berlin, Germany) followed by an enzymatic digestion with collagenase A and dispase II (both from Roche, Penzberg, Germany) at 37 °C. The material was filtered through a cell strainer (70 µm pores, BD Biosciences, Bedford, UK) and the isolated cell suspension was centrifuged at 118× g for 10 min. The cell sediment was re-suspended in cell culture medium. Human primary osteoblasts were cultivated in a special formulation of calcium-depleted Dulbecco's modified Eagle medium (DMEM, Pan Biotech GmbH, Aidenbach, Germany) containing 10% fetal calf serum (Pan Biotech GmbH), 1% penicillin/streptomycin, 1% amphotericin b, 1% HEPES buffer, and the osteogenic additives L-ascorbate-2-phosphate (50 µg/mL), β-glycerophosphate (10 mM), as well as dexamethasone (100 nM) (all: Sigma-Aldrich, Munich, Germany) at 37 °C and 5% CO_2. These specific conditions, in particular the addition of osteogenic factors dexamethasone, β-glycerophosphate and L-ascorbate-2-phosphate, led to the differentiation of the isolated pre-osteoblasts into mature osteoblasts in standard cell culture flasks (polystyrene) [26,27]. The osteoblastic phenotype was analyzed by alkaline phosphatase staining (with fuchsin+substrate-chromogen; DAKO, Hamburg, Germany). Cells from passage 3 were harvested for subsequent cell culture experiments as follows. Cells were washed with PBS, trypsinized, and centrifuged at $118\times g$. If not otherwise stated, 30,000 cells (in duplicate) were transferred into a well of a 24-well cell culture plate allowing cell adherence over 24 h at 37 °C and 5% CO_2. Afterwards, cells were exposed to different concentrations of metal salts. Untreated cells served as negative controls whereas osteoblasts treated with CoCr28Mo6 particles (particle concentration: 0.01 mg/mL) [28,29] were used as positive control. For the comparison of the influence of different culture media, matured cells from passage 3 were not only seeded in the above described Ca-depleted osteogenic medium but also in standard Dulbecco's modified Eagle medium (Gibco™ DMEM Glutamax, ThermoFisher Scientific, Waltham, MA, USA) with osteogenic additives L-ascorbate-2-phosphate (50 µg/mL), β-glycerophosphate (10 mM) and dexamethasone (100 nM) (all: Sigma-Aldrich) before treatment.

2.3. Cellular Activity

The cell activity of ion-exposed human osteoblasts was determined after 48 h of exposure. Osteoblasts (10,000 cells per well, in duplicates) were seeded in black 96-well cell culture plates (Thermo Fisher Scientific Inc., Waltham, MA, USA). After 24 h of adherence under standard cell culture conditions, cells were treated with metal salts or particles. Supernatants were removed after 48 h and the water soluble tetrazolium salt (WST-1) assay (Roche, Penzberg, Germany) was performed according to the manufacturer's recommendations. Cells were incubated with a defined volume of WST-1/medium (1:10 ratio) reagent for 30 min. Subsequently, supernatants were transferred in a 96-well cell culture plate and absorbance at 450 nm (reference wave length: 630 nm) was determined in a Tecan Infinite® 200 Pro microplate reader (Tecan Group AG, Maennedorf, Switzerland). Afterwards, the same cells were used to quantify cell numbers using the CyQUANT® NF Cell Proliferation Assay Kit (Invitrogen, Thermo Fisher Scientific Inc., Waltham, MA, USA) according to the recommendations of the manufacturer. This combined approach was possible as the CyQUANT® kit measures the emission of fluorescence at 520 nm after the lysis of the cells and the subsequent binding of a proprietary green fluorescent dye, CyQUANT® GR dye, to cellular nucleic acids. The fluorescence signal depends solely on the amount of DNA present in the sample. In order to relate the fluorescence signal to an actual cell number, a cell number calibration curve was prepared with defined cell numbers between 0 and 20,000 cells in duplicate prior to each experiment. The calibration curve was then used to determine the cell number of treated cells per well. Fluorescence intensity was measured with the Tecan Infinite® 200 Pro (Tecan Group AG, Maennedorf, Switzerland) microplate reader (excitation: 485 nm, emission: 535 nm). Cellular activity was calculated by dividing WST-1 results by the respective cell number as measured by CyQuant (see Supplementary Materials Figures S1 and S2). Since the activity values per cell were very small, results are presented as activity per one million cells for easier illustration of data.

Additionally, microscopic examinations of cell cultures were carried out after 48 h of incubation. Morphology of human osteoblasts was documented via light microscopy using a 200× magnification (Nikon ECLIPSE TS100, Nikon GmbH, Duesseldorf, Germany). The formation of actin filaments within cell structures was visualized by actin staining and DAPI counterstain following the procedure described by Klinder et al. (2018) [28]. Pictures for actin staining were taken with 400× magnification at 500 nm where actin stain fluoresced green. Cell nuclei were visualized with DAPI stain at 400 nm and showed a blue fluorescence. The respective pictures were taken from exactly the same spot and superimposed upon each other with the help of Adobe Photoshop CS6 image processing software Version 13.0.1 (Adobe Systems Software Ireland Ltd., Dublin, Ireland).

2.4. Gene Expression Analysis

RNA isolation was carried out using the peqGOLD Total RNA Kit (VWR International GmbH, Hanover, Germany) following the manufacturer's protocol. RNA was eluted into a fresh sterile tube using RNase free water and RNA concentration was measured using the Tecan Infinite® 200 (Tecan Group AG, Maennedorf, Switzerland) microplate reader and NanoQuant Plate™ with RNase free water as blank. The purity of the isolated RNA was assessed and median ratios at 260/280 nm of 2.10 (2.10–2.11), 2.10 (1.78–2.14), 2.13 (2.04–2.19), 2.11(1.90–2.13) and 2.09 (2.08–2.11) were recorded for untreated, Co-treated, Cr-treated, Ni-treated and particle-treated samples, respectively. After RNA isolation, a reverse transcriptase polymerase chain reaction (RT-PCR) was used to transcribe the RNA into cDNA. Here, the High Capacity cDNA Reverse Transcription Kit (Applied Biosystems, Foster City, CA, USA) was used. A master mix was prepared as described in the manufacturer's protocol. The specific amount of each RNA sample containing 200 ng RNA was calculated using the results from concentration measurement and added up to 10 µL with RNase free water in PCR tubes. Subsequently, 10 µL of master mix was added and mixed well. The samples were placed in a thermocycler (Analytik Jena, Jena, Germany) and the following RT-PCR protocol was used: 10 min at 25 °C, 120 min at 37 °C, 15 s at 85 °C. Afterwards, samples were diluted in additional 20 µL RNase free water and stored at −20 °C.

To determine the expression level of differentiation- and inflammation-associated genes, the cDNA of treated and untreated cells was used to perform a semiquantitative real-time (q-PCR) with SybrGreen. For the PCR reaction, the 2× innuMIX qPCR MasterMix SyGreen (Analytik Jena, Jena, Germany) was used following the manufacturer's protocol. To that master mix, 0.5 µL of forward and reverse primer (12 µM), respectively, as well as 3 µL of Aqua dest. were added. For each sample, 1 µL template cDNA (in duplicates) was pipetted onto the bottom of a 96-well PCR plate and filled up with 9 µL of the mentioned master mix. The used primer sequences of osteogenic (Col1A1, ALP) and pro-osteolytic mediators (IL-6, IL-8) are listed in Table 1. Distilled water, instead of cDNA, served as negative control. The plate was sealed with adhesive foil and placed in the qTower 2.0 (Analytik Jena, Jena, Germany). Gene expression analysis was done under the following conditions: initial activation time of 2 min at 95 °C, 40 times of rotation of denaturation for 5 s at 95 °C and annealing/elongation for 25 s at 60–65 °C. A cycle of threshold (Ct) of 28 was set as limit. The relative expression of each gene compared to the housekeeping gene hypoxanthine guanine phosphoribosyl transferase (HPRT) was calculated using the equation: $\Delta Ct = Ct_{target} - Ct_{HPRT}$. The relative amount of target mRNA of cells treated with metal salts and controls was calculated using $2^{(-\Delta\Delta Ct)}$ with $\Delta\Delta Ct_{treatment} = \Delta Ct_{treated} - \Delta Ct_{control}$.

Table 1. Overview of primer sequences for qRT-PCR.

Primer	Forward (5′-3′)	Reverse (5′-3′)
Alkaline phosphatase (ALP)	cattgtgaccaccacgagag	ccatgatcacgtcaatgtcc
Collagen 1 (Col1A1)	acgaagacatcccaccaatc	agatcacgtcatcgcacaac
Hypoxanthine guanine phosphoribosyltransferase (HPRT)	ccctggcgtcgtgattagtg	tcgagcaagacgttcagtcc
Interleukin 6 (IL-6)	tggattcaatgaggagacttgcc	ctggcatttgtggttgggtc
Interleukin 8 (IL-8)	tctgtgtgaaggtgcagttttg	atttctgtgttggcgcagtg
Toll-like receptor 4 (TLR 4)	ggtcagacggtgatagcgag	tttacgggccaagtctccacg

2.5. Protein Analysis

The protein contents of bone remodeling markers (pro-collagen type 1 (C1CP), osteoprotegerin (OPG), receptor activator of nuclear factor κb ligand (RANKL)) and pro-inflammatory mediators (interleukin (IL) 6 and 8) were determined in the supernatant of control and ion-exposed osteoblasts. For this purpose, the supernatants were collected and stored at −20 °C prior to quantification. C1CP was determined using the C1CP ELISA (Quidel, Marburg, Germany) according to the manufacturer's recommendations. Absorbance was measured at 405 nm (reference wave length: 630 nm) using the Tecan Infinite® 200 Pro (Tecan Group AG, Maennedorf, Switzerland) microplate reader. A standard curve was prepared to calculate protein concentration in samples. OPG and RANKL were determined via LEGENDplex™ (BioLegend, San Diego, CA, USA) using fluorescence-labeled beads. These beads are conjugated with the specific antibody on its surface. The samples were incubated with the antibody-conjugated beads, thus forming capture bead-analyte-detection antibody sandwiches. Afterwards, streptavidin-phycoerythrin was added, which bound to the biotinylated detection antibodies, providing fluorescent signal intensities in proportion to the amount of bound analytes. Fluorescence intensities (excitation: 575 nm, emission: 660 nm) in samples were analyzed on a flow cytometer (FACSAria™ IIIu, BD Biosciences). The concentrations of OPG and RANKL were quantified using a standard curve generated in the same assay as well as LEGENDplex™ Data Analysis Software v8 (BioLegend, San Diego, CA, USA).

Soluble proteins of IL6 and IL8 were quantified via eBioscience™ Human IL-6/IL-8 ELISA Ready-SET-Go!™ Kits (both: ThermoFisher Scientific, Waltham, MA, USA) according to the instructions of the manufacturer. Absorbance was measured at 405 nm (reference wave length: 630 nm) using the Tecan Infinite® 200 Pro (Tecan Group AG, Maennedorf, Switzerland) microplate reader. Sample concentrations were calculated using a standard curve, respectively.

Finally, all protein contents within the samples were normalized to the overall protein content which was quantified by the Qubit Protein Assay Kit and Qubit 1.0 (both: Invitrogen) according to the manufacturer's instructions.

2.6. Data Analysis and Illustration

Data analysis and illustration were performed by GraphPadPRISM v.7.02 (GraphPad Inc., San Diego, CA, USA). Results are shown as box plots. Boxes depict interquartile ranges, horizontal lines within boxes depict medians, and whiskers depict maximum and minimum values. For cell culture experiments, human osteoblasts were used in duplicates with a minimum of four independent donors. While gene expression results are depicted as percentage of $2^{(-\Delta\Delta Ct)}$ for a better visualization of the changes with the untreated control (0 μM) set as 100%, the underlying statistical analysis was performed with the ΔCt values to allow the statistical comparison to the untreated control (0 μM). For the statistical analysis of protein data the values of the specific protein amount normalized to total protein content were used.

Statistical comparisons regarding the influence metal salts at different concentrations on cellular activity, gene expression, and protein synthesis were performed with repeated measures (RM) two-way analysis of variance (ANOVA) with "type of metal salt" and "concentration" as variables. Tukey's multiple comparisons test was used for post hoc testing. Results after particle exposure (positive control) were compared to untreated samples (negative control) by either paired t-test or Wilcoxon matched-pairs test depending on normal distribution of data according to Shapiro-Wilk testing. The effects of the different cell culture media were analyzed with RM two-way ANOVA with "type of media" and "concentration" as variables. Post hoc testing was performed with Bonferroni's multiple comparison test. Significances were set to a p-value less than 0.05. Further details of statistical tests are indicated in the results section and the figure legends.

3. Results

3.1. Influence of Metal Ions on Cellular Activity

Evaluation of cell activity was tested via cell number determination and WST-1 assay (see Supplementary Materials Figures S1 and S2). In Figure 1 the metabolic activity of ion-exposed osteoblasts per million cells is depicted in comparison to negative (untreated) and positive (CoCr28Mo6 particles) controls. Cellular activity was influenced by the concentration ($p = 0.0006$) and the type of metal salt ($p = 0.0002$) with a strong interaction between both factors ($p < 0.0001$). Post hoc analysis showed that in comparison to untreated cells, the lowest concentration (10 µM) of Co ions led to decreased cell activity ($p = 0.0430$), while a concentration of 100 µM Co ions resulted in significantly enhanced cell activity levels compared to untreated control and the lower concentrations of 10 µM and 50 µM ($p = 0.0048$, $p < 0.0001$ and $p < 0.0001$, respectively). Moreover, exposure to 100 µM of Co ions and treatment with CoCr28Mo6 particles resulted in similar activity levels. However, when further increasing the concentration of cobalt salt to 500 µM, cellular activity decreased again compared to 100 µM ($p = 0.0430$). Metabolic activity of human osteoblasts exposed to Cr ions was significantly lower than untreated controls but this effect was not concentration-dependent ($p = 0.0046$, $p = 0.0007$, $p = 0.0023$ and $p = 0.0004$ of 10, 50, 100 and 500 µM all compared to untreated control, respectively). Exposure to the highest Ni(2+) concentration (500 µM) led to significantly reduced metabolic activity compared to all lower concentrations of Ni salt, as well as to untreated cells ($p < 0.0001$, $p = 0.0005$, $p < 0.0001$ and $p < 0.0001$ of 0, 10, 50 and 100 µM all compared to 500 µM, respectively).

When comparing the different metal salts, exposure to Cr(3+) (50 µM, 100 µM and 500 µM) resulted in significantly reduced activity levels compared to Co ions ($p = 0.0156$, $p < 0.0001$ and $p < 0.0001$, respectively). For Ni ions there were still significantly decreased metabolism rates for concentrations 100 µM and 500 µM detected when compared to Co ($p < 0.001$ and $p < 0.0001$ at 100 µM and 500 µM, respectively).

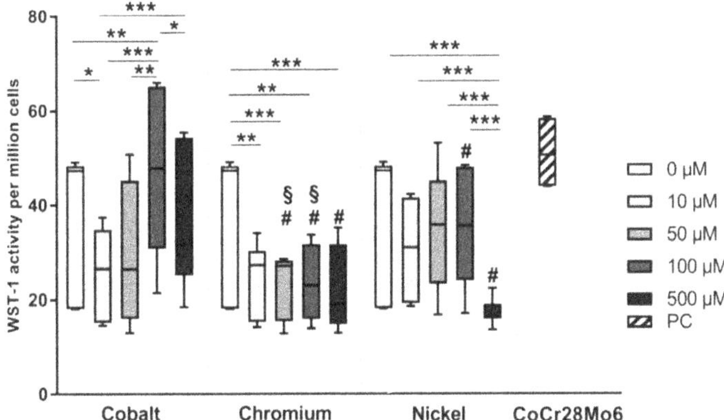

Figure 1. Cell activity of human osteoblasts after exposure to metal salts. Untreated cells served as controls (0 µM) while treatment with CoCr28Mo6 particles (0.01 mg/mL) was used as positive control (PC). Osteoblasts were treated with different concentrations of Co(2+), Cr(3+) and Ni(2+) over 48 h. Afterwards metabolic activity was determined via water soluble tetrazolium salt (WST-1) assay followed by cell number analysis using CyQUANT NF Cell Proliferation Assay. Data are shown as metabolic activity per million cells, depicted as box plots (n = 7). Significance was calculated with concentration-dependent differences: * $p < 0.05$, ** $p < 0.01$, *** $p < 0.001$; differences to cobalt: # $p < 0.05$; differences to nickel: § $p < 0.05$.

The evaluation of cell morphology after exposure to metal salts was carried out with light microscopy and actin staining. Light microscopy revealed a tendency to morphological changes after treatment with Co and Ni ions (Figure 2B,D). Compared to untreated and Cr(3+)-exposed osteoblasts, cells seemed to be more fusiform without clearly formed filopodia for cell connections. Additionally, the actin stain of cells revealed a decrease in cell number after treatment with the bivalent ions Co(2+) and Ni(2+). While actin filaments were clearly visible in the control (Figure 2E) and Cr(3+)-exposed cells (Figure 2G), a weakening of the fluorescence signal was observed in the Co(2+) group (Figure 2F). The treatment with Ni ions not only led to a reduction in cell number (Figure S1) but also seemed to affect the cytoskeleton of the cells. Partly, cells were completely negative for actin fluorescence staining with only the counterstained nucleus visible (indicated by arrows in Figure 2H) or the cells were only stained along the cell membrane with no visible intracellular network structure.

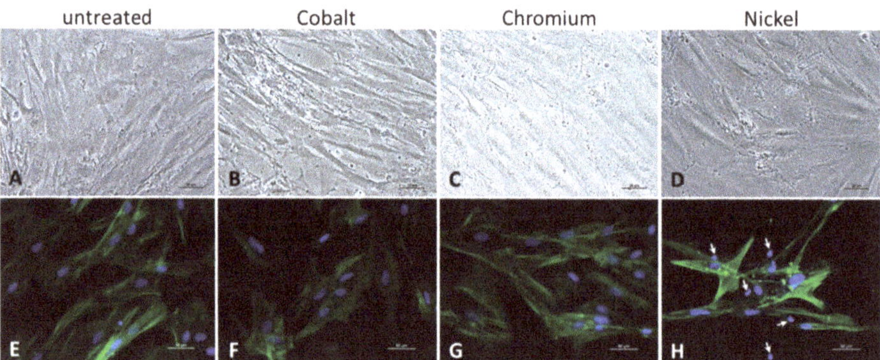

Figure 2. Determination of cell morphology of ion-exposed human osteoblasts (500 µM) after 48 h of incubation. (**A–D**): Phase contrast microscopy (scale bar: 20 µm); (**E–H**): Fluorescence staining of actin filaments and cell nuclei (scale bar: 50 µm).

3.2. Influence of Cell Culture Medium on Cell Activity after Exposure to Metal Salts

Since the expansion and long-term culture of osteoblasts in vitro is rather hampered by calcium phosphate deposition and mineralization, we have been using calcium depleted medium in our cell culture for several years. However, a calcium-free environment is far from the in vivo situation in human bone. In order to better mimic the in vivo situation and to determine differential effects of calcium-depleted and calcium-containing medium conditions on activity of ion-exposed human osteoblasts, the standard osteogenic cell culture medium was compared with standard Dulbecco's modified Eagle medium (Figure 3). The main differences between both media are the glucose concentration as well as the calcium and glutamine content. While the osteogenic medium is characterized by a low glucose concentration (1 g/L as well as calcium and glutamine depletion, DMEM contains high glucose (4.5 g/L), calcium chloride (0.26 g/L) and L-alanyl-L-glutamine (0.86 g/L). Since the respective bivalent ions Co(2+) or Ni(2+) would compete with Ca(2+) in the cell culture medium, our main hypothesis was that cellular activity was less affected in Ca-enriched medium after treatment with bivalent metal salts. We assumed that increased calcium levels in the cell culture allow osteoblasts to take up Ca(2+) rather than Co(2+) or Ni(2+), which may affect cell activity positively. With regard to our results, similar values of metabolic activity were detectable for untreated cells and ion-exposed cells in concentrations between 10 µM and 100 µM for Co(2+) and Ni(2+). At the highest ion concentration of 500 µM, a significant difference (Co: $p = 0.0061$; Ni: $p = 0.004$) between both media was determined with lower cell activity rates for Ca(2+) depleted medium. In Cr(3+)-exposed osteoblast, no differences between both media were detectable.

3.3. Expression of TLR4

Downstream processes are initiated by implant debris recognition via toll-like receptor (TLR) 4 known as a relevant receptor in aseptic implant loosening [30]. To determine *TLR4* gene expression in osteoblasts, cells were treated with 100 µM and 500 µM metal salt solutions since cell activity studies indicated significant differences between these concentrations. RM two-way ANOVA revealed significant differences for "type of metal salt" ($p = 0.0132$) and "concentration" ($p = 0.0224$) with a strong interaction between both factors ($p < 0.0001$). This was confirmed by post hoc testing (Figure 4). The positive control (CoCr28Mo6) as well as Co ions at the concentration of 500 µM, but not at 100 µM, significantly induced *TLR4* gene expression compared to the untreated control ($p = 0.0239$ and $p < 0.0001$, respectively) after 48 h. Co ions at 500 µM significantly enhanced *TLR4* transcripts compared to Cr ions ($p < 0.0001$) and Ni ions ($p = 0.0006$) at the same concentration. Meanwhile, Ni(2+) at 500 μM also showed an upregulation of mRNA when compared to Cr(3+) at 500 μM ($p = 0.0122$)—this effect did not reach significance when compared to the untreated control ($p = 0.1749$).

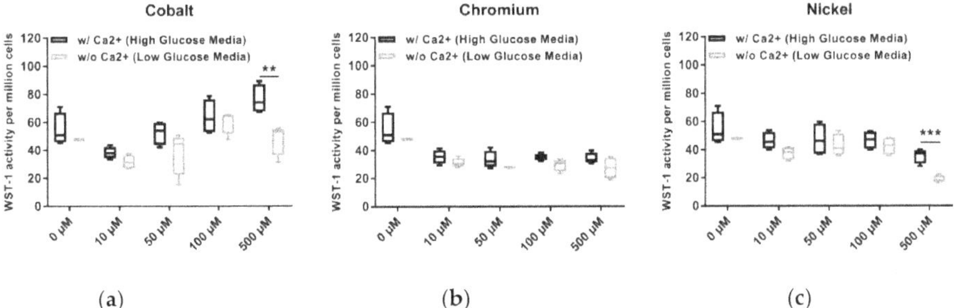

(a)   (b)   (c)

Figure 3. Comparison of cell activity of human osteoblasts after exposure to metal salts in Ca(2+) containing (high glucose medium) and Ca(2+) depleted (low glucose medium). Untreated cells served as negative control (0 μM). Osteoblasts were treated with different concentrations of (**a**) Co(2+), (**b**) Cr(3+) and (**c**) Ni(2+) over 48 h. Afterwards metabolic activity was determined via water soluble tetrazolium salt (WST-1) assay followed by cell number analyses using CyQUANT NF Cell Proliferation Assay. Data are shown as metabolic activity per million cells, depicted as box plots (n = 4). Data distribution and significance were calculated with Bonferroni test and two-way ANOVA. Medium-dependent differences: ** $p < 0.01$, *** $p < 0.001$.

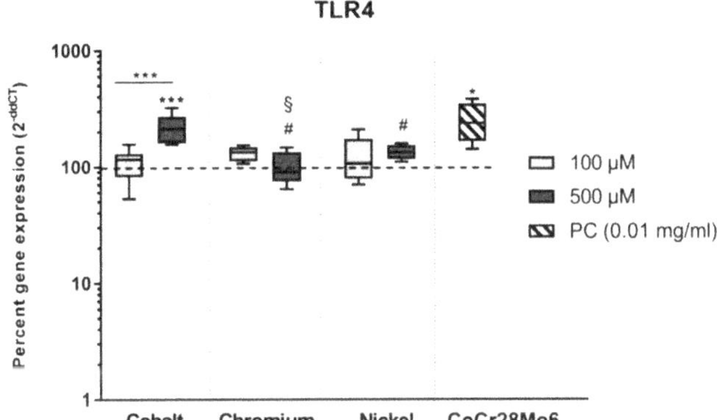

Figure 4. Relative transcript abundance of *TLR4* following exposure to metal salts. Gene expression levels were determined via semiquantitative real-time PCR in human osteoblasts treated with the respective metal salt concentration or particles (PC). Gene expression results are depicted as median and minimum/maximum values (n = 6) of the percentage of $2^{(-\Delta\Delta Ct)}$ related to the untreated control (100%). Significances between groups were calculated with RM two-way ANOVA using ΔCT values. Post hoc testing was performed with Tukey's multiple comparison test: * $p < 0.05$, *** $p < 0.001$; differences to cobalt: # $p < 0.05$; differences to nickel: § $p < 0.05$.

3.4. Osteogenic Differentiation and Bone Remodeling after Exposure to Co-, Cr- and Ni-Ions

The osteogenic differentiation capacity of human osteoblasts was analyzed after treatment with metal salts (100 μM and 500 μM). Collagen type 1, the main component of bone tissue, was clearly affected in human osteoblasts by bivalent ions (Figure 5, RM two-way ANOVA for gene expression: "type of metal salt" $p = 0.0002$, "concentration" $p = 0.0024$ and interaction of both $p < 0.0001$ [5A] and RM two-way ANOVA for protein synthesis: "type of metal salt" $p = 0.0020$, "concentration" $p = 0.0002$ and interaction of both $p < 0.0001$ [5C]). In detail, Co(2+) reduced both transcripts as well as protein levels in a concentration-dependent manner. Similar results were observed after exposure to Ni(2+), however, especially for gene expression, the reduction was not as pronounced as for Co ions. Exposure to 500 µM Co ions resulted in significantly lower mRNA transcripts than after exposure to Ni ions ($p < 0.001$). These results were comparable to *Col1A1* mRNA and protein expression rates of particle-treated human osteoblasts (positive control). Treatment with Cr(3+) did not affect transcript abundance and only slightly reduced protein levels ($p = 0.0054$ to untreated control). Besides collagen type 1, alkaline phosphatase (ALP) as another osteogenic differentiation marker was analyzed on transcript level and also showed significant variations with regard to the two tested variables (RM two-way ANOVA "type of metal salt" $p < 0.0001$, "concentration" $p = 0.0143$, and interaction of both $p < 0.0001$ [5B]). CoCr28Mo6 particles and Co ions significantly reduced *ALP* mRNA in osteoblasts after two days of exposure ($p = 0.0010$, $p = 0.0014$ and $p < 0.0001$ for particles, 100 µM and 500 µM to untreated control, respectively). The effect of Co(2+) was clearly concentration-dependent. Neither Ni nor Cr salts showed a significant influence on *ALP* gene expression and therefore differed significantly from Co at both tested concentrations (Cr to Co: $p < 0.0001$ at 100 and 500 µM; Ni to Co: $p = 0.0003$ and $p < 0.0001$ at 100 and 500 µM, respectively).

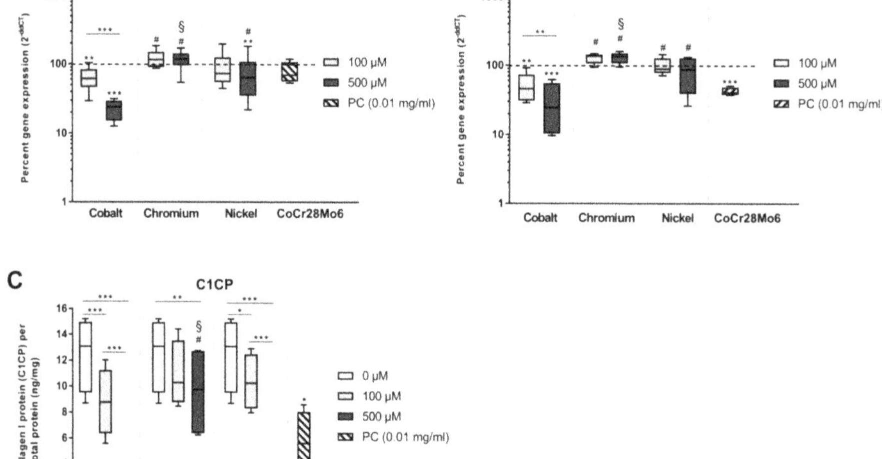

Figure 5. Relative transcript abundance of osteoblastic differentiation markers following exposure to metal salts. Gene expression levels of *Col1A1* and *ALP* by q-PCR (**A**,**B**) were determined in human osteoblasts treated with the respective metal salt concentration or particles (PC). Gene expression results are depicted as median and minimum/maximum values (n = 6) of the percentage of $2^{(-\Delta\Delta Ct)}$ related to the untreated control (100%). (**C**) Cell culture supernatants were used to evaluate the concentration of released pro-collagen 1 protein (C1CP per total protein content) by ELISA. Data are shown as median and minimum/maximum values (n = 4). Significances between groups were calculated with RM two-way ANOVA using ΔCT values for gene expression and values of the specific protein amount normalized to total protein content for protein expression. Post hoc testing was performed with Tukey's multiple comparison test: * $p < 0.05$, ** $p < 0.01$, *** $p < 0.001$; differences to cobalt: # $p < 0.05$; differences to nickel: § $p < 0.05$.

OPG is released by human osteoblasts to counteract activity of RANKL which directly influence osteoclastogenesis. Both bone remodeling markers are synthesized by human osteoblasts. *RANKL* and *OPG* mRNA were neither detected in treated nor in untreated osteoblasts as Ct values were higher than the threshold of 28 cycles. However, on protein level, both bone remodeling marker were detectable. Treatment with Co(2+) and Ni(2+) at 500 μM led to a reduced soluble OPG protein content in the cell culture supernatants (Figure 6A) while RANKL protein was clearly upregulated (Figure 6B). Exposure to Cr(3+) reduced both, OPG and RANKL protein biosynthesis, in human osteoblasts. However, the effect was only significant for OPG synthesis and seems rather due to the generally reduced cellular activity (see Section 3.1) after incubation with Cr ions. The observed reduction of OPG synthesis by CoCr28Mo6 particles did not reach significance ($p = 0.0568$) and RANKL protein synthesis was not affected by particles. It is noteworthy that despite the clearly counteracting trends on OPG and RANKL synthesis by Co and Ni ions the amount of OPG released by osteoblasts still far exceeded the amount of RANKL produced (median values ranging from 9.54 to 73.30 pg/mg total protein for OPG and from 0.06 to 0.32 pg/mg total protein for RANKL).

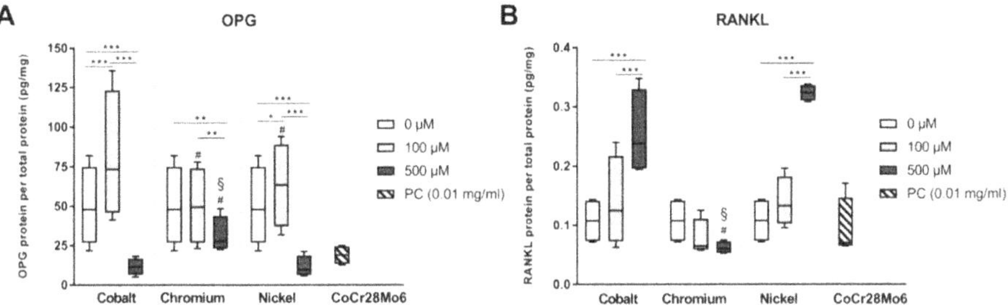

Figure 6. Protein expression of markers for bone remodeling following exposure to metal salts. Cell culture supernatants were used to evaluate the concentrations of released osteoprotegerin (OPG, **A**) and receptor activator of nuclear factor κb ligand (RANKL, **B**) by LEGENDplex™. Data are shown as median and minimum/maximum values (n = 4). Significances between groups were calculated with RM two-way ANOVA using values of the specific protein amount normalized to total protein content. Post hoc testing was performed with Tukey's multiple comparison test: * $p < 0.05$, ** $p < 0.01$, and *** $p < 0.001$; differences to cobalt: # $p < 0.05$; differences to nickel: § $p < 0.05$.

3.5. Induction of Inflammation

In our previous in vitro studies, we could show that IL6 and IL8 are the most relevant pro-inflammatory mediators released by human osteoblasts in response to wear and corrosion products. In this study, *IL6* and *IL8* mRNA and protein expression levels were evaluated after exposure to Co(2+), Ni(2+) and Cr(3+) (Figure 7). When analyzing the gene expression profiles, *IL6* mRNA expression was only induced by Ni(2+) at 500 μM ($p = 0.0006$) while *IL8* mRNA expression levels were elevated by exposure to 500 μM Co, 500 μM Ni and to CoCr28Mo6 particles ($p < 0.0001$, $p < 0.0001$ and $p = 0.0276$, respectively) compared to the untreated control. The release of IL6 into the cell culture supernatant seemed independent from an induction of the gene expression as significant, concentration-dependent increases of IL6 and IL8 protein were determined in the cell culture supernatants after exposure to the bivalent ions Co and Ni. Interestingly, CoCr28Mo6 particles, which were used as a positive control, significantly reduced IL6 and IL8 release into the supernatant after 48 h compared to the untreated control (IL6: $p = 0.0031$, IL8: $p = 0.0197$). Again, incubation with Cr(3+) led to reduced protein levels of IL6 and IL8 in the supernatants, which is in accordance with the observation of a reduced cellular activity and lower protein synthesis for OPG and RANKL as reported in Sections 3.1 and 3.4. In general, higher amounts of IL8 were released by osteoblasts than for IL6 with a maximum of 27.68 pg/mg total protein for IL6 and a maximum of 141.00 pg/mg total protein for IL8.

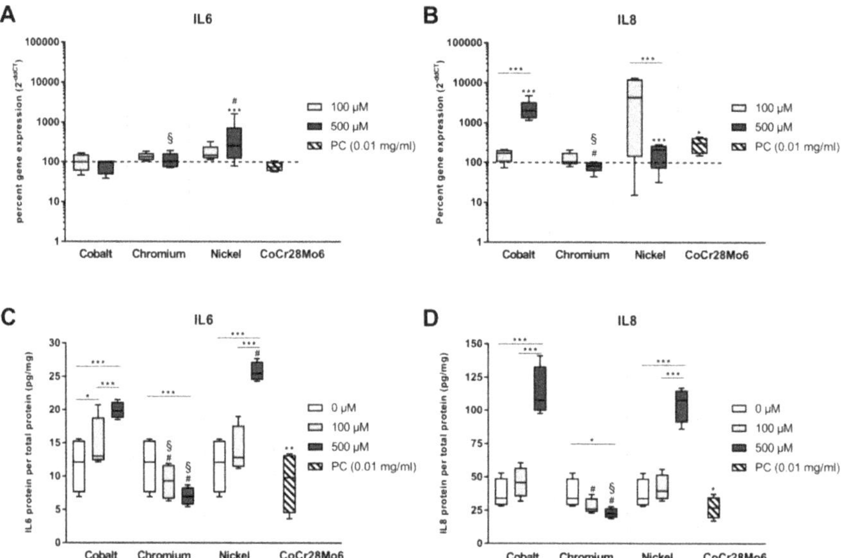

Figure 7. Expression of inflammatory mediators following exposure to metal salts. Gene expression levels of *IL6* and *IL8* by q-PCR (**A**,**B**) were determined in human osteoblasts treated with the respective metal salt concentration or particles (PC). Gene expression results are depicted as median and minimum/maximum values (n = 6) of the percentage of $2^{(-\Delta\Delta Ct)}$ related to the untreated control (100%). Cell culture supernatants were used to evaluate the concentration of released IL6 and IL8 protein (per total protein content) by ELISA (**C**,**D**). Protein data are shown as median and minimum/maximum values (n = 4). Significances between groups were calculated with RM two-way ANOVA using ΔCT values for gene expression and values of the specific protein amount normalized to total protein content for protein expression. Post hoc testing was performed with Tukey's multiple comparisons test: * $p < 0.05$, ** $p < 0.01$, *** $p < 0.001$; differences to cobalt: # $p < 0.05$; differences to nickel: § $p < 0.05$.

4. Discussion

The corrosion of metal implants can lead to the release of free ions, which affect cellular behavior either locally or systemically. In this study, we analyzed the effects of different Co and Cr ions on osteoblastic survival, metabolism, and inflammation potential and compared the results with CoCr28Mo6 particle-exposed cells in order to assess the grade of toxicity. Since Ni is a common component in metal alloys, we additionally treated cells with Ni ions. The main purpose here was to compare the effects of different bivalent ions (Ni and Co) on the above mentioned cell processes.

For the experiments, we used Co(2+), Cr(3+) and Ni(2+) ions derived from metal salts. The tested concentrations were similar to those found in periprosthetic soft tissue [18] but lower than those reported for mineralized bone tissue [22]. The authors explained the relative high values in the mineralized bone tissue with the fact that the tissue was derived from the immediate vicinity, i.e., within a few millimeters, of the implant. This already highlights the difficulty to determine the relevant concentrations for in vitro-testing as these might differ depending on the distance from the implant but also depend on the conditions that aid corrosion. Also, when analyzing the in vivo concentrations, there is no differentiation whether the metals occur in form of ions or particles. Thus, our tested concentrations can only represent a rough estimate of the in vivo situation. Within the applied concentration range we were able to show that Co(2+) had a concentration-dependent influence on cell activity. Although the cell number remained (Figure S1) almost stable, an increase of metabolic activity (Figure S2) was initially detectable between 10 μM and 100 μM ion treatment. This effect can be explained by the fact that Co(2+) is able to surmount the cell membrane via membrane transport systems similar to Ca(2+).

Intracellular, cobalt can then trigger hypoxia-like reactions in the cell, which lead to the upregulation of HIF-associated genes resulting in enhanced glycolytic reactions [31]. However, the highest Co(2+) concentration (500 µM) seemed to be more cytotoxic, not least because of a decline in cell number (Figure S1). The cytotoxicity of Co(2+) has been associated with enhanced oxidative stress, the formation of ROS, and DNA damage [2,32]. Consequentially, apoptosis or necrosis occurs [2,32] which was also detectable in our previous study [21]. Additionally, the toxic effects might be further explained by the inhibition of Ca(2+) entry and Ca(2+) signaling, leading to the intracellular competition of Co(2+) and Ca(2+) with intracellular Ca(2+) binding proteins affecting cellular behavior [6]. The results of our comparative analysis of Ca-containing and –depleted media might reflect the effect described by Scharf et al. (2014) [6]. Osteoblasts in Ca-containing medium showed increased metabolic activity compared to cells incubated in calcium-depleted medium after exposure to 500 µM Co(2+). We assume that the increased glucose content of the calcium-containing medium did not cause the increased metabolic activity, since no differences in cell activity were detected both in untreated cells and at the lower Co(2+) concentrations. As Co(2+) cannot be actively transported out of the cell, intracellular accumulation and cytotoxicity occurs [31]. Thus, in consideration of our results we rather hypothesize that, despite the increased presence of Co(2+), the cells prefer the Ca influx in Ca-enriched medium, resulting in a reduced intracellular Co(2+) accumulation, fewer cytotoxic effects and restored/increased metabolic activity. Bivalent Ni ions might affect osteoblasts via the same mechanism, but are however more cytotoxic than Co(2+). This was reflected by our observation of reduced cell numbers, especially at 500 µM and a loss of the actin filament structure.

In comparison to bivalent metal ions, the cell number was not affected by Cr(3+). However, a decrease in cell metabolism was detectable without concentration-dependent differences. This effect can be explained by the fact that Cr(3+) ions cannot enter the cell membrane [33] and therefore accumulate at the cell membrane [34], which may have an effect on cell metabolism. On the other hand, Magone et al. (2015) described the possibility of Cr(3+) oxidation to Cr(4+) in order to enter the cell membrane [7]. Intracellular, Cr(4+) conjugates with proteins resulting in reduction of Cr(4+) to Cr(3+) [7,32]. Although Shrivastava et al. (2002) mentioned that cell degeneration, DNA damage, and ROS accumulation is assigned to be the result of Cr(4+) exposure [23], other studies have revealed a release of ROS in the presence of Cr(3+) not least because of the rapid Cr (4+) reduction [6,7,32]. Additionally, cellular damage and the induction of apoptotic pathways by Cr(3+) were also described by Rudolf and Cervinka (2009) [35].

Exposure to CoCr28Mo6 particles resulted in unaffected cell activity compared to untreated cells. Compared to ion exposure, particle treatment led to an activity level similar to 100 µM Co(2+) or to obviously enhanced levels compared to Cr(3+). One simple explanation for this result can be the differing mechanism for the uptake and intracellular availability of ions and particles which directly affect cellular activity. As mentioned before, ions can enter the cell membrane via different ion-mediated transporter systems while particles have to be recognized by receptors or actively incorporated via phagocytosis [29,32]. On the other hand, the corrosion of particles within the cell culture media, or by cell-mediated mechanisms including phagocytosis, cannot be excluded. Thus, the released ions from the particles would have the same effect as Co(2+) ions from the metal salts on triggering the previously mentioned hypoxia-like reactions. However, the corrosion-mediated release of ions in the cell culture medium has not been examined yet, but will be the subject of further investigations.

A limitation for the determination of metabolic activity via WST-1 might be that this assay is influenced by the presence of superoxide, which is associated with phagocytosis and oxidative burst [36]. Therefore, the results for metabolic activity can also reflect the presence of ROS, as indeed WST-1 was used for the measurement of superoxide production [36]. However, WST-1 was performed in ion-free media, so we assumed that most of the results were caused by the reduction of formazan via the mitochondrial respiratory chain. Future experiments will establish whether the changes observed in metabolic activity after particle and metal ion exposure are linked to ROS production in our cells.

Intracellular downstream processes, including ROS production and oxidative burst, can be induced by a pathogen or substance binding on surface receptors. Here, toll-like receptors (TLR) recognize a wide spectrum of exogenous and endogenous danger signals [30] and therefore play a central role in the induction of intracellular cascades, especially in macrophages. Since TLR 4 is known to be the most relevant receptor in aseptic implant loosening [30,37], we looked at TLR4 sensitization in human osteoblasts after exposure to metal salts and controls. We initially assumed that CoCr28Mo6 particles primarily induce TLR4 expression and further downstream processes [38] while ions enter the cell membrane via channels and initiate immunological reactions. However, our results indicated that metal ions, especially Co(2+), have almost the same impact on *TLR4* gene expression as particles. It seems likely that the increased release of IL6 and IL8 can be directly linked to TLR4 activation. This is also supported by *TLR4* and IL6/IL8 results of Cr(3+) and Ni(2+) exposed cells. Lawrence and co-workers already showed, that Co(2+) activates TLR 4 which is in turn associated with enhanced chemokine release [39,40]. In contrast, Samelko et al. (2016) reported that CoCr particles did not preferentially activate TLR4 induced inflammation [41]. This is supported by our previous [28] as well as our current findings of reduced IL6 and IL8 secretion after particle exposure. Here, the release of pro-osteolytic mediators might be induced by other downstream mechanism such as inflammasome activation [41,42]. However, an impact of metal ions on inflammasome activation cannot be excluded as a strong link between ion channel expression and inflammasome activation has been described [43]. Thus, to finally prove the link between our findings of ion-mediated IL6 and IL8 release and TLR4 activation, further studies have been carried out, especially for the quantification of TLR4 on the surface of osteoblastic cells.

Osteoblasts exposed to metal ions and CoCr28Mo6 particles showed reduced capacity of differentiation which was almost in accordance to our previous work [21,28]. Interestingly, Drynda et al. (2018) reported in their study that Cr(3+) instead of Co(2+) led to a concentration-dependent inhibition (10 μM to 250 μM) of osteoblastic mineralization [22]. This fact could not be proven in our work, since we found significantly reduced osteogenic differentiation at higher Co(2+) concentration, whereas Cr(3+) had significantly less (Col1 protein) or no effects (*Col1A1, ALP*). Our present data show that not only CoCr28Mo6 particles affect extracellular matrix production and hence bone formation processes, but additionally the presence of bivalent metal ions can inhibit collagen type 1 synthesis. Bone degradation and osteolysis can be further advanced by altered OPG/RANKL ratios as these affect osteoclastogenesis and bone resorption. Here, it should be noted that OPG prevents bone loss while RANKL promotes osteoclastic resorption [7]. Our data indicated that mainly bivalent ions have a significant impact on OPG/RANKL ratios. Although OPG protein concentrations were clearly elevated compared to RANKL, concentration-dependent effects were found: while OPG was significantly downregulated at 500 μM, RANKL protein was significantly upregulated. As a result, OPG/RANKL ratio at 500 μM was only about one tenth of the ratio at 100 μM. This was in contrast to the study of Zijlstra et al. (2012) who revealed a time-dependent effect of Co and Cr ions on *OPG/RANKL* mRNA ratio without concentration-dependent differences [44]. However, they only calculated ratios via gene expression results and not on protein data, which is important to demonstrate osteolytic reactions in the periprosthetic tissue. In our study, we were not able to detect *OPG* and *RANKL* transcripts although we used clearly lower metal ion concentrations. Nonetheless, our protein data can provide important information about the effects of ion exposure on bone remodeling. In this context, co-cultures should be carried out in the future to prove the effects of osteoclastogenesis and further bone resorption as already shown in our previous study [21].

5. Conclusions

Our data show that especially Co(2+) as a bivalent ion has a significant influence on osteoblastic activity, differentiation, and inflammatory processes. In this work, we were further able to show clear differences between Co(2+) and Cr(3+), and thus to assess the toxicity of both ions. We were able to demonstrate an increased sensitization of human osteoblasts resulting in unbalanced bone remodeling as shown by lower collagen type production, decreased OPG/RANKL ratio, as well as the induction of inflammation. Moreover, corrosion products induced cellular downstream processes faster, i.e., after a short-time exposure of only 48 h, compared to CoCr28Mo6 particles where significant effects were shown after 96 h [28].

Supplementary Materials: The following are available online at http://www.mdpi.com/1996-1944/12/17/2771/s1, Figure S1: Cell numbers of human osteoblasts after exposure to metal salts. Untreated cells served as controls (0 μM). Osteoblasts were treated with different concentrations of Co(2+), Cr(3+) and Ni(2+) over 48 h. Afterwards cell number was determined via CyQUANT NF Cell Proliferation Assay. Data are depicted as box plots (n = 7). Significance was calculated with concentration-dependent differences: * $p < 0.05$, ** $p < 0.01$, *** $p < 0.001$; differences to cobalt: $^{\#}$ $p < 0.05$; differences to nickel: § $p < 0.05$, Figure S2: Metabolic activity of human osteoblasts after exposure to metal salts. Untreated cells served as controls (0 μM). Osteoblasts were treated with different concentrations of Co(2+), Cr(3+) and Ni(2+) over 48 h. Afterwards metabolic activity was determined via water soluble tetrazolium salt (WST-1) assay. Data are depicted as box plots (n = 7). Significance was calculated with concentration-dependent differences: * $p < 0.05$, *** $p < 0.001$; differences to cobalt: $^{\#}$ $p < 0.05$; differences to nickel: § $p < 0.05$.

Author Contributions: All authors were fully engaged in the study and in producing the manuscript. A.J.-H. and A.K. projected the study. M.-L.S. and A.S. performed the experiments with osteoblasts and evaluated the data with support of A.J.-H. and A.K. A.K. and M.-L.S. made the statistical analysis. A.J.-H. and A.K. wrote the primary manuscript with support of K.P., B.M.-H., T.F. and R.B. R.B. provided all laboratory equipment. A.J.-H., K.P., T.F. and B.M.-H. organized the research funding. All authors ensured the accuracy of the data and the analyses and reviewed the manuscript in its current state.

Funding: This work was financially supported by Rostock University Medical Center (Joint Project: KOBE) and the AFOR Foundation.

Acknowledgments: We gratefully acknowledge Katharina Ekat and Doris Hansmann for their technical support.

Conflicts of Interest: The authors declare no conflict of interest.

References

1. Matusiewicz, H. Potential release of in vivo trace metals from metallic medical implants in the human body: From ions to nanoparticles—A systematic analytical review. *Acta Biomater.* **2014**, *10*, 2379–2403. [CrossRef]

2. Sansone, V.; Pagani, D.; Melato, M. The effects on bone cells of metal ions released from orthopaedic implants: A review. *Clin. Cases Miner Bone Metab.* **2013**, *10*, 34–40. [CrossRef]

3. Hussenbocus, S.; Kosuge, D.; Solomon, L.B.; Howie, D.W.; Oskouei, R.H. Head-Neck Taper Corrosion in hip Arthroplasty. *Biomed. Res. Int.* **2015**, *2015*, 758123. [CrossRef]

4. Cooper, H.J.; Della Valle, C.J.; Berger, R.A.; Tetreault, M.; Paprosky, W.G.; Sporer, S.M.; Jacobs, J.J. Corrosion at the head-neck taper as a cause for adverse local tissue reactions after total hip arthroplasty. *J. Bone Joint Surg. Am.* **2012**, *94*, 1655–1661. [CrossRef]

5. Sabella, S.; Carney, R.P.; Brunetti, V.; Malvindi, M.A.; Al-Juffali, N.; Vecchio, G.; James, S.M.; Bakr, O.M.; Cingolani, R.; Stellacci, F.; et al. A general mechanism for intracellular toxicity of metal-containing nanoparticles. *Nanoscale* **2014**, *6*, 7052–7061. [CrossRef]

6. Scharf, B.; Clement, C.C.; Zolla, V.; Perino, G.; Yan, B.; Elci, S.G.; Purdue, E.; Goldring, S.; Macaluso, F.; Cobelli, N.; et al. Molecular analysis of chromium and cobalt-related toxicity. *Sci. Rep.* **2014**, *4*, 5729. [CrossRef]

7. Magone, K.; Luckenbill, D.; Goswami, T. Metal ions as inflammatory initiators of osteolysis. *Arch. Orthop. Trauma Surg.* **2015**, *135*, 683–695. [CrossRef]

8. Gilbert, J.L.; Sivan, S.; Liu, Y.; Kocagöz, S.B.; Arnholt, C.M.; Kurtz, S.M. Direct in vivo inflammatory cell-induced corrosion of CoCrMo alloy orthopedic implant surfaces. *J. Biomed. Mater. Res. A* **2015**, *103*, 211–223. [CrossRef]

9. Toh, W.Q.; Tan, X.; Bhowmik, A.; Liu, E.; Tor, S.B. Tribochemical characterization and tribocorrosive behavior of CoCrMo alloys: A Review. *Materials* **2017**, *11*, 30. [CrossRef]

10. Gibon, E.; Amanatullah, D.F.; Loi, F.; Pajarinen, J.; Nabeshima, A.; Yao, Z.; Hamadouche, M.; Goodman, S.B. The biological response to orthopaedic implants for joint replacement: Part I: Metals. *J. Biomed. Mater. Res. B Appl. Biomater.* **2016**, *105*, 2162–2173. [CrossRef]

11. Hannemann, F.; Hartmann, A.; Schmitt, J.; Lützner, J.; Seidler, A.; Campbell, P.; Delaunay, C.P.; Drexler, H.; Ettema, H.B.; García-Cimbrelo, E.; et al. European multidisciplinary consensus statement on the use and monitoring of metal-on-metal bearings for total hip replacement and hip resurfacing. *Orthop. Traumatol. Surg. Res.* **2013**, *99*, 263–271. [CrossRef]

12. Test ID: COS. Cobalt, Serum. Available online: https://www.mayocliniclabs.com/test-catalog/Clinical+and+Interpretive/80084 (accessed on 8 August 2019).

13. Steens, W.; Loehr, J.F.; von Foerster, G.; Katzer, A. Chronic cobalt poisoning in endoprosthetic replacement. *Orthopade* **2006**, *35*, 860–864. [CrossRef]

14. Oldenburg, M.; Wegner, R.; Baur, X. Severe cobalt intoxication due to prosthesis wear in repeated total hip arthroplasty. *J. Arthroplasty* **2009**, *24*, 825.e15–825.e20. [CrossRef]

15. Machado, C.; Appelbe, A.; Wood, R. Arthroprosthetic cobaltism and cardiomyopathy. *Heart Lung Circ.* **2012**, *21*, 759–760. [CrossRef]

16. Fox, K.A.; Phillips, T.M.; Yanta, J.H.; Abesamis, M.G. Fatal cobalt toxicity after total hip arthroplasty revision for fractured ceramic components. *Clin Toxicol. (Phila)* **2016**, *54*, 874–877. [CrossRef]

17. Peters, R.M.; Willemse, P.; Rijk, P.C.; Hoogendoorn, M.; Zijlstra, W.P. Fatal Cobalt toxicity after a non-metal-on-metal total hip arthroplasty. *Case Rep. Orthop.* **2017**, 9123684. [CrossRef]

18. Kuba, M.; Gallo, J.; Pluhácek, T.; Hobza, M.; Milde, D. Content of distinct metals in periprosthetic tissues and pseudosynovial joint fluid in patients with total joint arthroplasty. *J. Biomed. Mater. Res. B Appl. Biomater.* **2019**, *107*, 454–462. [CrossRef]

19. Kim, C.H.; Choi, Y.H.; Jeong, M.Y.; Chang, J.S.; Yoon, P.W. Cobalt intoxication heart failure after revision total hip replacement for ceramic head fracture: A case report. *Hip Pelvis.* **2016**, *28*, 259–263. [CrossRef]

20. Andrews, R.E.; Shah, K.M.; Wilkinson, J.M.; Gartland, A. Effects of cobalt and chromium ions at clinically equivalent concentrations after metal-on-metal hip replacement on human osteoblasts and osteoclasts: Implications for skeletal health. *Bone* **2011**, *49*, 717–723. [CrossRef]

21. Jonitz-Heincke, A.; Tillmann, J.; Klinder, A.; Krueger, S.; Kretzer, J.P.; Høl, P.J.; Paulus, A.C.; Bader, R. The impact of metal ion exposure on the cellular behavior of human osteoblasts and PBMCs: In vitro analyses of osteolytic processes. *Materials (Basel)* **2017**, *10*, 734. [CrossRef]

22. Busse, B.; Hahn, M.; Niecke, M.; Jobke, B.; Püschel, K.; Delling, G.; Katzer, A. Allocation of nonbirefringent wear debris: Darkfield illumination associated with PIXE microanalysis reveals cobalt deposition in mineralized bone matrix adjacent to CoCr implants. *J. Biomed. Mater. Res. A* **2008**, *87*, 536–545. [CrossRef]

23. Drynda, S.; Drynda, A.; Feuerstein, B.; Kekow, J.; Lohmann, C.H.; Bertrand, J. The effects of cobalt and chromium ions on transforming growth factor-beta patterns and mineralization in human osteoblast-like MG63 and SaOs-2 cells. *J. Biomed. Mater. Res. A* **2018**, *106*, 2105–2115. [CrossRef]

24. Hallab, N.J.; Vermes, C.; Messina, C.; Roebuck, K.A.; Glant, T.T.; Jacobs, J.J. Concentration- and composition-dependent effects of metal ions on human MG-63 osteoblasts. *J. Biomed. Mater. Res.* **2002**, *60*, 420–433. [CrossRef]

25. Lochner, K.; Fritsche, A.; Jonitz, A.; Hansmann, D.; Mueller, P.; Mueller-Hilke, B.; Bader, R. The potential role of human osteoblasts for periprosthetic osteolysis following exposure to wear particles. *Int. J. Mol. Med.* **2011**, *28*, 1055–1063. [CrossRef]

26. Brauer, A.; Pohlemann, T.; Metzger, W. Osteogenic differentiation of immature osteoblasts: Interplay of cell culture media and supplements. *Biotech. Histochem.* **2016**, *91*, 161–169. [CrossRef]

27. Skottke, J.; Gelinsky, M.; Bernhardt, A. In vitro Co-culture model of primary human osteoblasts and osteocytes in collagen gels. *Int. J. Mol. Sci.* **2019**, *20*, 1998. [CrossRef]

28. Klinder, A.; Seyfarth, A.; Hansmann, D.; Bader, R.; Jonitz-Heincke, A. Inflammatory response of human peripheral blood mononuclear cells and osteoblasts incubated with metallic and ceramic submicron particles. *Front. Immunol.* **2018**, *9*, 831. [CrossRef]

29. Jonitz-Heincke, A.; Tillmann, J.; Ostermann, M.; Springer, A.; Bader, R.; Hol, P.J.; Cimpan, M.R. Label-free monitoring of uptake and toxicity of endoprosthetic wear particles in human cell cultures. *Int. J. Mol. Sci.* **2018**, *19*, 3486. [CrossRef]

30. Nich, C.; Goodman, S.B. The role of macrophages in the biological reaction to wear debris from joint replacements. *J. Long Term Eff. Med. Implants.* **2014**, *24*, 259–265. [CrossRef]

31. Simonsen, L.O.; Harbak, H.; Bennekou, P. Cobalt metabolism and toxicology—A brief update. *Sci. Total Environ.* **2012**, *432*, 210–215. [CrossRef]

32. Bijukumar, D.R.; Segu, A.; Souza, J.C.M.; Li, X.; Barba, M.; Mercuri, L.G.; J Jacobs, J.; Mathew, M.T. Systemic and local toxicity of metal debris released from hip prostheses: A review of experimental approaches. *Nanomedicine* **2018**, *14*, 951–963. [CrossRef]

33. Shrivastava, R.; Upreti, R.K.; Seth, P.K.; Chaturvedi, U.C. Effects of chromium on the immune system. *FEMS Immunol. Med. Microbiol.* **2002**, *34*, 1–7. [CrossRef]

34. Keegan, G.M.; Learmonth, I.D.; Case, C.P. Orthopaedic metals and their potential toxicity in the arthroplasty patient: A review of current knowledge and future strategies. *J. Bone Joint Surg. Br.* **2007**, *89*, 567–573. [CrossRef]

35. Rudolf, E.; Cervinka, M. Trivalent chromium activates Rac-1 and Src and induces switch in the cell death mode in human dermal fibroblasts. *Toxicol. Lett.* **2009**, *188*, 236–242. [CrossRef]

36. Berridge, M.V.; Herst, P.M.; Tan, A.S. Tetrazolium dyes as tools in cell biology: New insights into their cellular reduction. *Biotechnol. Annu. Rev.* **2005**, *11*, 127–152.

37. Gu, Q.; Shi, Q.; Yang, H. The role of TLR and chemokine in wear particle-induced aseptic loosening. *J. Biomed. Biotechnol.* **2012**, 596870. [CrossRef]

38. Potnis, P.A.; Dutta, D.K.; Wood, S.C. Toll-like receptor 4 signaling pathway mediates proinflammatory immune response to cobalt-alloy particles. *Cell Immunol.* **2013**, *282*, 53–65. [CrossRef]

39. Lawrence, H.; Mawdesley, A.E.; Holland, J.P.; Kirby, J.A.; Deehan, D.J.; Tyson-Capper, A.J. Targeting Toll-like receptor 4 prevents cobalt-mediated inflammation. *Oncotarget* **2016**, *7*, 7578–7585. [CrossRef]

40. Lawrence, H.; Deehan, D.J.; Holland, J.P.; Anjum, S.A.; Mawdesley, A.E.; Kirby, J.A.; Tyson-Capper, A.J. Cobalt ions recruit inflammatory cells in vitro through human Toll-like receptor 4. *Biochem. Biophys. Rep.* **2016**, *7*, 374–378. [CrossRef]

41. Samelko, L.; Landgraeber, S.; McAllister, K.; Jacobs, J.; Hallab, N.J. Cobalt alloy implant debris induces inflammation and bone loss primarily through danger signaling, Not TLR4 activation: Implications for DAMP-ening implant related inflammation. *PLoS ONE* **2016**, *11*, e0160141. [CrossRef]

42. Bitar, D.; Parvizi, J. Biological response to prosthetic debris. *World J. Orthop.* **2015**, *6*, 172–189. [CrossRef] [PubMed]

43. Santoni, G.; Cardinali, C.; Morelli, M.B.; Santoni, M.; Nabissi, M.; Amantini, C. Danger- and pathogen-associated molecular patterns recognition by pattern-recognition receptors and ion channels of the transient receptor potential family triggers the inflammasome activation in immune cells and sensory neurons. *J. Neuroinflamm.* **2015**, *12*, 21. [CrossRef] [PubMed]

44. Zijlstra, W.P.; Bulstra, S.K.; van Raay, J.J.; van Leeuwen, B.M.; Kuijer, R. Cobalt and chromium ions reduce human osteoblast-like cell activity in vitro, reduce the OPG to RANKL ratio, and induce oxidative stress. *J. Orthop. Res.* **2012**, *30*, 740–747. [CrossRef] [PubMed]

Article

Leaf Age-Dependent Effects of Foliar-Sprayed CuZn Nanoparticles on Photosynthetic Efficiency and ROS Generation in *Arabidopsis thaliana*

Ilektra Sperdouli [1,2], Julietta Moustaka [1,3], Orestis Antonoglou [4],
Ioannis-Dimosthenis S. Adamakis [1,5], Catherine Dendrinou-Samara [4] and Michael Moustakas [1,*]

[1] Department of Botany, Aristotle University of Thessaloniki, GR-54124 Thessaloniki, Greece
[2] Institute of Plant Breeding and Genetic Resources, Hellenic Agricultural Organisation–Demeter, Thermi, GR-57001 Thessaloniki, Greece
[3] Department of Plant and Environmental Sciences, University of Copenhagen, Thorvaldsensvej 40, DK-1871 Frederiksberg C, Denmark
[4] Laboratory of Inorganic Chemistry, Department of Chemistry, Aristotle University of Thessaloniki, 54124 Thessaloniki, Greece
[5] Department of Botany, Faculty of Biology, National and Kapodistrian University of Athens, 157 84 Athens, Greece
*   Correspondence: moustak@bio.auth.gr; Tel.: +30-2310-99-8335

Received: 26 June 2019; Accepted: 3 August 2019; Published: 6 August 2019

Abstract: Young and mature leaves of *Arabidopsis thaliana* were exposed by foliar spray to 30 mg L^{-1} of CuZn nanoparticles (NPs). The NPs were synthesized by a microwave-assisted polyol process and characterized by dynamic light scattering (DLS), X-ray diffraction (XRD), and transmission electron microscopy (TEM). CuZn NPs effects in *Arabidopsis* leaves were evaluated by chlorophyll fluorescence imaging analysis that revealed spatiotemporal heterogeneity of the quantum efficiency of PSII photochemistry (Φ_{PSII}) and the redox state of the plastoquinone (PQ) pool (q_p), measured 30 min, 90 min, 180 min, and 240 min after spraying. Photosystem II (PSII) function in young leaves was observed to be negatively influenced, especially 30 min after spraying, at which point increased H_2O_2 generation was correlated to the lower oxidized state of the PQ pool. Recovery of young leaves photosynthetic efficiency appeared only after 240 min of NPs spray when also the level of ROS accumulation was similar to control leaves. On the contrary, a beneficial effect on PSII function in mature leaves after 30 min of the CuZn NPs spray was observed, with increased Φ_{PSII}, an increased electron transport rate (ETR), decreased singlet oxygen ($^1O_2$) formation, and H_2O_2 production at the same level of control leaves. An explanation for this differential response is suggested.

Keywords: bimetallic nanoparticles; hydrogen peroxide; mature leaves; non-photochemical quenching; photoprotective mechanism; photosynthetic heterogeneity; plastoquinone pool; redox state; spatiotemporal heterogeneity; young leaves

---

1. Introduction

Both zinc (Zn) and copper (Cu) are essential elements for plant growth [1]. Zn deficiency results in a rapid inhibition of plant growth and development, while several physiological processes are impaired [1–3]. Zinc scarcity in arable soils [4] is a major problem worldwide [2], which is mainly due to the low Zn soil solubility resulting in Zn unavailability to plant roots [5]. Adequate Zn supply is suggested to improve productivity and nutrients in crops [6]. Low Zn concentrations in soils can be improved by adding Zn fertilizers, but this is a costly and ineffective policy [7]. However, to increase grain Zn concentrations, foliar Zn application can be applied [7–9]. As Cu is also an essential element for all organisms, Cu-based fertilizers and fungicides have been widely used in agriculture as well [10,11].

Nanoparticles (NPs) in agriculture are used to reduce the amount of sprayed chemical products by the smart delivery of active ingredients, diminish nutrient losses in fertilization, and increase yields through optimized water and nutrient management [12–15]. Nevertheless, NPs are experiencing problems in reaching the market as novel products in agriculture, making agriculture still a negligible area for nanotechnology due mainly to the high production costs required in high volumes and to doubtful practical profits and lawmaking uncertainties [15]. In addition, the major concern regarding the use of NPs is their possible phytotoxicity, which is closely related to their chemical composition, structure, size, and surface area [12].

Since plant production is driven by photosynthesis, it is possible to estimate the fate of plant growth and development by evaluating photosynthetic function [16]. The process of photosynthesis converts light energy into chemical energy by the collaboration of photosystem I (PSI), and photosystem II (PSII), which work in coordination [16–18]. Chlorophyll fluorescence analysis has been widely used as a highly sensitive indicator of photosynthetic efficiency [19–26]. The obtained information can be interpreted to acquire knowledge about the state of the photosynthetic machinery and the effects of environmental pressure on plants [27,28]. Nevertheless, photosynthetic functioning is not uniform at the leaf area particularly under abiotic stress circumstances, which makes conventional chlorophyll fluorescence measurements non-characteristic of the physiological status of the entire leaf [29]. This disadvantage overcomes chlorophyll fluorescence imaging analysis and permits the detection of spatiotemporal heterogeneity at the total leaf surface [30–35].

The synthesis of hydrophilic CuZn NPs is challenging and has been hardly reported before [36,37]. We have previously evaluated the phytotoxicity of 15 mg L^{-1} and 30 mg L^{-1} of CuZn NPs sprayed on tomato plants by determining their effects on the light reactions of photosynthesis [38]. The evaluated CuZn NPs displayed minimal ionic dissolution (<10%), a significant amount of biocompatible polyol surface coating (32%), and high crystallinity, factors that minimize their toxic effects [39]. While no significant effects in PSII functionality were noticed with 15 mg L^{-1} of NPs, the application of 30 mg L^{-1} of CuZn NPs resulted in a reduced plastoquinone (PQ) pool that gave rise to H_2O_2 generation [38]. A reduced PQ pool reflects an imbalance between energy supply and demand [40,41]—or, in other words, excess excitation energy [42–44]. Young leaves have the ability to dissipate the excess excitation energy by non-photochemical quenching (NPQ) more efficiently than mature leaves [45], and this is sufficient in scavenging reactive oxygen species (ROS) [42,46,47]. Based on these reports, we hypothesized that exposing young and mature leaves to 30 mg L^{-1} of CuZn NPs will result in differential effects on them, with young leaves retaining a more oxidized PQ pool and less H_2O_2 accumulation. *Arabidopsis thaliana* was selected as the plant material to test our hypothesis, since it has been widely used as a model system in understanding the physiological mechanisms of higher plants [31,48].

2. Materials and Methods

2.1. Synthesis of CuZn NPs

Based on our previous results, a modified synthesis of CuZn NPs was followed using a commercial microwave accelerated reaction system, the Model MARS 6-240/50-CEM [38]. Equal amounts of $Zn(NO_3)_2 \cdot 4H_2O$ (2.0 mmol) and $Cu(NO_3)_2 \cdot 3H_2O$ (2.0 mmol) were mixed and dissolved in 20 mL of triethylene glycol (TrEG). After centrifugation at 2800 g, the supernatant liquids were discarded, and the black-brown precipitate was washed three times with ethanol. NPs were re-dispersed in water via sonication assistance to form stable aqueous suspensions at concentrations <200 mg L^{-1}.

2.2. Characterization of CuZn NPs

Primary particle size and morphology was determined by a conventional transmission electron microscope (TEM) (JEOL JEM 1010, Tokyo, Japan) [38].

The crystal structure was investigated through X-ray diffraction (XRD) performed on a Philips PW 1820 diffractometer (Amsterdam, The Netherlands) at a scanning rate of 0.050/3 s, in the 2θ range from $10°$ to $90°$, with monochromatized Cu Kα radiation ($\lambda = 1.5406$ nm) [38].

The hydrodynamic size of CuZn NPs was determined by dynamic light scattering (DLS) measurements, which were carried out at 25 °C utilizing a Nano ZS Zetasizer (Malvern Instrument, Worcestershire, UK) apparatus [33].

2.3. Plant Material and Exposure to CuZn NPs

Arabidopsis thaliana ecotype Columbia (Col-0) seeds obtained from the Nottingham Arabidopsis Stock Centre (NASC) were sown on a soil and peat mixture in a controlled-environment growth chamber under $22 \pm 1/18 \pm 1$ °C day/night temperature with a 16-hour day at 130 ± 20 μmol photons m^{-2} s^{-1} light intensity and $50 \pm 5/60 \pm 5\%$ day/night humidity. Four and six-week-old Arabidopsis plants were spayed with 30 mg L^{-1} of CuZn NPs. Rosette leaf 8 from four-week-old (young, immature leaf) and six-week-old (mature to senescing leaf) were selected for photosynthetic measurements 30 min, 90 min, 180 min, and 240 min after the foliar spray with 30 mg L^{-1} of CuZn NPs, while control plants were sprayed with distilled water.

2.4. Chlorophyll Fluorescence Imaging Analysis

Chlorophyll fluorescence analysis was conducted with an Imaging-PAM Chlorophyll Fluorometer (Walz, Effeltrich, Germany) as described in detail previously [27]. Chlorophyll fluorescence parameters were measured in dark-adapted (15 min) leaves (rosette leaf 8) from four and six-week-old Arabidopsis plants sprayed with distilled water (control), or 30 mg L^{-1} of CuZn NPs. In each leaf, eight to 11 areas of interest (AOI) that covered the whole leaf area were selected for analysis. Four to five leaves from different plants were measured at each treatment at the actinic light intensity of 140 μmol photons m^{-2} s^{-1}. By using the Imaging Win software (Heinz Walz GmbH, Effeltrich, Germany), we measured the effective quantum yield of photochemistry in PSII (Φ_{PSII}), the quantum yield of regulated non-photochemical energy loss in PSII (Φ_{NPQ}), the quantum yield of non-regulated energy loss in PSII (Φ_{NO}), the photochemical quenching (q_p) that is a measure of the redox state of the PQ pool, the non-photochemical quenching (NPQ), and the relative PSII electron transport rate (ETR).

Representative results of the effective quantum yield of photochemical energy conversion in PSII (Φ_{PSII}) and the redox state of PQ pool (q_p) are also shown as color-coded images, after 5 min of illumination with 140 μmol photons m^{-2} s^{-1}.

2.5. Imaging of ROS

The presence of ROS was detected in young and mature leaves of *Arabidopsis thaliana* by staining with 25 μM of 2′,7′-dichlorofluorescein diacetate (H$_2$DCF-DA, Sigma, St. Louis, MO, USA) in the dark, as described before [41]. After 30 min of incubation, the leaves were observed with a Zeiss AxioImager Z.2 fluorescence microscope (Jena, Germany) equipped with an MRc5 Axiocam using the AxioVision SE64 4.8.3 software according to the manufacturer's instructions.

2.6. Statistical Analyses

Four to five leaves from different plants were analyzed for each treatment, and the differences between chlorophyll fluorescence parameters were separated by paired t-test at a level of $p < 0.05$ with the StatView software (computer package, Abacus Concepts, Inc., Berkley, CA, USA) [23].

3. Results

3.1. Characterization of the Synthesized CuZn NPs

The crystal structure of the NPs was investigated through X-ray diffraction (XRD) (Figure 1a). The observed peaks at 35.84°, 42.4°, 49.78°, and 73.47° are attributed to γ-brass (JCPDS no. 5-6566) and α-brass (JCPDS no. 50-1333 and no. 65-6567), while no significant changes were observed from the previous reported by us CuZn NPs [38]. The composition analysis of the NPs by inductively coupled plasma (ICP) indicated a 52%/48% copper/zinc proportion, respectively, and thus an overall composition of α-Cu$_{47}$Zn$_{29}$/γ-Cu$_9$Zn$_{15}$ based on the X-ray diffractions. TEM images of CuZn NPs (Figure 2) revealed small, spherical nanoparticles in the range of 20 nm to 30 nm in contrast to the formation of nanoclusters [38]. The different nanoarchitecture is attributed to the half amount of polyol (TrEG) that has been used in the present synthesis.

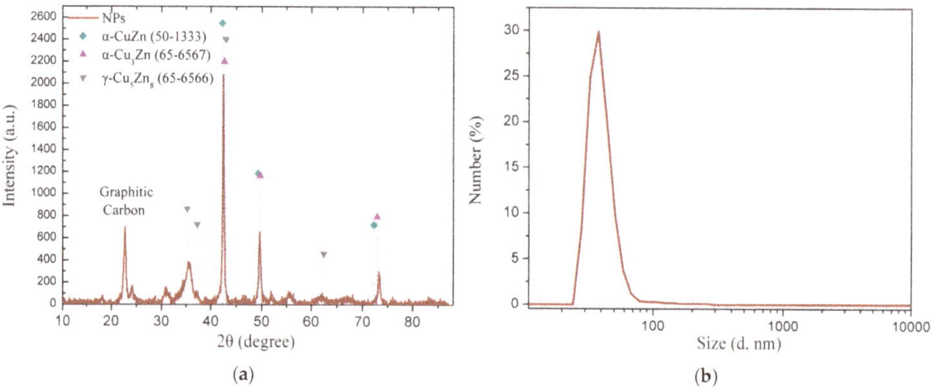

Figure 1. The X-ray diffraction (XRD) patterns of CuZn nanoparticles (NPs) (**a**), and the size distribution (diameter in nm) of the aqueous suspensions of CuZn NPs evaluated by dynamic light scattering (DLS) numbers measurements (**b**).

CuZn NPs are hydrophilic, and thus readily disperse in water. The hydrodynamic diameter provided by DLS number measurements (Figure 1b) was 35 nm, matching well to the size provided by TEM and indicated monodispersity. Additionally, the amount of leached ions in a 30 mg L^{-1} aqueous suspension of CuZn NPs after 24 h of incubation was found to be 1.8 mg L^{-1} for Cu and 2.2 mg L^{-1} for Zn, respectively.

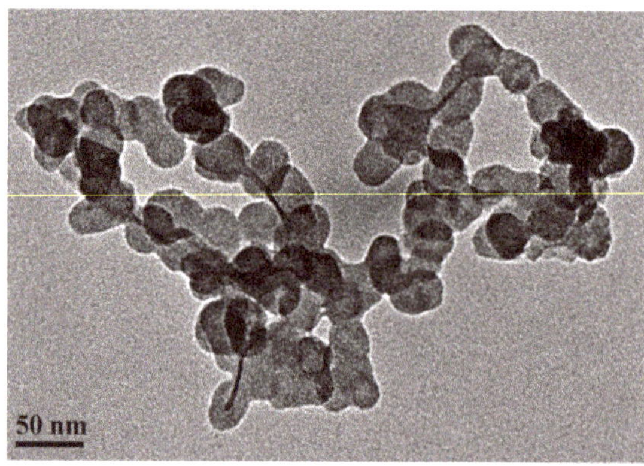

Figure 2. Size and morphology of CuZn NPs determined by transmission electron microscopy.

3.2. Changes in Lght Energy Partitioning at PSII in Young and Mature Leaves After Exposure to CuZn NPs

We estimated the light energy partitioning at PSII, that is, Φ_{PSII}, Φ_{NPQ} and Φ_{NO}, which sum to one. The quantum yield of photochemical energy conversion (Φ_{PSII}) at 30 min, 90 min, and 180 min after the foliar spray with 30 mg L^{-1} of CuZn NPs presented a significant decrease in young leaves, while in mature leaves, it increased significantly compared to controls (Figure 3a). Φ_{PSII} recovered to control values in young leaves 240 min after the spray, while at the same time remaining significantly higher than controls in mature leaves (Figure 3a). Φ_{PSII} in mature leaves at 30 min, 90 min, 180 min, and 240 min after spraying with CuZn NPs was significantly higher than young leaves (Figure 3a).

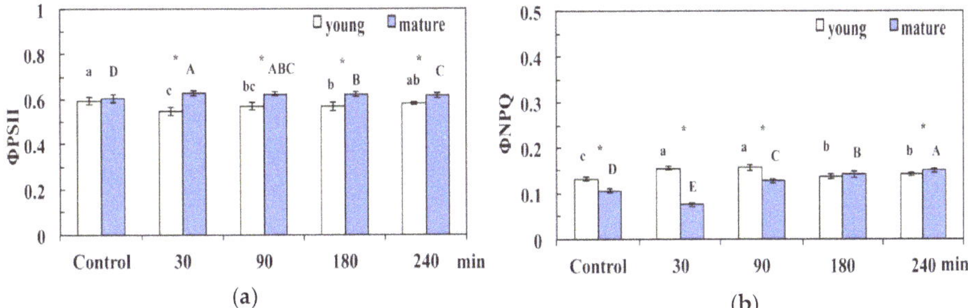

Figure 3. Changes in the quantum efficiency of photosystem II (PSII) photochemistry (Φ_{PSII}) (**a**), and the quantum yield of regulated non-photochemical energy loss in PSII (Φ_{NPQ}) (**b**); of *Arabidopsis thaliana* young and mature leaves measured (at 140 µmol photons m^{-2} s^{-1}) 30 min, 90 min, 180 min, and 240 min after the foliar spay with 30 mg L^{-1} of CuZn NPs or distilled water (control). Error bars on columns are standard deviations based on four to five leaves from different plants. Columns with different letters (lowercase for young leaves and capitals for mature) are statistically different ($p < 0.05$). An asterisk (*) represents a significantly different mean of the same time treatment between young and mature leaves ($p < 0.05$).

The quantum yield of regulated non-photochemical energy loss (Φ_{NPQ}) at 30 min, 90 min, 180 min, and 240 min after the CuZn NPs spray increased significantly, compared to control, in young leaves (Figure 3b). Φ_{NPQ} 30 min after the NPs spray decreased in mature leaves, while it increased significantly afterwards, compared to control (Figure 3b). Φ_{NPQ} in young control leaves, and at 30 min and 90 min after the CuZn NPs spray was significantly higher than that in mature leaves (Figure 3b). In comparison, Φ_{NPQ} was significantly higher in mature leaves 240 min after the spray with CuZn NPs (Figure 3b).

The quantum yield of non-regulated energy loss (Φ_{NO}), which is a loss process due to PSII inactivity, increased significantly in young leaves 30 min after the CuZn NPs spray compared to control, while it remained unchanged in mature leaves, where it decreased significantly later on (90 min, 180 min, and 240 min after the foliar spray) (Figure 4). In young leaves, Φ_{NO} 90 min after the foliar spray with NPs decreased to control values, and increased later on (180 min), but retained control values 240 min after spraying with the NPs (Figure 4). Φ_{NO} in mature control leaves was significantly higher than that in young leaves, but at 90 min, 180 min, and 240 min after the foliar spray with 30 mg L^{-1} of CuZn NPs, it decreased significantly compared to young leaves and control values (Figure 4).

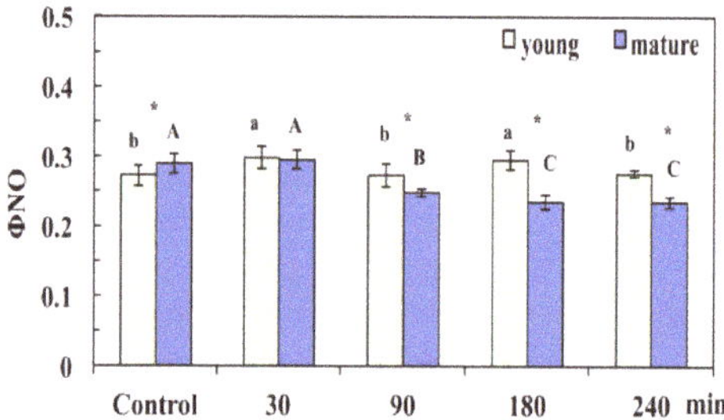

Figure 4. Changes in the quantum yield of non-regulated energy dissipation in PSII (Φ_{NO}) of *Arabidopsis thaliana* young and mature leaves measured (at 140 µmol photons m^{-2} s^{-1}) 30 min, 90 min, 180 min, and 240 min after the foliar spay with 30 mg L^{-1} of CuZn NPs or distilled water (control). Error bars on columns are standard deviations based on four to five leaves from different plants. Columns with different letter (lower case for young leaves and capitals for mature) are statistically different ($p < 0.05$). An asterisk (*) represents a significantly different mean of the same time treatment between young and mature leaves ($p < 0.05$).

3.3. Changes in the Photoprotective Energy Dissipation and the Electron Transport Rate in Young and Mature Leaves After Exposure to CuZn NPs

The non-photochemical quenching (NPQ) increased significantly at 30 min and 90 min after the CuZn NPs spray in young leaves, compared to the control, while it decreased 180 min after spraying, and increased again significantly 240 min after spraying (Figure 5a). NPQ in mature leaves decreased 30 min after spraying with NPs, but later on (90 min, 180 min, and 240 min after the foliar spray), it increased compared to control values (Figure 5a). NPQ was significantly higher in young leaves compared to mature and control leaves and 30 min and 90 min after the CuZn NPs spray, but significantly lower than in mature leaves at 180 min and 240 min after the foliar spray (Figure 5a).

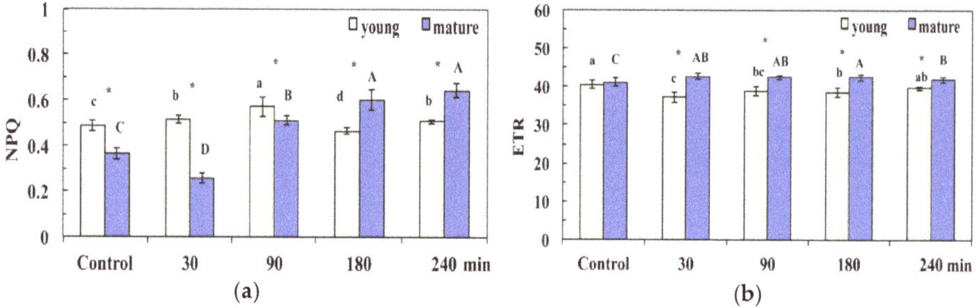

Figure 5. Changes in the non-photochemical fluorescence quenching (NPQ) (**a**), and the relative PSII electron transport rate (ETR) (**b**); of *Arabidopsis thaliana* young and mature leaves measured (at 140 µmol photons m^{-2} s^{-1}) 30 min, 90 min, 180 min, and 240 min after the foliar spay with 30 mg L^{-1} of CuZn NPs or distilled water (control). Error bars on columns are standard deviations based on four to five leaves from different plants. Columns with different letter (lower case for young leaves and capitals for mature) are statistically different ($p < 0.05$). An asterisk (*) represents a significantly different mean of the same time treatment between young and mature leaves ($p < 0.05$).

The relative electron transport rate at PSII (ETR) decreased significantly in young leaves 30 min, 90 min, and 180 min after the foliar spray with 30 mg L^{-1} of CuZn NPs, while at the same time it increased significantly in mature leaves compared to controls (Figure 5b). ETR recovered to control values in young leaves 240 min after the spray, while it remained significantly higher than controls in mature leaves (Figure 5b). The ETR in mature leaves at 30 min, 90 min, 180 min, and 240 min after the spray with CuZn NPs was significantly higher than that in young leaves (Figure 5b).

3.4. Changes in the Redox State of Plastoquinone (PQ) Pool in Young and Mature Leaves After Exposure to CuZn NPs

The redox state of PQ pool (q_p), which is a measure of the fraction of open PSII reaction centers, decreased significantly at 30 min and 180 min after the CuZn NPs spray in young leaves, compared to control; in contrast, it was at control values 90 min and 240 min after spraying (Figure 6).

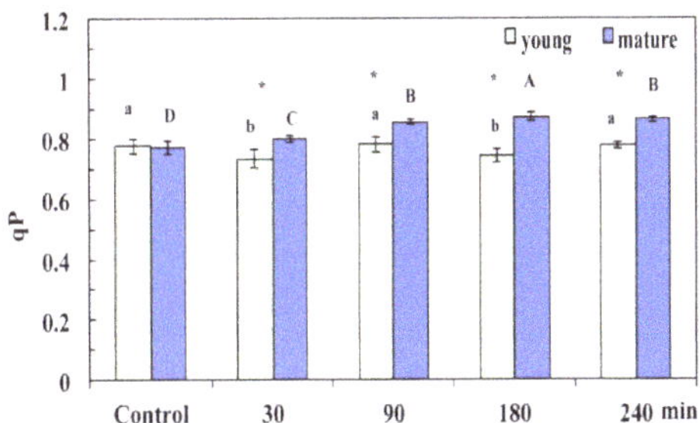

Figure 6. Changes in the photochemical fluorescence quenching, which is the relative reduction state of the plastoquinone (PQ) pool, reflecting the fraction of open PSII reaction centers (q_p) of young and mature *Arabidopsis thaliana* leaves measured (at 140 µmol photons m^{-2} s^{-1}) 30 min, 90 min, 180 min, and 240 min after the foliar spay with 30 mg L^{-1} of CuZn NPs or distilled water (control). Error bars on columns are standard deviations based on four to five leaves from different plants. Columns with different letter (lower case for young leaves and capitals for mature) are statistically different ($p < 0.05$). An asterisk (*) represents a significantly different mean of the same time treatment between young and mature leaves ($p < 0.05$).

At all the sampling periods (30 min, 90 min, 180 min, and 240 min after the NPs spray), the mature leaves were in a more oxidized state than control (Figure 6).

3.5. Spatiotemporal Heterogeneity of the Quantum Efficiency of PSII Photochemistry and the Redox State of Plastoquinone (PQ) Pool in Young and Mature Leaves After Exposure to CuZn NPs

The quantum yield of photochemical energy conversion (Φ_{PSII}) in control young leaves showed a spatial heterogeneity, with higher values in the midrib of the leaves than in the lamina (Figure 7a). A spatiotemporal heterogeneity of Φ_{PSII} in young leaves was evident 30 min after the foliar spray with 30 mg L^{-1} of CuZn NPs with lower values in the distal (tip) leaf area (Figure 7b). The spatiotemporal heterogeneity of Φ_{PSII} in young leaves 90 min after the foliar spray was still evident due to an increase of whole leaf Φ_{PSII} values (Figure 7c), and became amplified 180 min after spraying (Figure 7d). Then, 240 min after spraying with CuZn NPs, Φ_{PSII} increased to the control whole leaf values, showing also a spatial heterogeneity, with higher Φ_{PSII} values in the area where lower values were previously scored (distal leaf area) (Figure 7e).

Mature control leaves (Figure 8a) presented less spatial heterogeneity in Φ_{PSII} compared to control young leaves (Figure 7a), with higher values in the distal (tip) leaf area (Figure 8a). Higher values also occurred in the same area 30 min after the foliar spray with 30 mg L^{-1} CuZn NPs, which also caused the whole leaf Φ_{PSII} to increase (Figure 8b). The spatiotemporal heterogeneity of Φ_{PSII} in mature leaves was also evident 90 min after spraying with NPs (Figure 8c), but become less apparent 180 min after the foliar spray (Figure 8d). At 240 min after the foliar spray, Φ_{PSII} decreased in mature leaves in the whole leaf area, but remained higher than in the control leaves (Figure 8e).

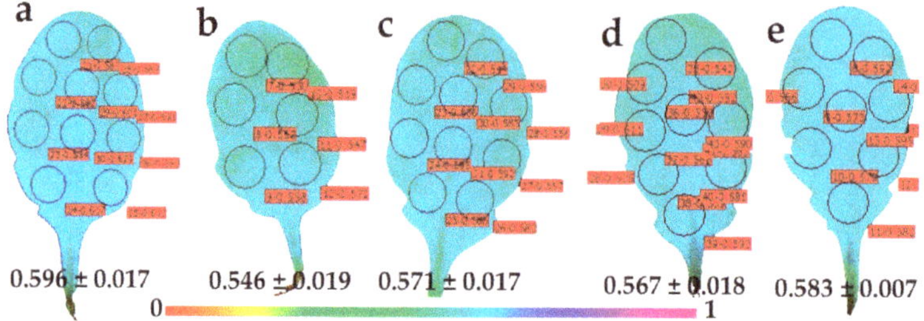

0.596 ± 0.017   0.546 ± 0.019   0.571 ± 0.017   0.567 ± 0.018   0.583 ± 0.007

Figure 7. Representative chlorophyll fluorescence images of the effective quantum yield of PSII photochemistry (Φ_{PSII}) of *Arabidopsis thaliana* young leaves after 5 min of illumination at 140 μmol photons m^{-2} s^{-1}. Leaves were measured after the foliar spray with distilled water (control) (**a**), or 30 min (**b**), 90 min (**c**), 180 min (**d**) and 240 min (**e**) after the foliar spay with 30 mg L^{-1} of CuZn NPs. The color code depicted at the bottom of the images ranges from values 0.0 to 1.0. The areas of interest (AOI) are shown in each image. The average Φ_{PSII} value of all the AOI for the whole leaf is shown.

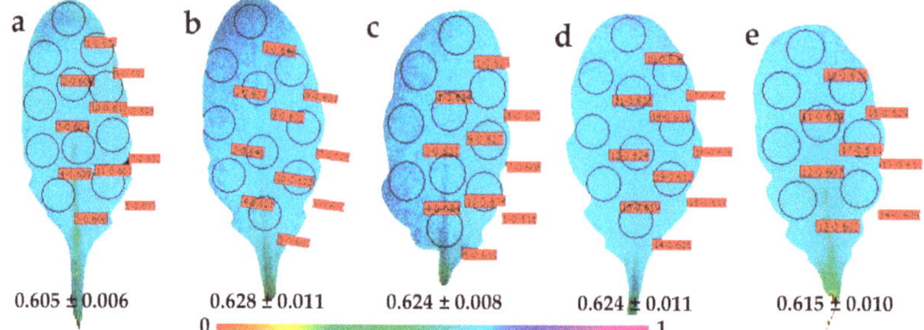

0.605 ± 0.006   0.628 ± 0.011   0.624 ± 0.008   0.624 ± 0.011   0.615 ± 0.010

Figure 8. Representative chlorophyll fluorescence images of the effective quantum yield of PSII photochemistry (Φ_{PSII}) of *Arabidopsis thaliana* mature leaves after 5 min of illumination at 140 μmol photons m^{-2} s^{-1}. Leaves were measured after the foliar spray with distilled water (control) (**a**), or 30 min (**b**), 90 min (**c**), 180 min (**d**) and 240 min (**e**) after the foliar spay with 30 mg L^{-1} of CuZn NPs. The color code depicted at the bottom of the images ranges from values 0.0 to 1.0. The areas of interest (AOI) are shown in each image. The average Φ_{PSII} value of all the AOI for the whole leaf is shown.

Images of the redox state of the PQ pool (q_P) of control young leaves showed a spatial heterogeneity, with higher values in the proximal (base) midrib of leaves (Figure 9a), as observed in the images of Φ_{PSII} (Figure 7a). At 30 min after the foliar spray with 30 mg L^{-1} of CuZn NPs, a spatiotemporal heterogeneity of q_P in young leaves was noticed, with lower values in the distal (tip) leaf area (Figure 9b) and significantly lower whole leaf q_P values than those of the young control leaves (Figure 9a). At 90 min after the foliar spray with CuZn NPs, the q_P images of young leaves (Figure 9c) were similar to the images of control young leaves (Figure 9a). At 180 min after the foliar spray with CuZn NPs, the whole leaf q_P values in young leaves decreased (Figure 9d), resembling the images 90 min after the foliar spray (Figure 9c) with less evident heterogeneity. At 240 min after spraying with NPs, whole leaf q_P values (Figure 9e) resemble those of control young leaves (Figure 9a).

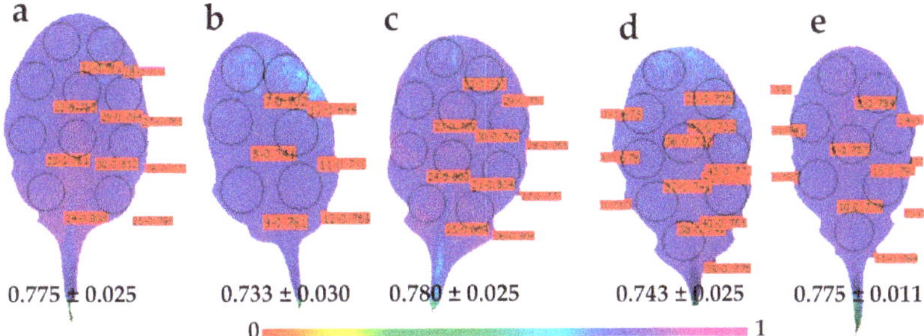

Figure 9. Representative chlorophyll fluorescence images of the relative reduction state of the plastoquinone (PQ) pool, that is, the photochemical fluorescence quenching, reflecting the fraction of open PSII reaction centers (q_p), of *Arabidopsis thaliana* young leaves after 5 min of illumination at 140 µmol photons m^{-2} s^{-1}. Leaves were measured after the foliar spray with distilled water (control) (**a**), or 30 min (**b**), 90 min (**c**), 180 min (**d**), and 240 min (**e**) after the foliar spay with 30 mg L^{-1} of CuZn NPs. The color code depicted at the bottom of the images ranges from values 0.0 to 1.0. The areas of interest (AOI) are shown in each image. The average q_p value of all the AOI for the whole leaf is shown.

Images of the redox state of the PQ pool (q_P) of control mature leaves showed leaf homogeneity rather than leaf heterogeneity (Figure 10a). At 30 min after the foliar spray with 30 mg L^{-1} of CuZn NPs, a slight heterogeneity of q_P was observed in mature leaves, with increased q_P values in the whole leaf area (Figure 10b). Later on (90 min, 180 min, and 240 min after spraying with CuZn NPs), a further increase of q_P values compared to the control mature leaves was observed in the whole leaf area (Figure 10c–e).

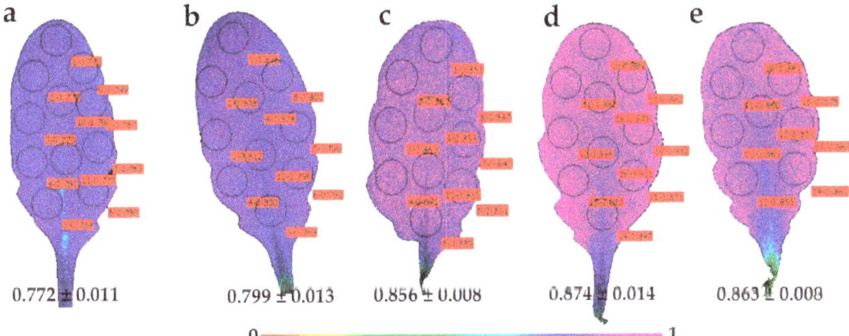

Figure 10. Representative chlorophyll fluorescence images of the relative reduction state of the plastoquinone (PQ) pool—that is, the photochemical fluorescence quenching, reflecting the fraction of open PSII reaction centers (q_p) of *Arabidopsis thaliana* mature leaves after 5 min of illumination at 140 µmol photons m^{-2} s^{-1}. Leaves were measured after the foliar spray with distilled water (control) (**a**), or 30 min (**b**), 90 min (**c**), 180 min (**d**) and 240 min (**e**) after the foliar spay with 30 mg L^{-1} of CuZn NPs. The color code depicted at the bottom of the images ranges from values 0.0 to 1.0. The areas of interest (AOI) are shown in each image. The average q_p value of all the AOI for the whole leaf is shown.

3.6. ROS Generation in Young and Mature Leaves After Exposure to CuZn NPs

ROS generation was quantified in young (Figure 11a–e) and mature (Figure 11f–j) *A. thaliana* leaves by the fluorescent probe DCF-DA. In both young (Figure 11a) and mature (Figure 11f) control leaves, no notable quantities of H_2O_2 could be observed. At 30 min after the foliar spray with CuZn NPs, the highest H_2O_2 generation was noticed in young leaves (Figure 11b), accompanying the lower measured q_P values (Figure 9b). At the same time in mature leaves (Figure 11g), the level of ROS accumulation was similar to the control values (Figure 11f). At 90 min after the CuZn NPs spray, almost no H_2O_2 could be detected in young leaves (Figure 11c). In mature leaves, no H_2O_2 could be detected at 90 min, 180 min, and 240 min after the CuZn NPs spray, either (Figure 11h–j). At 180 min after spraying with CuZn NPs, a high H_2O_2 production (but substantially less than 30 min after spraying) was observed in young leaves (Figure 11d). At 240 min after spraying with CuZn NPs, the level of ROS accumulation in young leaves (Figure 11e) was similar to that of the control (Figure 11a).

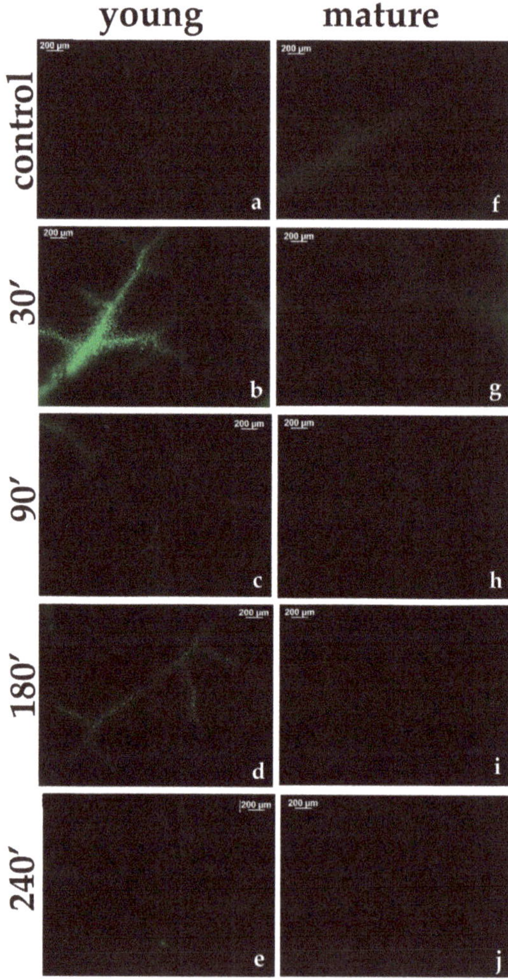

Figure 11. Representative patterns of reactive oxygen species (ROS) (H_2O_2) production in *Arabidopsis thaliana* young (**a–e**) and mature (**f–j**) leaves, as indicated by the fluorescence of H_2DCF-DA. The H_2O_2 generation after the foliar spray with distilled water (control) in a young leaf (**a**) and mature leaf (**f**); or 30 min after foliar spay with 30 mg L^{-1} of CuZn NPs in a young leaf (**b**) and mature leaf (**g**); 90 min after foliar spay with 30 mg L^{-1} of CuZn NPs in a young leaf (**c**) and mature leaf (**h**); 180 min after foliar spay with 30 mg L^{-1} of CuZn NPs in a young leaf (**d**) and mature leaf (**i**); and 240 min after foliar spay with 30 mg L^{-1} of CuZn NPs in a young leaf (**e**), and mature leaf (**j**). Scale bare: 200 µm. A higher H_2O_2 content is indicated by the light green color.

4. Discussion

Inorganic NPs are emerging as novel agrochemicals due to their unique characteristics and high surface energy, which make them effective in lower doses compared to conventional inorganic ionic formulations. For instance, bulk brass has been utilized in the healthcare industry, while ionic forms of zinc and copper such as Bordeaux mixture, sulfate, and chloride salts are used in agrochemistry, but with adverse environmental effects and toxicity. However, due to the low water solubility, these ionic forms of agrochemicals are applied in relatively large amounts in order to effectively control the phytopathogens when the spores vegetate, which is through causing the secretion of malic acid and amino acids and subsequently dissolving them [49]. As a consequence, the limit between plant protection and phytotoxicity is still a matter of discussion. A need exists for new products that are going to have high biological activity and less metal in the formulation. Under these perspectives, hydrophilic CuZn NPs retain the desired characteristics of bulk brass, while forming stable aqueous suspensions with minimal ionic dissolution that are effective in low doses.

Young leaves can utilize only a fraction of absorbed irradiance in photochemical reactions via CO_2 assimilation, since light capture ability develops earlier than CO_2 assimilation capacity [45,50,51]. When the absorbed light is not used in photochemistry, in order to avoid photodamage, the excess excitation energy has to be safely removed by a photoprotective mechanism called non-photochemical quenching (NPQ) [52,53]. Consequently, the ability to dissipate excess excitation energy by NPQ is higher in young leaves than in mature leaves [45,47], which means that under control growth conditions, NPQ is significantly higher in young leaves compared to mature leaves (Figure 5a).

Heterogeneity in PSII photochemistry has been frequently reported to depend on the leaf age [41–43,45,47,48]. We observed changes in light energy partitioning related to leaf age under control growth conditions mostly related to Φ_{NPQ} and Φ_{NO}. Control young leaves had higher Φ_{NPQ} than mature leaves (Figure 3b), and without any significant difference in Φ_{PSII} (Figure 3a), it resulted in significantly lower Φ_{NO} (Figure 4). However, 90 min, 180 min, and 240 min after the NPs spray, Φ_{NO} increased in young leaves compared to mature leaves (Figure 4) due to a decreased photochemical energy conversion (Φ_{PSII}) (Figure 3a) that could not be compensated by the increased Φ_{NPQ} (Figure 3b). Φ_{NO} consists of chlorophyll fluorescence internal conversions and intersystem crossing, which leads to the formation of singlet oxygen ($^1O_2$) via the triplet state of chlorophyll (^{3}chl*) [41,54–56], thus suggesting increased $^1O_2$ formation in young leaves compared to mature leaves. NPQ is one of the most important photoprotective mechanisms in plants [25,41,57,58]. The enhancement of NPQ that reflects the dissipation of excess excitation energy in the form of harmless heat in young leaves (Figure 5a) at 30 min after spraying with CuZn NPs, could not protect young leaves from ROS generation at 30 min after the NPs spray (Figure 11b). However, the increase of NPQ in young leaves 90 min after the CuZn NPs spray (Figure 5a) was effective at retaining the same redox state of the PQ pool with control leaves and reducing H_2O_2 production at 90 min after the NPs spray to control levels (Figure 11c). An effective photoprotection can be attained only if NPQ is adjusted in such a way that no changes occur in the redox state of the PQ pool [41,59]. Otherwise, an imbalance between energy supply and demand occurs, indicating excess excitation energy [57–59]. Under such circumstances, the generation of H_2O_2 occurs (Figure 11b), which can be diffused through the leaf veins to act as a long-distance signaling molecule [38,41,60–62]. The intracellular ROS signaling pathways are initiated by the redox state of the PQ pool that regulates photosynthetic gene expression, comprising also a mechanism of plant acclimation [38,63,64]. The redox state of the PQ pool is of unique significance for antioxidant defense and signaling [65]. It has been shown recently that ROS generation is influenced also by the circadian system [66,67]. We postulate that ROS generation at 30 min after the NPs spray (Figure 11b) possibly served as the signaling molecule to contribute to a more oxidized state of the PQ pool at 90 min after the NPs spray (Figure 6, Figure 9c), resulting in a H_2O_2 production similar to the control leaf level (Figure 11c).

The foliar spay of *Arabidopsis thaliana* young and mature leaves with 30 mg L^{-1} of CuZn NPs revealed a spatiotemporal heterogeneity of Φ_{PSII} and q_p measured (at 140 µmol photons m^{-2} s^{-1})

30 min, 90 min, 180 min, and 240 min after spraying (Figures 7–10). Young leaves show a higher spatial heterogeneity (Figures 7 and 9) compared to mature leaves (Figures 8 and 10). Nevertheless, PSII function was not uniform for both leaf types, making conventional chlorophyll fluorescence instruments not suitable for abiotic stress studies and pointing out the advantages of using chlorophyll fluorescence imaging analysis in the recognition of spatial heterogeneity at the leaf surface [29–35]. The response of cells to the same stress condition is not uniform, with some cells behaving more vulnerably than others [68].

In contrast to previous reports that young leaves acclimatize better to environmental changes and can maintain a better ROS homeostasis [43,45,69], mature leaves responded better. Thus, in disagreement with our hypothesis, the PSII photochemistry of young leaves seem to be negatively influenced when exposed to 30 mg L^{-1} of CuZn NPs. Young leaves could overcome the negative effects on the function of PSII only after 240 min of the NPs spray, at which point the level of ROS accumulation was also similar to that of control young leaves. On the contrary, a beneficial effect was observed in the PSII function of mature leaves 30 min after the CuZn NPs spray, which was through an increased quantum efficiency of PSII photochemistry (Φ_{PSII}), an increased electron transport rate (ETR), an increased fraction of open PSII reaction centers (q_p), decreased $^1O_2$ formation, and no notable changes in H_2O_2 generation.

Zinc and Cu are important micronutrients that are required for plant growth and development [1,38], but when they are in excess, they can cause toxicity effects on plant growth and development, affecting photosynthetic function [3,70]. In young leaves with sufficient Zn and Cu concentrations, the spray with 30 mg L^{-1} of CuZn NPs resulted in an excess supply of them, causing negative effects on PSII function and increased ROS production. Increased H_2O_2 generation in young leaves after 30 min of spraying with 30 mg L^{-1} of CuZn NPs (Figure 11b) was correlated to a lower oxidized state of the PQ pool (Figure 9b). Zinc is involved in a wide variety of physiological processes, playing catalytic, regulatory, and structural roles with several crucial functions in the cell [1,71–73], but excess Zn has to be detoxified in roots by sequestration to protect the sensitive photosynthetic leaf tissues [3,72]. Zn phytotoxicity varies extensively, depending on combinations with other heavy metals, the environmental conditions, the plant species, and the plant age [72], as well as by the leaf age, as shown here.

Leaf senescence turns leaves from units with a primary assimilation role into centers of nutrient mobilization [74,75]. During leaf senescence, new metabolic pathways are activated and others are de-activated, with nutrient and material remobilization, followed by a declining photosynthesis [74,75]. The *A. thaliana* rosette leaf 8 from six-week-old plants is a mature to senescing leaf; thus, nutrient remobilization occurs, resulting in nutrient deficiency. Spraying these leaves with 30 mg L^{-1} of CuZn NPs restores Zn and Cu deficiency and improves photosynthetic efficiency. Zinc is known to contribute to the repair processes of PSII by turning over the photodamaged D1 protein [72,76]. Copper is vital for photosynthesis, and more than half of Cu is found in chloroplasts participating in the light reactions [77]. The foliar spraying of Cu NPs induced stress tolerance by stimulating antioxidant mechanisms [78]. However, this nutrient remobilization explanation has yet to be established.

Although plants are producers and play a key role in the ecosystem, the impact of NPs upon them is not well studied [79]. In order to understand the uptake, transport, and also bioaccumulation of NPs in plants after foliar exposure, different qualitative and quantitative methods are still being developed with an unclear comparability of results among the different techniques [80,81]. Among the different techniques, inductively coupled plasma mass spectroscopy (ICP-MS) is one of the most reliable methods for the detection of NPs, offering a range of advantages in high detection limits and high sensitivity for many elements [79,81–83].

Previously, both the positive and harmful impacts of NPs on terrestrial and aquatic plants have been established, which are mainly due to the concentration, size, and specific surface area of NPs, the exposure methodology, and the plant species that was examined [38,79,81,84–86]. In the root uptake of NPs, translocation to the above-ground parts takes place in a unidirectional pathway through xylem vessels, while in the foliar uptake of NPs, translocation takes place in bidirectional pathways throughout the plant by the phloem [85]. The efficiency of uptake and translocation, and the effects of NPs on growth metabolism and photosynthesis vary from plant to plant [84]. Some studies report that the foliar application of NPs considerably increases the chlorophyll content in plants and results in a higher amount of light energy capture and photosynthesis enhancement [84], while another study reports that the root uptake of NPs decreases Φ_{PSII} and q_p, and increases Φ_{NO} due to an ineffective photoprotective mechanism (NPQ) resulting from a significant decrease of the PsbS protein, which is the key regulator of the energy dissipation process [87]. Nevertheless, the investigation on NPs is still at an initial phase; more laborious work is required in order to understand their impact on the physiological, biochemical, and molecular mechanisms in plants [79,84].

The present study demonstrates that considering the leaf developmental stage is important for understanding the mechanisms underlying leaf growth responses to environmental stresses [41–43,88]; thus, it must be taken into account in environmental stress studies in order to compare leaves of the same developmental stage [41–43,48,89]. Leaves of distinct ages differentially control stress responses, and plant responses against biotic and abiotic stresses are balanced in a leaf age-dependent manner [90].

Author Contributions: C.D.-S. and M.M. conceived and designed the experiments and analyzed the data; I.S., J.M., O.A. and I.-D.S.A. performed the experiments and analyzed the data; M.M. wrote the paper; all the authors review and approved the manuscript.

Funding: This research received no external funding.

Conflicts of Interest: The authors declare no conflict of interest.

References

1. Marschner, H. *Mineral Nutrition of Higher Plants*, 2nd ed.; Academic Press: London, UK, 1995.
2. Cakmak, I. Possible roles of zinc in protecting plant cells from damage by reactive oxygen species. *New Phytol.* **2000**, *146*, 185–205. [CrossRef]
3. Moustakas, M.; Bayçu, G.; Gevrek-Kürüm, N.; Moustaka, J.; Csatári, I.; Rognes, S.E. Spatiotemporal heterogeneity of photosystem II function during acclimation to zinc exposure and mineral nutrition changes in the hyperaccumulator *Noccaea caerulescens*. *Environ. Sci. Pollut. Res.* **2019**, *26*, 6613–6624. [CrossRef] [PubMed]
4. Alloway, B.J. Soil factors associated with zinc deficiency in crops and humans. *Environ. Geochem. Health* **2009**, *31*, 537–548. [CrossRef] [PubMed]
5. Ahmad, P.; Alyemeni, M.N.; Ahanger, M.A.; Wijaya, L.; Alam, P.; Kumar, A.; Ashraf, M. Upregulation of antioxidant and glyoxalase systems mitigates NaCl stress in *Brassica juncea* by supplementation of zinc and calcium. *J. Plant Interact.* **2018**, *13*, 151–162. [CrossRef]
6. Rameshraddy; Pavithra, G.J.; Rajashekar Reddy, B.H.; Mahesh, S.; Geetha, K.N.; Shankar, A.G. Zinc oxide nano particles increases Zn uptake, translocation in rice with positive effect on growth, yield and moisture stress tolerance. *Ind. J. Plant Physiol.* **2017**, *22*, 287–294. [CrossRef]
7. Doolette, C.L.; Read, T.L.; Li, C.; Scheckel, K.G.; Donner, E.; Kopittke, P.M.; Schjoerring, J.K.; Lombi, E. Foliar application of zinc sulphate and zinc EDTA to wheat leaves: Differences in mobility, distribution, and speciation. *J. Exp. Bot.* **2018**, *69*, 4469–4481. [CrossRef] [PubMed]
8. Cakmak, I. Enrichment of cereal grains with zinc: Agronomic or genetic biofortification? *Plant Soil* **2008**, *302*, 1–17. [CrossRef]
9. Prasad, R.; Shivay, Y.S.; Kumar, D. Agronomic biofortification of cereal grains with iron and zinc. *Adv. Agron.* **2014**, *125*, 55–91.
10. Garcia, P.C.; Rivero, R.M.; Ruiz, J.M.; Romero, L. The role of fungicides in the physiology of higher plants: Implications for defense responses. *Bot. Rev.* **2003**, *69*, 162–172. [CrossRef]
11. Zhao, L.; Huang, Y.; Hu, J.; Zhou, H.; Adeleye, A.S.; Keller, A.A. ^{1}H NMR and GC-MS based metabolomics reveal defense and detoxification mechanism of cucumber plant under nano-Cu stress. *Environ. Sci. Technol.* **2016**, *50*, 2000–2010. [CrossRef]
12. Ruttkay-Nedecky, B.; Krystofova, O.; Nejdl, L.; Adam, V. Nanoparticles based on essential metals and their phytotoxicity. *J. Nanobiotechnol.* **2017**, *15*, 33. [CrossRef] [PubMed]

13. Servin, A.D.; White, J.C. Nanotechnology in agriculture: Next steps for understanding engineered nanoparticle exposure and risk. *NanoImpact* **2016**, *1*, 9–12. [CrossRef]

14. Liu, R.; Lal, R. Potentials of engineered nanoparticles as fertilizers for increasing agronomic productions. *Sci. Total Environ.* **2015**, *514*, 131–139. [CrossRef] [PubMed]

15. Parisi, C.; Vigani, M.; Rodríguez-Cerezo, E. Agricultural nanotechnologies: What are the current possibilities? *Nano Today* **2015**, *10*, 124–127. [CrossRef]

16. Moustaka, J.; Ouzounidou, G.; Sperdouli, I.; Moustakas, M. Photosystem II is more sensitive than photosystem I to Al^{3+} induced phytotoxicity. *Materials* **2018**, *11*, 1772. [CrossRef] [PubMed]

17. Anderson, J.M. Changing concepts about the distribution of photosystems I and II between grana-appressed and stroma-exposed thylakoid membranes. *Photosynth. Res.* **2002**, *73*, 157–164. [CrossRef]

18. Apostolova, E.L.; Dobrikova, A.G.; Ivanova, P.I.; Petkanchin, I.B.; Taneva, S.G. Relationship between the organization of the PSII supercomplex and the functions of the photosynthetic apparatus. *J. Photochem. Photobiol. B* **2006**, *83*, 114–122. [CrossRef]

19. Krause, G.H.; Weis, E. Chlorophyll fluorescence and photosynthesis: The basics. *Annu. Rev. Plant Physiol. Plant Mol. Biol.* **1991**, *42*, 313–349. [CrossRef]

20. Murchie, E.H.; Lawson, T. Chlorophyll fluorescence analysis: A guide to good practice and understanding some new applications. *J. Exp. Bot.* **2013**, *64*, 3983–3998. [CrossRef]

21. Kalaji, H.M.; Oukarroum, A.A.; Alexandrov, V.; Kouzmanova, M.; Brestic, M.; Zivcak, M.; Samborska, I.A.; Cetner, M.D.; Allakhverdiev, S.I. Identification of nutrient deficiency in maize and tomato plants by in vivo chlorophyll *a* fluorescence measurements. *Plant Physiol. Biochem.* **2014**, *81*, 16–25. [CrossRef]

22. Guidi, L.; Calatayud, A. Non-invasive tools to estimate stress-induced changes in photosynthetic performance in plants inhabiting Mediterranean areas. *Environ. Exp. Bot.* **2014**, *103*, 42–52. [CrossRef]

23. Moustaka, J.; Ouzounidou, G.; Bayçu, G.; Moustakas, M. Aluminum resistance in wheat involves maintenance of leaf Ca^{2+} and Mg^{2+} content, decreased lipid peroxidation and Al accumulation, and low photosystem II excitation pressure. *BioMetals* **2016**, *29*, 611–623. [CrossRef] [PubMed]

24. Kalaji, H.M.; Jajoo, A.; Oukarroum, A.; Brestic, M.; Zivcak, M.; Samborska, I.A.; Cetner, M.D.; Łukasik, I.; Goltsev, V.; Ladle, R.J. Chlorophyll a fluorescence as a tool to monitor physiological status of plants under abiotic stress conditions. *Acta Physiol. Plant.* **2016**, *38*, 1–11. [CrossRef]

25. Kitao, M.; Tobita, H.; Kitaoka, S.; Harayama, H.; Yazaki, K.; Komatsu, M.; Agathokleous, E.; Koike, T. Light energy partitioning under various environmental stresses combined with elevated CO_2 in three deciduous broadleaf tree species in Japan. *Climate* **2019**, *7*, 79. [CrossRef]

26. Guidi, L.; Lo Piccolo, E.; Landi, M. Chlorophyll fluorescence, photoinhibition and abiotic stress: Does it make any difference the fact to be a C3 or C4 species? *Front. Plant Sci.* **2019**, *10*, 174. [CrossRef] [PubMed]

27. Moustaka, J.; Panteris, E.; Adamakis, I.D.S.; Tanou, G.; Giannakoula, A.; Eleftheriou, E.P.; Moustakas, M. High anthocyanin accumulation in poinsettia leaves is accompanied by thylakoid membrane unstacking, acting as a photoprotective mechanism, to prevent ROS formation. *Environ. Exp. Bot.* **2018**, *154*, 44–55. [CrossRef]

28. Sperdouli, I.; Moustakas, M. Differential blockage of photosynthetic electron flow in young and mature leaves of *Arabidopsis thaliana* by exogenous proline. *Photosynthetica* **2015**, *53*, 471–477. [CrossRef]

29. Sperdouli, I.; Moustakas, M. Spatio-temporal heterogeneity in *Arabidopsis thaliana* leaves under drought stress. *Plant Biol.* **2012**, *14*, 118–128. [CrossRef]

30. Gorbe, E.; Calatayud, A. Applications of chlorophyll fluorescence imaging technique in horticultural research: A review. *Sci. Hortic.* **2012**, *138*, 24–35. [CrossRef]

31. Moustaka, J.; Moustakas, M. Photoprotective mechanism of the non-target organism *Arabidopsis thaliana* to paraquat exposure. *Pest. Biochem. Physiol.* **2014**, *111*, 1–6. [CrossRef]

32. Agathokleous, E.; Kitao, M.; Harayama, H. On the nonmonotonic, hormetic photoprotective response of plants to stress. *Dose-Response* **2019**, *17*, 1–3. [CrossRef] [PubMed]

33. Moustakas, M.; Malea, P.; Haritonidou, K.; Sperdouli, I. Copper bioaccumulation, photosystem II functioning and oxidative stress in the seagrass *Cymodocea nodosa* exposed to copper oxide nanoparticles. *Environ. Sci. Pollut. Res.* **2017**, *24*, 16007–16018. [CrossRef] [PubMed]

34. Bayçu, G.; Moustaka, J.; Gevrek-Kürüm, N.; Moustakas, M. Chlorophyll fluorescence imaging analysis for elucidating the mechanism of photosystem II acclimation to cadmium exposure in the hyperaccumulating plant *Noccaea caerulescens*. *Materials* **2018**, *11*, 2580. [CrossRef] [PubMed]

35. Moustakas, M.; Malea, P.; Zafeirakoglou, A.; Sperdouli, I. Photochemical changes and oxidative damage in the aquatic macrophyte *Cymodocea nodosa* exposed to paraquat-induced oxidative stress. *Pest. Biochem. Physiol.* **2016**, *126*, 28–34. [CrossRef] [PubMed]

36. Cokoja, M.; Parala, H.; Schroter, M.K.; Birkner, A.; van den Berg, M.W.E.; Klementiev, K.V.; Grunert, W.; Fischer, R.A. Nanobrass colloids: Synthesis by Co-Hydrogenolysis of [CpCu(PMe3)] with [ZnCp*2] and investigation of the oxidation behaviour of a/b-CuZn nanoparticles. *J. Mater. Chem.* **2006**, *16*, 2420–2428. [CrossRef]

37. Schütte, K.; Meyer, H.; Gemel, C.; Barthel, J.; Fischer, R.A.; Janiak, C. Synthesis of Cu, Zn and Cu/Zn brass alloy nanoparticles from metal amidinate precursors in ionic liquids or propylene carbonate with relevance to methanol synthesis. *Nanoscale* **2014**, *6*, 3116–3126. [CrossRef] [PubMed]

38. Antonoglou, O.; Moustaka, J.; Adamakis, I.D.; Sperdouli, I.; Pantazaki, A.; Moustakas, M.; Dendrinou-Samara, C. Nanobrass CuZn nanoparticles as foliar spray non phytotoxic fungicides. *ACS Appl. Mater. Interfaces* **2018**, *10*, 4450–4461. [CrossRef] [PubMed]

39. Chang, Y.N.; Zhang, M.; Xia, L.; Zhang, J.; Xing, G. The toxic effects and mechanisms of CuO and ZnO nanoparticles. *Materials* **2012**, *5*, 2850–2871. [CrossRef]

40. Takahashi, S.; Badger, M.R. Photoprotection in plants: A new light on photosystem II damage. *Trends Plant Sci.* **2011**, *16*, 53–60. [CrossRef]

41. Moustaka, J.; Tanou, G.; Adamakis, I.D.; Eleftheriou, E.P.; Moustakas, M. Leaf age dependent photoprotective and antioxidative mechanisms to paraquat-induced oxidative stress in *Arabidopsis thaliana*. *Int. J. Mol. Sci.* **2015**, *16*, 13989–14006. [CrossRef]

42. Sperdouli, I.; Moustakas, M. A better energy allocation of absorbed light in photosystem II and less photooxidative damage contribute to acclimation of *Arabidopsis thaliana* young leaves to water deficit. *J. Plant Physiol.* **2014**, *171*, 587–593. [CrossRef] [PubMed]

43. Sperdouli, I.; Moustakas, M. Leaf developmental stage modulates metabolite accumulation and photosynthesis contributing to acclimation of *Arabidopsis thaliana* to water deficit. *J. Plant Res.* **2014**, *127*, 481–489. [CrossRef] [PubMed]

44. Agathokleous, E.; Mouzaki-Paxinou, A.C.; Saitanis, C.J.; Paoletti, E.; Manning, W.J. The first toxicological study of the antiozonant and research tool ethylene diurea (EDU) using a *Lemna minor* L. bioassay: Hints to its mode of action. *Environ. Pollut.* **2016**, *213*, 996–1006. [CrossRef] [PubMed]

45. Sperdouli, I.; Moustakas, M. Differential response of photosystem II photochemistry in young and mature leaves of *Arabidopsis thaliana* to the onset of drought stress. *Acta Physiol. Plant.* **2012**, *34*, 1267–1276. [CrossRef]

46. Jung, S. Variation in antioxidant metabolism of young and mature leaves of *Arabidopsis thaliana* subjected to drought. *Plant Sci.* **2004**, *166*, 459–466. [CrossRef]

47. Jiang, C.D.; Li, P.M.; Gao, H.Y.; Zou, Q.; Jiang, G.M.; Li, L.H. Enhanced photoprotection at the early stages of leaf expansion in field-grown soybean plants. *Plant Sci.* **2005**, *168*, 911–919. [CrossRef]

48. Bielczynski, L.W.; Lacki, M.K.; Hoefnagels, I.; Gambin, A.; Croce, R. Leaf and plant age affects photosynthetic performance and photoprotective capacity. *Plant Physiol.* **2017**, *175*, 1634–1648. [CrossRef]

49. Montag, J.; Schreiber, L.; Schonherr, J. An in vitro study on the post infection activities of copper hydroxide and copper sulfate against conidia of *Venturia inaequalis*. *J. Agric. Food Chem.* **2006**, *54*, 893–899. [CrossRef]

50. Balakumar, T.; Vincent, V.H.B.; Paliwal, K. On the interaction of UV-B radiation (280–315 nm) with water stress in crop plants. *Physiol. Plant.* **1993**, *87*, 217–222. [CrossRef]

51. Dillenburg, L.R.; Sullivan, J.H.; Teramura, A.H. Leaf expansion and development of photosynthetic capacity and pigments in *Liquidambar styraciflua*. *Am. J. Bot.* **1995**, *82*, 433–440. [CrossRef]

52. Demmig-Adams, B.; Adams, W.W., III. Photoprotection and other responses of plants to high light stress. *Annu. Rev. Plant Physiol. Plant Mol. Biol.* **1992**, *43*, 599–626. [CrossRef]

53. Demmig-Adams, B.; Adams, W.W., III; Barker, D.H.; Logan, B.A.; Bowling, D.R.; Verhoeven, A.S. Using chlorophyll fluorescence to assess the fraction of absorbed light allocated to thermal dissipation of excess excitation. *Physiol. Plant.* **1996**, *98*, 253–264. [CrossRef]

54. Klughammer, C.; Schreiber, U. Complementary PS II quantum yields calculated from simple fluorescence parameters measured by PAM fluorometry and the Saturation Pulse method. *PAM Appl. Notes* **2008**, *1*, 27–35.

55. Kasajima, I.; Ebana, K.; Yamamoto, T.; Takahara, K.; Yano, M.; Kawai-Yamada, M.; Uchimiya, H. Molecular distinction in genetic regulation of nonphotochemical quenching in rice. *Proc. Natl. Acad. Sci. USA* **2011**, *108*, 13835–13840. [CrossRef] [PubMed]

56. Gawroński, P.; Witoń, D.; Vashutina, K.; Bederska, M.; Betliński, B.; Rusaczonek, A.; Karpiński, S. Mitogen-activated protein kinase 4 is a salicylic acid-independent regulator of growth but not of photosynthesis in Arabidopsis. *Mol. Plant* **2014**, *7*, 1151–1166. [CrossRef] [PubMed]

57. Müller, P.; Li, X.P.; Niyogi, K.K. Non-photochemical quenching. A response to excess light energy. *Plant Physiol.* **2001**, *125*, 1558–1566. [CrossRef] [PubMed]

58. Ruban, A.V. Nonphotochemical chlorophyll fluorescence quenching: Mechanism and effectiveness in protecting plants from photodamage. *Plant Physiol.* **2016**, *170*, 1903–1916. [CrossRef] [PubMed]

59. Lambrev, P.H.; Miloslavina, Y.; Jahns, P.; Holzwarth, A.R. On the relationship between non-photochemical quenching and photoprotection of photosystem II. *Biochim. Biophys. Acta* **2012**, *1817*, 760–769. [CrossRef]

60. Wilson, K.E.; Ivanov, A.G.; Öquist, G.; Grodzinski, B.; Sarhan, F.; Huner, N.P.A. Energy balance, organellar redox status, and acclimation to environmental stress. *Can. J. Bot.* **2006**, *84*, 1355–1370. [CrossRef]

61. Mittler, R.; Vanderauwera, S.; Suzuki, N.; Miller, G.; Tognetti, V.B.; Vandepoele, K.; Gollery, M.; Shulaev, V.; Van Breusegem, F. ROS signaling: The new wave? *Trends Plant Sci.* **2011**, *16*, 300–309. [CrossRef]

62. Baxter, A.; Mittler, R.; Suzuki, N. ROS as key players in plant stress signaling. *J. Exp. Bot.* **2014**, *65*, 1229–1240. [CrossRef] [PubMed]

63. Bräutigam, K.; Dietzel, L.; Kleine, T.; Ströher, E.; Wormuth, D.; Dietz, K.J.; Radke, D.; Wirtz, M.; Hell, R.; Dörmann, P.; et al. Dynamic plastid redox signals integrate gene expression and metabolism to induce distinct metabolic states in photosynthetic acclimation in Arabidopsis. *Plant Cell* **2009**, *21*, 2715–2732. [CrossRef] [PubMed]

64. Dietz, K.J.; Pfannschmidt, T. Novel regulators in photosynthetic redox control of plant metabolism and gene expression. *Plant Physiol.* **2011**, *155*, 1477–1485. [CrossRef] [PubMed]

65. Borisova-Mubarakshina, M.M.; Vetoshkina, D.V.; Ivanov, B.N. Antioxidant and signaling functions of the plastoquinone pool in higher plants. *Physiol. Plant.* **2019**, *166*, 181–198. [CrossRef] [PubMed]

66. Lai, A.G.; Doherty, C.J.; Mueller-Roeber, B.; Kay, S.A.; Schippers, J.H.M.; Dijkwel, P.P. CIRCADIAN CLOCK-ASSOCIATED 1 regulates ROS homeostasis and oxidative stress responses. *Proc. Natl. Acad. Sci. USA* **2012**, *109*, 17129–17134. [CrossRef] [PubMed]

67. Jones, M.A. Retrograde signalling as an informant of circadian timing. *New Phytol.* **2019**, *221*, 1749–1753. [CrossRef] [PubMed]

68. Stasolla, C.; Huang, S.; Hill, R.D.; Igamberdiev, A.U. Spatio-temporal expression of phytoglobin—A determining factor in the NO specification of cell fate. *J. Exp. Bot.* **2019**. [CrossRef]

69. Pinheiro, C.; Chaves, M.M. Photosynthesis and drought: Can we make metabolic connections from available data? *J. Exp. Bot.* **2011**, *62*, 869–882. [CrossRef]

70. Bayçu, G.; Gevrek-Kürüm, N.; Moustaka, J.; Csatári, I.; Rognes, S.E.; Moustakas, M. Cadmium-zinc accumulation and photosystem II responses of *Noccaea caerulescens* to Cd and Zn exposure. *Environ. Sci. Pollut. Res.* **2017**, *24*, 2840–2850. [CrossRef]

71. Doncheva, S.; Stoyanova, Z.; Velikova, V. Influence of succinate on zinc toxicity of pea plants. *J. Plant Nutr.* **2001**, *24*, 789–804. [CrossRef]

72. Tsonev, T.; Lidon, F.J.C. Zinc in plants—An overview. *Emir. J. Food Agric.* **2012**, *24*, 322–333.

73. Caldelas, C.; Weiss, D.J. Zinc homeostasis and isotopic fractionation in plants: A review. *Plant Soil* **2017**, *411*, 17–46. [CrossRef]

74. Hoch, W.A.; Zeldin, E.L.; McCown, B.H. Physiological significance of anthocyanins during autumnal leaf senescence. *Tree Physiol.* **2001**, *21*, 1–8. [CrossRef]

75. Ougham, H.J.; Morris, P.; Thomas, H. The colors of autumn leaves as symptoms of cellular recycling and defenses against environmental stresses. *Curr. Top. Dev. Biol.* **2005**, *66*, 135–160.

76. Bailey, S.; Thompson, E.; Nixon, P.J.; Horton, P.; Mullineaux, C.W.; Robinson, C.; Mann, N.H. A critical role for the *Var2 FtsH* homologue of *Arabidopsis thaliana* in the photosystem II repair cycle in vivo. *J. Biol. Chem.* **2002**, *277*, 2006–2011. [CrossRef]

77. Hänsch, R.; Mendel, R.R. Physiological functions of mineral micronutrients (Cu, Zn, Mn, Fe, Ni, Mo, B, Cl). *Curr. Opin. Plant Biol.* **2009**, *12*, 259–266. [CrossRef]

78. Pérez-Labrada, F.; López-Vargas, E.R.; Ortega-Ortiz, H.; Cadenas-Pliego, G.; Benavides-Mendoza, A.; Juárez-Maldonado, A. Responses of tomato plants under saline stress to foliar application of copper nanoparticles. *Plants* **2019**, *8*, 151. [CrossRef]

79. Rastogi, A.; Zivcak, M.; Sytar, O.; Kalaji, H.M.; He, X.; Mbarki, S.; Brestic, M. Impact of metal and metal oxide nanoparticles on plant: A critical review. *Front. Chem.* **2017**, *5*, 78. [CrossRef]

80. Zhao, F.J.; Moore, K.L.; Lombi, E.; Zhu, Y.G. Imaging element distribution and speciation in plant cells. *Trends Plant Sci.* **2014**, *19*, 183–192. [CrossRef]

81. Deng, Y.; Petersen, E.J.; Challis, K.E.; Rabb, S.A.; Holbrook, R.D.; Ranville, J.F.; Nelson, B.C.; Xing, B. Multiple method analysis of TiO_2 nanoparticle uptake in rice (*Oryza sativa* L.) plants. *Environ. Sci. Technol.* **2017**, *51*, 10615–10623. [CrossRef]

82. Wu, B.; Becker, J.S. Imaging techniques for elements and element species in plant science. *Metallomics* **2012**, *4*, 403–416. [CrossRef]

83. Hanć, A.; Piechalak, A.; Tomaszewska, B.; Barałkiewicz, D. Laser ablation inductively coupled plasma mass spectrometry in quantitative analysis and imaging of plant's thin sections. *Int. J. Mass Spectrom.* **2014**, *363*, 16–22. [CrossRef]

84. Kataria, S.; Jain, M.; Rastogi, A.; Zivcak, M.; Brestic, M.; Liu, S.; Tripathi, D.K. Role of nanoparticles on photosynthesis: Avenues and applications. In *Nanomaterials in Plants, Algae and Microorganisms: Concepts and Controversies*; Tripathi, D.K., Ahmad, P., Sharma, S., Chauhan, D.K., Dubey, N.K., Eds.; Elsevier, Academic Press: Cambridge, MA, USA, 2019; Volume 2, pp. 103–127.

85. Hussain, I.; Singh, A.; Singh, N.B.; Singh, A.; Singh, P. Plant-nanoceria interaction: Toxicity, accumulation, translocation and biotransformation. *S. Afr. J. Bot.* **2019**, *121*, 239–247. [CrossRef]

86. Malea, P.; Charitonidou, K.; Sperdouli, I.; Mylona, Z.; Moustakas, M. Zinc uptake, photosynthetic efficiency and oxidative stress in the seagrass *Cymodocea nodosa* exposed to ZnO nanoparticles. *Materials* **2019**, *12*, 2101. [CrossRef]

87. Chen, Y.; Wu, N.; Mao, H.; Zhou, J.; Su, Y.; Zhang, Z.; Zhang, H.; Yuan, S. Different toxicities of nanoscale titanium dioxide particles in the roots and leaves of wheat seedlings. *RSC Adv.* **2019**, *9*, 19243–19252. [CrossRef]

88. Pantin, F.; Simonneau, T.; Muller, B. Coming of leaf age: Control of growth by hydraulics and metabolics during leaf ontogeny. *New Phytol.* **2012**, *196*, 349–366. [CrossRef]

89. Majer, P.; Hideg, É. Developmental stage is an important factor that determines the antioxidant responses of young and old grapevine leaves under UV irradiation in a green-house. *Plant Physiol. Biochem.* **2012**, *50*, 15–23. [CrossRef]

90. Berens, M.L.; Wolinska, K.W.; Spaepen, S.; Ziegler, J.; Nobori, T.; Nair, A. Balancing trade-offs between biotic and abiotic stress responses through leaf age-dependent variation in stress hormone cross-talk. *Proc. Natl. Acad. Sci. USA* **2019**, *116*, 2364–2373. [CrossRef]

Article

Zinc Uptake, Photosynthetic Efficiency and Oxidative Stress in the Seagrass *Cymodocea nodosa* Exposed to ZnO Nanoparticles

Paraskevi Malea [1], Katerina Charitonidou [1,2], Ilektra Sperdouli [1,3], Zoi Mylona [1] and Michael Moustakas [1,*]

[1] Department of Botany, Aristotle University of Thessaloniki, GR-54124 Thessaloniki, Greece
[2] School of Agricultural Sciences, University of Thessaly, GR-38446 Volos, Greece
[3] Institute of Plant Breeding and Genetic Resources, Hellenic Agricultural Organisation-Demeter, Thermi, GR-57001 Thessaloniki, Greece
* Correspondence: moustak@bio.auth.gr; Tel.: +30-2310-99-8335

Received: 22 May 2019; Accepted: 27 June 2019; Published: 29 June 2019

Abstract: We characterized zinc oxide nanoparticles (ZnO NPs) by dynamic light scattering (DLS) measurements, and transmission electron microscopy (TEM), while we evaluated photosystem II (PSII) responses, Zn uptake kinetics, and hydrogen peroxide (H_2O_2) accumulation, in *C. nodosa* exposed to 5 mg L^{-1} and 10 mg L^{-1} ZnO NPs for 4 h, 12 h, 24 h, 48 h and 72 h. Four h after exposure to 10 mg L^{-1} ZnO NPs, we noticed a disturbance of PSII functioning that became more severe after 12 h. However, after a 24 h exposure to 10 mg L^{-1} ZnO NPs, we observed a hormetic response, with both time and dose as the basal stress levels needed for induction of the adaptive response. This was achieved through the reduced plastoquinone (PQ) pool, at a 12 h exposure, which mediated the generation of chloroplastic H_2O_2; acting as a fast acclimation signaling molecule. Nevertheless, longer treatment (48 h and 72 h) resulted in decreasing the photoprotective mechanism to dissipate excess energy as heat (NPQ) and increasing the quantum yield of non-regulated energy loss (Φ_{NO}). This increased the formation of singlet oxygen ($^1O_2$), and decreased the fraction of open reaction centers, mostly after a 72-h exposure at 10 mg L^{-1} ZnO NPs due to increased Zn uptake compared to 5 mg L^{-1}.

Keywords: adaptive response; hormetic response; hydrogen peroxide; marine angiosperms; non-photochemical quenching; photoprotective mechanism; plastoquinone pool; reactive oxygen species (ROS); redox state; zinc oxide nanoparticles

1. Introduction

The small size of nanoparticles (NPs) provides them with special physical and chemical properties that are not found in bulk materials allowing their utilization, among others, in agricultural products, catalysis, cosmetics, electronics, energy production, engineering, food industry, pharmaceutics and textiles [1–5].

Among the variety of metal NPs that are often used for marketable purposes, zinc oxide (ZnO) NPs are the most commonly used ones [6–8]. ZnO NPs, with their unique chemical and physical properties, such as high photostability, broad range of radiation absorption, high electrochemical coupling coefficient, and high chemical stability, are widely used in a diversity of applications, varying from paints to chemicals, from tires to ceramics, and from pharmaceuticals to agriculture [3,9]. ZnO NPs are specifically used in clothing, skin care products, anticancer medicines, sunscreens, coatings for solar cells, bottle coatings, and gas sensors [10].

The rapid expansion, due to their unique properties, and their release in the environment has raised considerable worries regarding manufactured NPs [11]. An essential aspect of the risk assessment of NPs is to understand their interactions with plants, a basic component of all ecosystems [11]. Thus, the extensive use of ZnO NPs has extended the requirements of research on their consequences on living organisms [6]. Environmental levels of ZnO-NPs were stated to be among 3.1–31 $\mu g\ kg^{-1}$ soil and 76–760 $\mu g\ L^{-1}$ water [12]. Manufactured NPs are unavoidably released into the soil and through streams, rivers, and sewage treatment they finally reach the sea [5]. Since NPs finally end in aquatic ecosystems, aquatic plants may be at higher risks than terrestrial. Thus, there is a need to evaluate the risks related to NP presence in aquatic ecosystems [12].

In photosynthesis, electron transport is mediated by photosystem I (PSI), and photosystem II (PSII) that work coordinately in the thylakoid membranes [13–15]. The most prone constituent of the photosynthetic apparatus to environmental stress is thought to be PSII [15–18]. Perturbations of PSII functionality resulted in declining the photosynthetic capacity, limiting the growth and development of plants, and reducing crop production [19,20]. Measuring PSII function by chlorophyll fluorescence imaging analysis is the most appropriate methodology to detect NPs-induced stress on plants [21–23] Metal oxides NPs alter photosynthetic efficiency by reducing energy transfer efficiency and quantum yield [5,21,23].

Reactive oxygen species (ROS) generated as byproducts in chloroplasts by the light reactions of photosynthesis are responsible for NPs induced toxicity [24], and thus the impact of NPs toxicity on plants can be also estimated by ROS production [22,25]. However, ROS generation can activate the plant's defense mechanisms in order to cope with the oxidative stress damage [15,26–29].

C. nodosa (Ucria) Ascherson is a perennial, fast-growing seagrass that colonizes shallow waters and degraded environments [5,30]. It grows in coastal areas in vicinity to anthropogenic actions and has been proposed a suitable bio-indicator species [31]. In this species, cellular, physiological and biochemical measurable responses induced by various chemical stressors have been proposed to monitor environmental quality [32,33].

The objectives of our study were to investigate the relationship between ZnO NPs effects in *C. nodosa* with Zn uptake in order to understand their impact on seagrasses. We wanted to test whether exposure of seagrasses to NPs will be a dose dependent response (concentration- and time-dependent) or a hormetic response. We evaluated ZnO NPs effects on *C. nodosa* PSII photochemistry by chlorophyll fluorescence imaging analysis, and detected ROS generation as a byproduct of ZnO NP's effects, linking also the ROS generation to PSII functionality.

2. Materials and Methods

2.1. Plant Material

C. nodosa (Ucria) Ascherson plants were collected from the Gulf of Thessaloniki, Aegean Sea (40°33 N, 22°58 E), by their maximum leaf biomass production, at 0.7 m–1.0 m depth [34].

2.2. Experimental Conditions and Exposure to Zinc Oxide Nanoparticles

C. nodosa plants (leaves, orthotropic and plagiotropic rhizomes and roots) were kept in seawater aquaria that have a salinity of 36.9 psu, pH of 7.9, and dissolved oxygen of 5.9 $mg\ L^{-1}$, using continuously aerated aquarium pumps as previously described [5].

Zinc oxide NPs with less than 50 nm particle sizes were purchased from Sigma-Aldrich (St. Louis, MO, USA). Stock solution of ZnO NPs in millique water (50 $mg\ L^{-1}$), after sonication for 30 min, was stored in the dark at 4 °C [5]. ZnO NPs concentrations in natural waters stated to be among 76–760 $\mu g\ L^{-1}$ [12] are below concentrations known to have environmental effects on aquatic organisms [35]. In preliminary experiments with 1 and 3 $mg\ L^{-1}$ ZnO NPs, no effect was detected on PSII functionality. Thus, we applied 5 $mg\ L^{-1}$ and 10 $mg\ L^{-1}$ ZnO NPs, which is 7–13 times more the maximum levels of ZnO-NPs reported for water environments [12].

C. nodosa plants were exposed to 5 and 10 mg L^{-1} for 0 (control), 4 h, 12 h, 24 h, 48 h and 72 h. The two ZnO NPs concentrations were prepared with filtered (0.45 μm GF/C Whatman) seawater immediately before use. Control and treatment solutions were changed every 24 h. *C. nodosa* intermediate leaf blades (about 300 mm length) were used for chlorophyll fluorescence imaging, H$_2$O$_2$ imaging, and for Zn uptake measurements.

2.3. Zinc Oxide Nanoparticles Characterization

Primary particle size, and the morphological and structural characteristics of ZnO NPs were investigated by transmission electron microscopy (TEM). Samples were prepared by drop-casting dispersions of the NPs onto carbon-coated Cu grids after a 3 min sonication with an ultrasonic (VibraCell 400 W, Sonics & Materials Inc., Newtown, CT, USA), applying a microtip probe under intensity settings 4 [36]. Finally, TEM images were obtained with a Jeol JEM 1010 microscope (Jeol, Tokyo, Japan) [37].

Dynamic light scattering (DLS) analysis was used to define the size-distribution profile of ZnO NPs (5 mg L^{-1}, 10 mg L^{-1} and 50 mg L^{-1}). Zeta (ζ) potential measurements were conducted to assess the surface charge of the particles as described previously [5]. All measurements were performed in Milli-Q water after brief sonication at 25 °C [5]. Results are presented as means (±SD) of three measurements.

2.4. Zinc Determination

Intermediate blades after wet digestion were processed following the methodology described previously [38,39]. Zinc concentrations were determined by flame atomic absorption spectrophotometry (AAnalyst 400 FAAS, Perkin-Elmer, Waltham, MA, USA) with the procedure described in detail before [38,39].

2.5. Zinc Leaf Uptake Kinetics

Zinc leaf uptake kinetics was fitted to the Michaelis-Menten equation: $(C_{max} \times t)/(K_m + t)$, as described in detail previously [5]. Briefly, C represents Zn leaf concentration reached in time t, K_m the time taken to reach half of the value of C_{max}, and C_{max} the maximum or saturation Zn concentration. The rate of the initial uptake ($C_{max}/2 \times K_m$), the time needed to get equilibrium (T_{eq}), the equilibrium concentration (C_{eq}) and the mean rate of uptake (V_c) were also estimated. Equilibrium concentration (C_{eq}) is a concentration where the hourly increase is less than 1% compared to the previous hour [38–41]. The time required to reach equilibrium (T_{eq}) was assessed as the time needed to get the C_{eq}, and the mean rate of uptake (V_c) was assessed as C_{eq}/T_{eq} [38,39]. Bioconcentration factor (BCF) was estimated as $(C_{eq} - C_i)/C_w$, where C_i is the initial Zn tissue concentration and C_w is the Zn concentration in water [38,39].

2.6. Chlorophyll Fluorescence Imaging Analysis

An Imaging-PAM Chlorophyll Fluorometer (Walz, Effeltrich, Germany) was used for photosynthetic efficiency measurements as previously described [5]. In *C. nodosa* dark-adapted (15 min) leaf samples, we selected six areas of interest, and the allocation of absorbed light energy to photochemistry (Φ_{PSII}), non-photochemical energy loss as heat (Φ_{NPQ}), and non-regulated energy loss (Φ_{NO}), were calculated as described previously [29]. Relative PSII electron transport rate (ETR), non-photochemical quenching (NPQ), and photochemical quenching (q_P), were also measured [42].

Color-coded images, acquired with 200 μmol photons m^{-2} s^{-1}, of Φ_{PSII}, Φ_{NPQ}, Φ_{NO}, and the redox state of plastoquinone (PQ) pool (q_P), are also presented.

2.7. Imaging of Hydrogen Peroxide Generation

For the estimation of H$_2$O$_2$ production, *C. nodosa* leaves were treated with 25 μM 2′,7′-dichlorofluorescein diacetate (Sigma) in the dark for 30 min, as described previously [43,44].

2.8. Statistical Analyses

Zinc leaf uptake kinetics data were analyzed using IBM Statistics SPSS® 24 (New York, NY, USA). The significant differences on the fluorescence variables, between control and different incubation time in each concentration and between different concentrations at the same exposure time were tested at the 5% level of probability using t-test analysis (IBM Statistics SPSS® 24). Modal analysis by using NORMSEP computer program was employed to estimate particle size (nm) distribution of ZnO NPs [45].

3. Results

3.1. Characterization of ZnO NPs

The size and morphology of ZnO NPs was measured in stock solution by TEM (Figure 1). Modal analysis was used in data emerged from TEM micrographs, in order to determine the ZnO particle diameter distribution (Figure 2). A percentage of 92.8% of the particles was generally in agreement with manufacturer's characteristics (size < 50 nm). The computed mean (±SD) NPs size group with the highest frequency was 20.44 nm ± 7.95 nm (Figure 2).

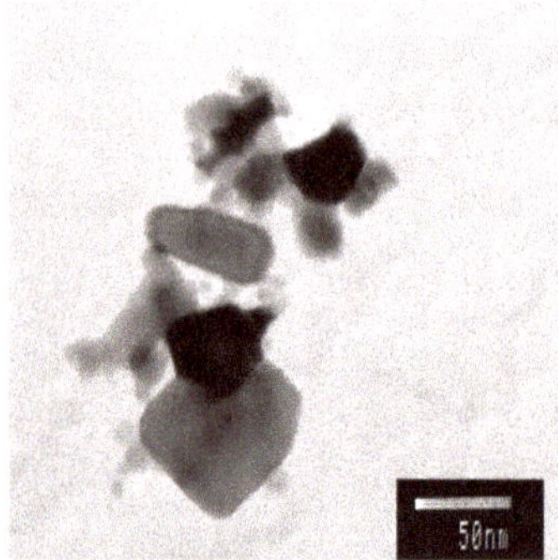

Figure 1. Transmission electron microscope (TEM) images of zinc oxide nanoparticles (ZnO NPs) stock solution (50 mg L^{-1}).

The hydrodynamic size of 5 mg L^{-1}, 10 mg L^{-1} and 50 mg L^{-1} (stock) ZnO NPs solutions, ranged from 220.6 nm to 225.0 nm (Table 1). The negative surface charge (ζ potential) ranged from −17.5 mV to −18.13 mV (Table 1). This similarity among the different NPs concentrations was indicative of the colloidal stability of the different populations. The size distribution by intensity of ZnO NPs is shown in Figure 3a,b.

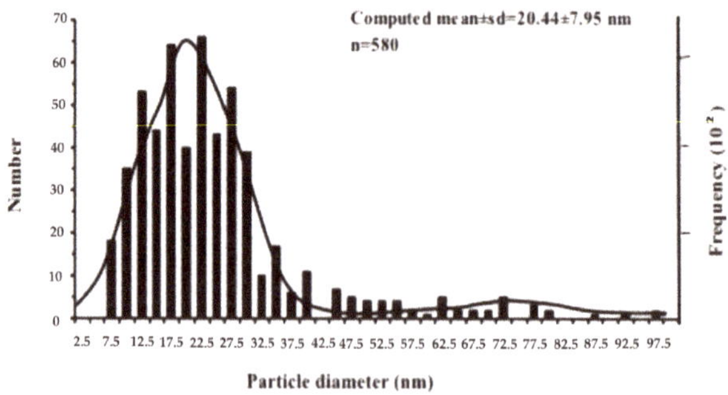

Figure 2. Distribution pattern of ZnO NPs from TEM micrographs.

Table 1. Hydrodynamic size (±SD) (diameter in nm) and zeta potential (±SD) (mV) values of the ZnO NPs (n = 3).

Formulation	Size	Zeta Potential
ZnO 50 mg L^{-1}	221.0 ± 7.2	−17.93 ± 0.11
ZnO 10 mg L^{-1}	225.0 ± 11.2	−17.50 ± 2.15
ZnO 5 mg L^{-1}	220.6 ± 5.0	−18.13 ± 0.55

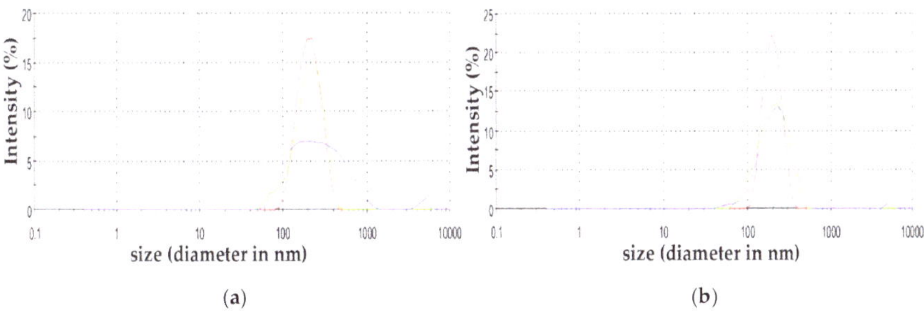

(a) (b)

Figure 3. Size distribution by intensity of 5 mg L^{-1} ZnO NPs (**a**); and 10 mg L^{-1} ZnO NPs (**b**). Different colours indicate the replications.

3.2. Zinc Leaf Uptake Kinetics

Leaf Zn uptake at both ZnO NPs concentration displayed a time dependent variation (Figure 4). The uptake kinetics at both exposure concentrations was fitted to the Michaelis-Menten equation (r^2:0.744 for 5 mg L^{-1}, and 0.681 for 10 mg L^{-1}, $p < 0.01$; Figure 4, Table 2). Zinc uptake of *C. nodosa* leaves increased more rapidly at the beginning of the experiment in the lower ZnO NPs solution (5 mg L^{-1}), while it showed a higher and more than doubled initial rate [C$_{max}$/(2 × K$_m$)], and a higher mean rate (V$_c$) in comparison to the 10 mg L^{-1} solution (Figure 4, Table 2). At 5 mg L^{-1} exposure concentration, the uptake reached the equilibrium concentration earlier (T$_{eq}$ = 28 h) than in the higher solution (42 h) (Table 2). However, both the maximum concentration (C$_{max}$) and the equilibrium concentration (C$_{eq}$) displayed their higher values at the higher ZnO NPs concentration (10 mg L^{-1}) (Figure 4, Table 2). During the experiment, the highest uptake was observed after 72 h of exposure, at both ZnO NP treatments (748.7 ± 29.7 µg g^{-1} dry wt at 5 mg L^{-1} and 1086.0 ± 33.7 µg g^{-1} dry wt at 10 mg L^{-1}) (Figure 4). Moreover, BCF value was higher at the lower exposure concentration (Table 2).

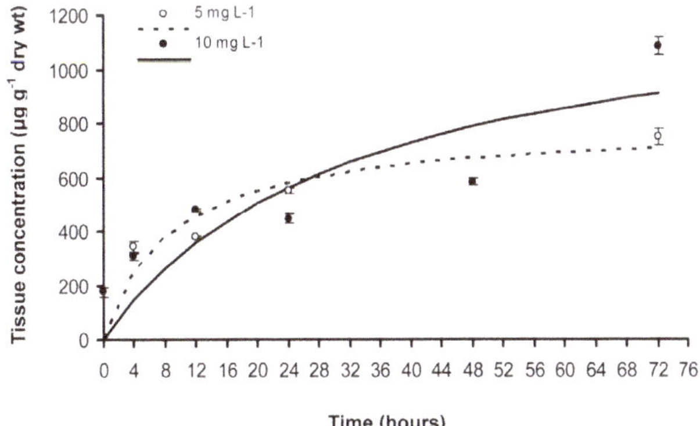

Figure 4. Kinetics of zinc uptake (µg g^{-1} dry weight) in *C. nodosa* leaf blades at 5 mg L^{-1} and 10 mg L^{-1} ZnO NPs ± SD (n = 3); dashed and bold lines are the uptake kinetics calculated using Michaelis-Menten equation.

Table 2. Kinetics of Zn accumulation in *C. nodosa* leaf blades exposed to 5 mg L^{-1} and 10 mg L^{-1} ZnO NPs.

Parameter	5 mg L^{-1}	10 mg L^{-1}
C_{max}	795.4 ($\pm$178.9)	1316.7 ($\pm$614.7)
K_m	8.8 ($\pm$6.7)	32.1 ($\pm$34.2)
$C_{max}/(2 \times K_m)$	44.950	20.507
r^2	0.744 *	0.681 *
C_{eq}	573.272	761.263
T_{eq}	28	42
V_c	20.473	18.125
BCF	79.588	58.593

* $p < 0.01$.

The fits correspond to a Michaelis-Menten equation: C = ($C_{max} \times t$)/($K_m + t$); C, in µg g^{-1} dry wt; K_m, in hours; t, in hours; C_{eq}, in µg g^{-1} dry wt; T_{eq}, in hours; V_c, concentration/hours. In parentheses, standard errors are given.

3.3. Allocation of Absorbed Light Energy in Leaf Blades of Cymodocea nodosa Exposed to ZnO NPs

The allocation of absorbed light energy at PSII in leaf blades of *C. nodosa* exposed to ZnO NPs for 4 h, 24 h, 48 h and 72 h is shown in Figures 5 and 6. Exposure to 5 and 10 mg L^{-1} ZnO NPs significantly decreased the quantum efficiency of PSII photochemistry (Φ_{PSII}) compared to control values, with the exception of the 24 h treatment that Φ_{PSII} increased at 10 mg L^{-1} ZnO NPs and did not differ compared to control values at 5 mg L^{-1} ZnO NPs (Figure 5a). Thus, at the 24-h treatment, Φ_{PSII} in 10 mg L^{-1} ZnO NPs was higher than in 5 mg L^{-1} (Figure 5a).

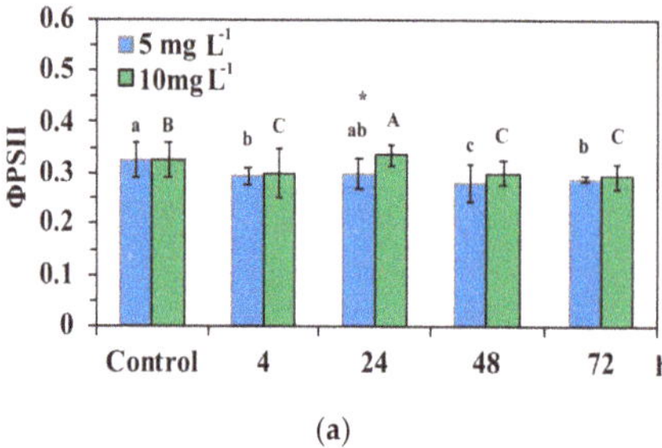
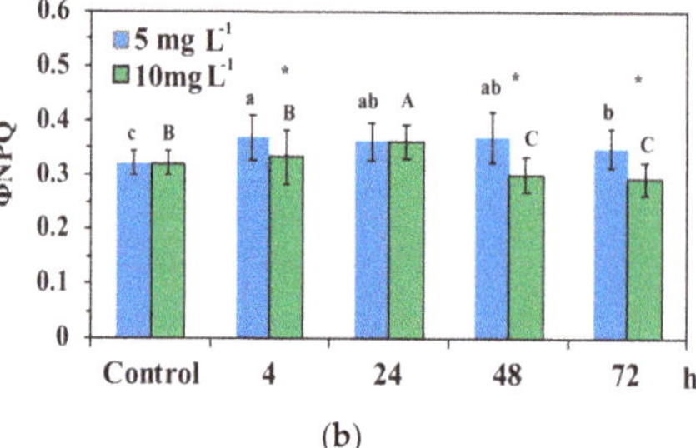

(a)   (b)

Figure 5. The quantum efficiency of PSII photochemistry (Φ_{PSII}) (**a**); and the quantum yield of regulated non-photochemical energy loss as heat (Φ_{NPQ}) (**b**), in control leaf blades of *C. nodosa* and in leaf blades exposed to 5 mg L^{-1} and 10 mg L^{-1} ZnO NPs for 4 h, 24 h, 48 h and 72 h. Columns with the same letter (lower case for 5 mg L^{-1} ZnO NPs and capitals for 10 mg L^{-1} ZnO NPs) are not statistically different ($p < 0.05$). An asterisk represents a significantly different mean of the same time treatment between 5 and 10 mg L^{-1} ZnO NPs ($p < 0.05$). Bars in columns represent standard deviation.

The quantum yield of regulated non-photochemical energy loss (Φ_{NPQ}) increased significantly at 5 mg L^{-1} ZnO NPs compared to control, while at 10 mg, L^{-1} increased after 24 h of exposure, but decreased at further exposure time (48 h and 72 h) (Figure 5b). Φ_{NPQ} after 4 h, 48 h and 72 h exposure to the low concentration was significantly higher than in the high concentration (Figure 5b).

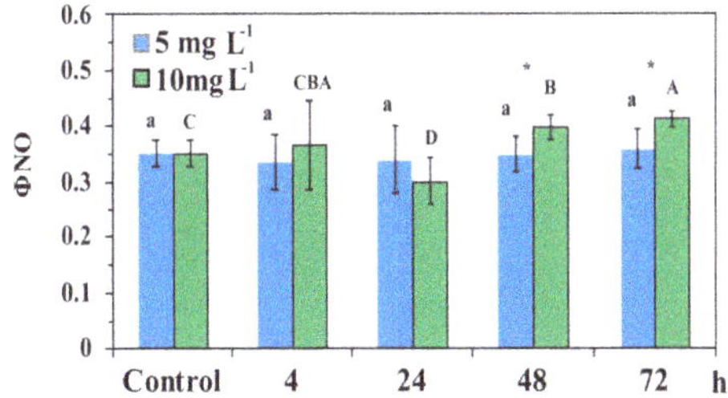

Figure 6. The quantum yield of non-regulated energy dissipated in PSII (non-regulated heat dissipation, a loss process due to PSII inactivity) (Φ_{NO}) in control leaf blades of *C. nodosa* (*Cymodocea nodosa*) and in leaf blades exposed to 5 mg L^{-1} and 10 mg L^{-1} ZnO NPs for 4 h, 24 h, 48 h and 72 h. Symbol explanations as in Figure 5.

The non-regulated energy loss (Φ_{NO}) did not differ compared to control values after exposure to 5 mg L^{-1} ZnO NPs (Figure 6), but decreased compared to the control values after 24 h of exposure to the high NPs concentration, and increased at further exposure time (48 h and 72 h) (Figure 6). At 48 h and 72 h treatment Φ_{NO} of *C. nodosa* leaf blades exposed to the high concentration was significantly higher than in the low concentration (Figure 6).

3.4. Electron Transport Rate and Non-Photochemical Quenching in Leaf Blades of Cymodocea nodosa Exposed to ZnO NPs

The relative electron transport rate (ETR) of PSII decreased significantly at both ZnO NPs concentrations, with the exception of the 24 h exposure, where the ETR increased at the high concentration and did not differ compared to control values at the low concentration (Figure 7a). With the 24 h treatment, the ETR of *C. nodosa* leaf blades exposed to the high concentration was significantly higher than in the low concentration (Figure 7a).

The non-photochemical quenching (NPQ) increased at the low concentration compared to control, while at the high concentration it increased after 24 h of exposure, but decreased significantly at further exposure time (48 h and 72 h) (Figure 7b). After 4 h, 48 h and 72 h treatment, the NPQ of *C. nodosa* leaf blades exposed to 5 mg L^{-1} was significantly higher than 10 mg L^{-1} (Figure 7b).

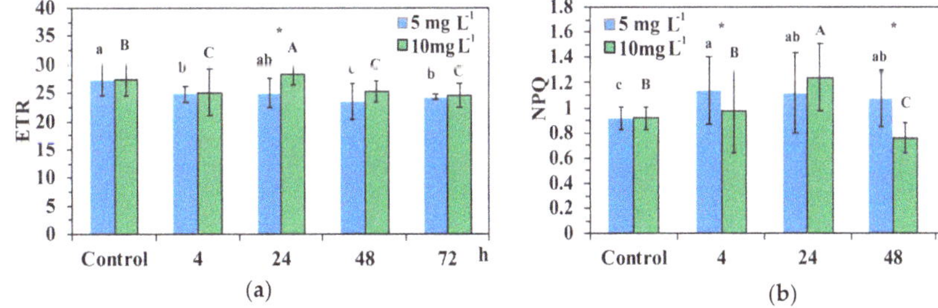

Figure 7. The relative electron transport rate of PSII (ETR) (a); and the non-photochemical quenching (NPQ) (b), in control leaf blades of *C. nodosa* and in leaf blades exposed to 5 mg L^{-1} and 10 mg L^{-1} ZnO NPs for 4 h, 24 h, 48 h and 72 h. Symbol explanations as in Figure 5.

3.5. The Redox State of Plastoquinone Pool of Cymodocea nodosa Leaf Blades Exposed to ZnO NPs

The redox state of plastoquinone (PQ) pool (a measure of PSII open reaction centers) (q_P), did not differ compared to control values after 4 h, 24 h, and 48 h exposure to 5 mg L^{-1} ZnO NPs but reduced after 72 h (Figure 8). PQ pool reduced compared to control values after 4 h, 48 h and 72 h exposure to

10 mg L^{-1} ZnO, but increased after a 24 h exposure (Figure 8). After a 72 h exposure of *C. nodosa* to ZnO NPs, the PQ pool was more oxidized at 5 mg L^{-1} than at 10 mg L^{-1}, but less than controls (Figure 8).

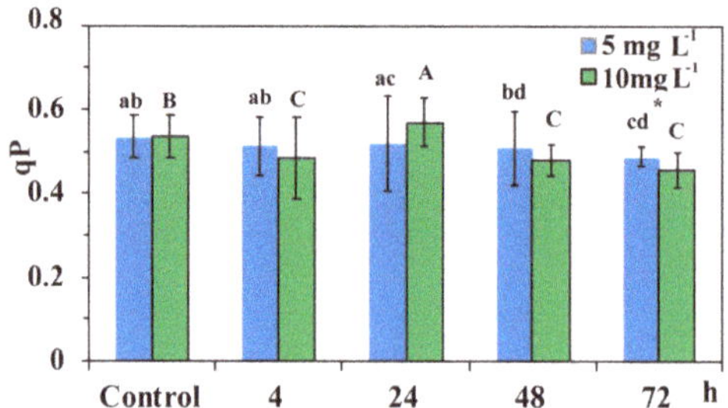

Figure 8. The photochemical quenching (q_P) in control leaf blades of *C. nodosa* and in leaf blades exposed to 5 mg L^{-1} and 10 mg L^{-1} ZnO NPs for 4 h, 24 h, 48 h and 72 h. Symbol explanations as in Figure 5.

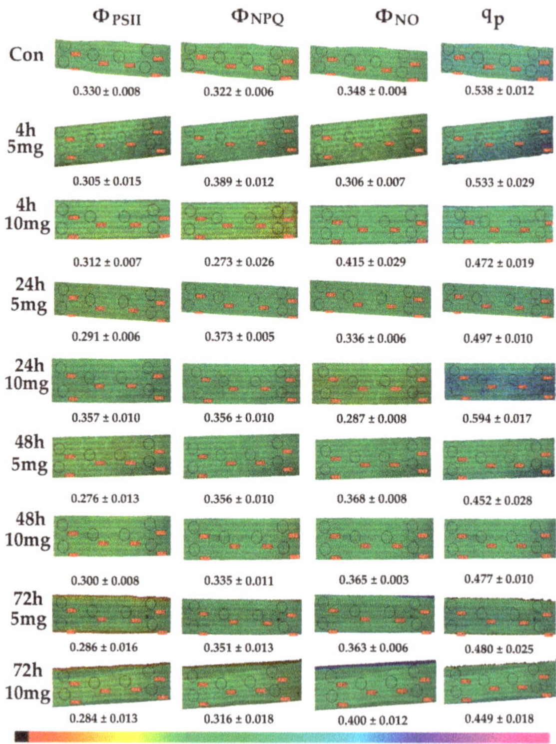

Figure 9. Representative chlorophyll fluorescence images at 200 µmol photons m^{-2} s^{-1} actinic light of Φ_{PSII}, Φ_{NPQ}, Φ_{NO}, and q_P; of *C. nodosa* control leaf blades and leaf blades exposed to 5 mg L^{-1} and 10 mg L^{-1} ZnO NPs for 4 h, 24 h, 48 h and 72 h. The colour code depicted at the bottom of the images ranges from black (pixel values 0.0) to purple (1.0). The six areas of interest are shown. Average values are presented for each photosynthetic parameter.

3.6. Chlorophyll Fluorescence Images of Cymodocea nodosa Leaf Blades Exposed to ZnO NPs

We did not detect any spatial heterogeneity of Φ_{PSII}, Φ_{NPQ}, Φ_{NO}, and q_P images in the control leaf blades of *C. nodosa* measured at 200 µmol photons m^{-2} s^{-1} actinic light (Figure 9). In addition, exposure to both NPs concentrations for 4 h, 24 h, 48 h and 72 h did not significantly alter these patterns (Figure 9). A temporal heterogeneity was observed at the images of q_P and Φ_{NPQ} (Figure 9). The lowest Φ_{PSII} with simultaneous low q_P values was observed after a 12 h exposure at the high concentration (Figure 10a). At the same time, the highest levels of H_2O_2 were detected (Figure 10b). Immediately after this, we noticed that the 24 h exposure to 10 mg L^{-1} ZnO NPs resulted in the highest Φ_{PSII} values with the highest q_P values, both of them being higher than the control values (Figure 9). A parallel decreased Φ_{NO} was detected (Figure 9).

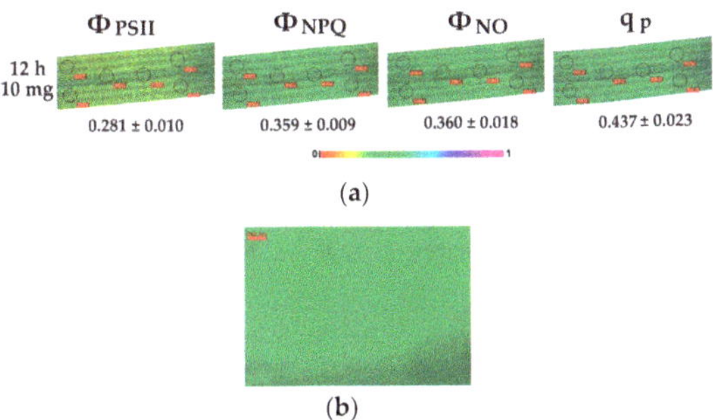

Figure 10. Representative chlorophyll fluorescence images after 5 min illumination at 200 μmol photons $m^{-2} s^{-1}$ actinic light of Φ_{PSII}, Φ_{NPQ}, Φ_{NO}, and q_P of *C. nodosa* leaf blades exposed to 10 mg L^{-1} ZnO NPs for 12 h. The colour code depicted at the bottom of the images ranges from 0.0 to 1.0. The average values are presented for each photosynthetic parameter (**a**). Below is the representative pattern of H_2O_2 production in a *C. nodosa* leaf blade exposed for 12 h to 10 mg L^{-1} ZnO NPs. Scale bare: 200 μm. Increased H_2O_2 content is indicated by light green colour (**b**).

3.7. Imaging of Hydrogen Peroxide Production After Exposure of Cymodocea nodosa Leaf Blades to ZnO NPs

No noteworthy quantities of H_2O_2 could be noticed in the control leaf blades of *C. nodosa* (Figure 11a). Exposure at the low concentration for 4 h did not result in any change to H_2O_2 production (Figure 11b), while the same exposure time at the high concentration, resulted in increased production of H_2O_2 (Figure 11c). After a 24-h exposure to 5 mg L^{-1} ZnO NPs, H_2O_2 levels were the same as the control ones (Figure 11d), while they also dropped at 10 mg L^{-1} and could not be detected at all (Figure 11e). Later, after a 48-h exposure to ZnO NPs, there was an increase in the accumulation of H_2O_2 being higher at 5 mg L^{-1} (Figure 11f), than at 10 mg L^{-1} ZnO NPs (Figure 11g). However, after a 72-h exposure to ZnO NPs the accumulation of H_2O_2 decreased at 5 mg L^{-1} (Figure 11h), but increased at 10 mg L^{-1} ZnO NPs (Figure 11i). The highest H_2O_2 accumulation was detected after a 12-h exposure to 10 mg L^{-1} ZnO NPs (Figure 10b) and the lowest after a 24-h exposure to 10 mg L^{-1} (Figure 11e).

4. Discussion

Previously, extensive literature survey has demonstrated both the positive and detrimental impacts of NPs on terrestrial and aquatic plants, which are due to size and type of NPs (especially their specific surface area) and the plant species [5,7,8,12,22,25]. Toxicity of ZnO NPs is determined to be due to the dissolution, release and uptake of free Zn ions, but specific nanoparticulate effects may be hard to unravel from effects due to free zinc ions [46,47]. Thus, ZnO NPs effects in *C. nodosa* were correlated to both applied ZnO NPs concentration and to Zn uptake.

In *C. nodosa* cellular, physiological and biochemical measurable responses (biomarkers) to metallic elements (e.g., Cd, Cr, Cu, Ni) have been proposed as early warning signals of alterations in seawater quality [32,33,38]. However, there has been little consideration of the seagrasses, especially *C. nodosa*, as the test material in evaluating the effects of metal oxide nanoparticles [5], despite the fact that coastal ecosystems are expected to be the destination of the majority of the nanoparticles, mainly ZnO and

TiO$_2$ NPs, discharged by industry [46]. NPs are released into aquatic environment either by direct uses or wastewater plant effluents [47–49].

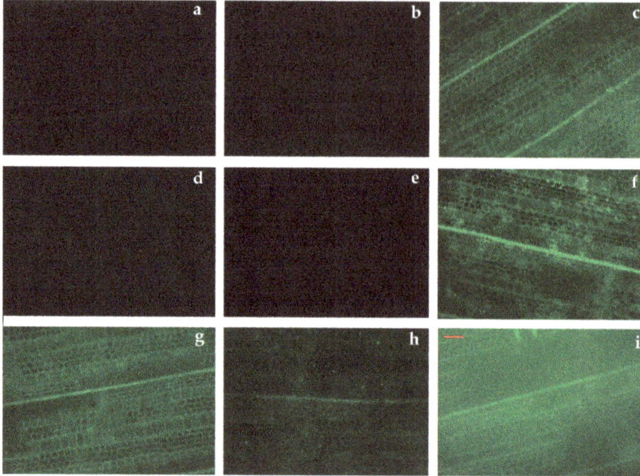

Figure 11. Representative patterns of H$_2$O$_2$ production in *C. nodosa* control leaf blades and after exposure to 5 mg L^{-1} and 10 mg L^{-1} ZnO NPs, as indicated by the fluorescence of H$_2$DCF-DA. The H$_2$O$_2$ real-time generation in control leaf blade (**a**); after a 4-h exposure to 5 mg L^{-1} (**b**); after a 4-h exposure to 10 mg L^{-1} (**c**); after a 24-h exposure to 5 mg L^{-1} (**d**); after a 24-h exposure to 10 mg L^{-1} (**e**); after a 48-h exposure to 5 mg L^{-1} (**f**); after a 48-h exposure to 10 mg L^{-1} (**g**); after a 72-h exposure to 5 mg L^{-1} (**h**); and after a 72-h exposure to 10 mg L^{-1} (**i**). Scale bare: 200 µm. A higher H$_2$O$_2$ content is indicated by light green colour.

Photosynthetic organisms via the process of photosynthesis, transform the light energy into chemical energy with the collaboration of PSII and PSI, while the most susceptible constituent of the photosynthetic apparatus to environmental stresses is believed to be PSII [15–18,50]. PSII functionality estimated by chlorophyll fluorescence imaging has been considered as the most suitable method to identify NPs toxicity effects on plants [5,21,22]. Exposure of plants to NPs can have positive or negative effects on the light reactions of photosynthesis [51].

At the beginning of the experiment, Zn uptake kinetics at 5 mg L^{-1} ZnO NPs with a more than twice initial rate and higher mean rate (V_c) than 10 mg L^{-1} (Figure 4, Table 2), resulted in a significant lower quantum efficiency of PSII photochemistry (Φ_{PSII}) compared to 10 mg L^{-1}, after a 24 h exposure (Figure 5a). However, at 5 mg L^{-1} exposure concentration, Zn uptake reached the equilibrium concentration earlier (T_{eq} = 28 h) than at 10 mg L^{-1} (T_{eq} = 42 h), resulting in a significantly higher regulated non-photochemical energy loss as heat (Φ_{NPQ}) after 48 h and 72 h exposure, compared to 10 mg L^{-1} (Figure 5b). This lower Φ_{NPQ} at 10 mg L^{-1} ZnO NPs, resulted in significantly higher non-regulated energy loss (Φ_{NO}) of *C. nodosa* leaf blades exposed to 10 mg L^{-1} for 48 h and 72 h (Figure 6), since there was no difference in Φ_{PSII} (Figure 5a). The significantly increased Φ_{NO} (Figure 6) implies higher singlet oxygen (^{1}O$_2$) production. Φ_{NO} consists of chlorophyll fluorescence internal conversions and intersystem crossing, indicative of ^{1}O$_2$ formation via the triplet state of chlorophyll (^{3}chl*) [42,52–54]. The increased ^{1}O$_2$ formation, mostly after 72 h exposure to 10mg L^{-1} ZnO NPs, was due to an increased Zn uptake compared to 5 mg L^{-1} (Figure 4).

A reduced photosynthetic efficiency, measured as the maximum quantum efficiency of PSII (F_v/F_m), and the redox state of plastoquinone (PQ) pool (q_P), was also observed previously due to increased Zn accumulation [7]. Furthermore, ZnO NPs treatments enhanced generation of H$_2$O$_2$ [7]. A relationship between closed reaction centers (q_P) and increased H$_2$O$_2$ generation was noticed (Figures 9–11). The PQ pool is considered as the component integrated in plant antioxidant defense [55], which at a 12 h of exposure mediated the generation of chloroplastic H_2O_2 [22], acting as a fast acclimation-signaling molecule [55,56].

After 12 h exposure of *C. nodosa* to 10 mg L^{-1} ZnO NPs, leaf Zn uptake was higher than in 5 mg L^{-1} (Figure 4), resulting in the lower Φ_{PSII} and the lowest q_P values (Figure 10a). At the same time, an increased accumulation of H_2O_2 was detected in the leaves of *C. nodosa* (Figure 10b). Closed reaction centers (q_P) indicate excess photon supply and associated ROS production [57–59]. This photosynthesis derived H_2O_2 is moving throughout leaf veins, serving as a signaling molecule and triggering a stress-defense response [22,35,44,55,56,60,61]. Thus, this stress-defense response triggered a significant increase of the fraction of open PSII reaction centers (q_P) (Figure 8) and a significant increase of Φ_{PSII} (Figure 5a) and ETR (Figure 7a) observed after 24 h exposure of *C. nodosa* to 10 mg L^{-1} ZnO NPs; while at the same time, H_2O_2 generation decreased to control levels (Figure 11e). However, we have to emphasize that antibacterial activity of ZnO and ZnO NPs is due to ROS generation [62,63], which also orchestrates a regulatory action at various plant developmental stages [64].

Hormesis has been extensively documented in plants, revealing that biphasic dose-responses occur commonly [65,66]. Thus, at low-level stress, plants are activating responses at the cellular and molecular level that enhance adaptation and plant tolerance [65,66]. The photoprotective mechanism of non-photochemical quenching (NPQ) which is closely related to ROS, follows a biphasic dose-response pattern typical of hormesis [67]. NPQ in *C. nodosa* leaf blades exposed to 10 mg L^{-1} ZnO NPs depicts a hormetic response. A hormetic response suggests that a basal stress level is needed for adaptive responses [68,69]. This basal stress level was 10 mg L^{-1} ZnO NPs and the time required for the induction of this mechanism was 24 h exposure.

Author Contributions: Conceptualization, P.M. and M.M.; methodology, P.M., I.S. and M.M.; software, P.M.; validation, P.M., K.C. and I.S.; formal analysis, P.M. and K.C.; investigation, P.M., K.C., I.S. and Z.M.; resources, P.M. and M.M.; data curation, P.M. and K.C.; writing—original draft preparation, P.M. and M.M; writing—review and editing, M.M; visualization, P.M., K.C. and M.M.; supervision, P.M. and M.M.; project administration, P.M. and M.M.; funding acquisition, P.M. and M.M.

Funding: This research received no external funding.

Conflicts of Interest: The authors declare no conflict of interest.

References

1. Maynard, A.; Aitken, R.J.; Butz, T.; Colvin, V.; Donaldson, K.; Oberdörster, G.; Philbert, M.A.; Ryan, J.; Seaton, A.; Stone, V.; et al. Safe handling of nanotechnology. *Nature* **2006**, *444*, 267–269. [CrossRef] [PubMed]

2. Chang, Y.N.; Zhang, M.; Xia, L.; Zhang, J.; Xing, G. The toxic effects and mechanisms of CuO and ZnO nanoparticles. *Materials* **2012**, *5*, 2850–2871. [CrossRef]

3. Kołodziejczak-Radzimska, A.; Jesionowski, T. Zinc oxide—From synthesis to application: A review. *Materials* **2014**, *7*, 2833–2881. [CrossRef]

4. Hong, J.; Rico, C.; Zhao, L.; Adeleye, A.S.; Keller, A.A.; Peralta-Videa, J.R.; Gardea-Torresdey, J.L. Toxic effects of copper-based nanoparticles or compounds to lettuce (*Lactuca sativa*) and alfalfa (*Medicago sativa*). *Environ. Sci. Proc. Impacts* **2015**, *17*, 177–185. [CrossRef] [PubMed]

5. Moustakas, M.; Malea, P.; Haritonidou, K.; Sperdouli, I. Copper bioaccumulation, photosystem II functioning and oxidative stress in the seagrass *Cymodocea nodosa* exposed to copper oxide nanoparticles. *Environ. Sci. Pollut. Res.* **2017**, *24*, 16007–16018. [CrossRef]

6. Ghosh, M.; Sinha, S.; Jothiramajayam, M.; Jana, A.; Nag, A.; Mukherjee, A. Cytogenotoxicity and oxidative stress induced by zinc oxide nanoparticle in human lymphocyte cells in-vitro and Swiss albino male mice in-vivo. *Food Chem. Toxicol.* **2016**, *97*, 286–296. [CrossRef]

7. Tripathi, D.K.; Mishra, R.K.; Singh, S.; Singh, S.; Vishwakarma, K.; Sharma, S.; Singh, V.P.; Singh, P.K.; Prasad, S.M.; Dubey, N.K.; et al. Nitric oxide ameliorates zinc oxide nanoparticles phytotoxicity in wheat seedlings: Implication of the ascorbate-glutathione cycle. *Front. Plant Sci.* **2017**, *8*, 1. [CrossRef]

8.   Pullagurala, V.L.R.; Adisa, I.O.; Rawat, S.; Kim, B.; Barrios, A.C.; Medina-Velo, I.A.; Hernandez-Viezcas, J.A.; Peralta-Videa, J.A.; Gardea-Torresdey, J.L. Finding the conditions for the beneficial use of ZnO nanoparticles towards plants—A review. *Environ. Pollut.* **2018**, *241*, 1175–1181. [CrossRef]

9.   Segets, D.; Gradl, J.; Taylor, R.K.; Vassilev, V.; Peukert, W. Analysis of optical absorbance spectra for the determination of ZnO nanoparticle size distribution, solubility, and surface energy. *ACS Nano* **2009**, *3*, 1703–1710. [CrossRef]

10.   Song, W.; Zhang, J.; Guo, J.; Zhang, J.; Ding, F.; Li, L.; Sun, Z. Role of the dissolved zinc ion and reactive oxygen species in cytotoxicity of ZnO nanoparticles. *Toxicol. Lett.* **2010**, *199*, 389–397. [CrossRef]

11.   Ma, X.; Geiser-Lee, J.; Deng, Y.; Kolmakov, A. Interactions between engineered nanoparticles (ENPs) and plants: Phytotoxicity, uptake and accumulation. *Sci. Total Environ.* **2010**, *408*, 3053–3061. [CrossRef] [PubMed]

12.   Rajput, V.D.; Minkina, T.M.; Behal, A.; Sushkova, S.N.; Mandzhieva, S.; Singh, R.; Gorovtsov, A.; Tsitsuashvili, V.S.; Purvis, W.O.; Ghazaryan, K.A.; et al. Effects of zinc-oxide nanoparticles on soil, plants, animals and soil organisms: A review. *Environ. Nanotechnol. Monit. Manag.* **2018**, *9*, 76–84. [CrossRef]

13.   Anderson, J.M. Changing concepts about the distribution of photosystems I and II between grana-appressed and stroma-exposed thylakoid membranes. *Photosynth. Res.* **2002**, *73*, 157–164. [CrossRef]

14.   Apostolova, E.L.; Dobrikova, A.G.; Ivanova, P.I.; Petkanchin, I.B.; Taneva, S.G. Relationship between the organization of the PSII supercomplex and the functions of the photosynthetic apparatus. *J. Photochem. Photobiol. B* **2006**, *83*, 114–122. [CrossRef]

15.   Moustaka, J.; Ouzounidou, G.; Sperdouli, I.; Moustakas, M. Photosystem II is more sensitive than photosystem I to Al^{3+} induced phytotoxicity. *Materials* **2018**, *11*, 1772. [CrossRef] [PubMed]

16.   Ekmekçi, Y.; Tanyolaç, D.; Ayhan, B. Effects of cadmium on antioxidant enzyme and photosynthetic activities in leaves of two maize cultivars. *J. Plant Physiol.* **2008**, *165*, 600–611. [CrossRef] [PubMed]

17.   Bayçu, G.; Moustaka, J.; Gevrek-Kürüm, N.; Moustakas, M. Chlorophyll fluorescence imaging analysis for elucidating the mechanism of photosystem II acclimation to cadmium exposure in the hyperaccumulating plant *Noccaea caerulescens*. *Materials* **2018**, *11*, 2580. [CrossRef]

18.   Moustakas, M.; Bayçu, G.; Gevrek-Kürüm, N.; Moustaka, J.; Csatári, I.; Rognes, S.E. Spatiotemporal heterogeneity of photosystem II function during acclimation to zinc exposure and mineral nutrition changes in the hyperaccumulator. *Noccaea caerulescens*. *Environ. Sci. Pollut. Res.* **2019**, *26*, 6613–6624. [CrossRef]

19.   Ding, Z.S.; Zhou, B.Y.; Sun, X.F.; Zhao, M. High light tolerance is enhanced by overexpressed PEPC in rice under drought stress. *Acta Agron. Sin.* **2012**, *38*, 285–292. [CrossRef]

20.   Zeng, L.D.; Li, M.; Chow, W.S.; Peng, C.L. Susceptibility of an ascorbate-deficient mutant of Arabidopsis to high-light stress. *Photosynthetica* **2018**, *56*, 427–432. [CrossRef]

21.   Falco, W.F.; Queiroz, A.M.; Fernandes, J.; Botero, E.R.; Falcão, E.A.; Guimarães, F.E.G.; M'Peko, J.C.; Oliveira, S.L.; Colbeck, I.; Caires, A.R.L. Interaction between chlorophyll and silver nanoparticles: A close analysis of chlorophyll fluorescence quenching. *J. Photochem. Photobiol. A* **2015**, *299*, 203–209. [CrossRef]

22.   Antonoglou, O.; Moustaka, J.; Adamakis, I.-D.; Sperdouli, I.; Pantazaki, A.; Moustakas, M.; Dendrinou-Samara, C. Nanobrass CuZn nanoparticles as foliar spray non phytotoxic fungicides. *ACS Appl. Mater. Interfaces* **2018**, *10*, 4450–4461. [CrossRef] [PubMed]

23.   Rico, C.M.; Hong, J.; Morales, M.I.; Zhao, L.; Barrios, A.C.; Zhang, J.Y.; Peralta-Videa, J.R.; Gardea-Torresdey, J.L. Effect of cerium oxide nanoparticles on rice: A study involving the antioxidant defense system and in vivo fluorescence imaging. *Environ. Sci. Technol.* **2013**, *47*, 5635–5642. [CrossRef] [PubMed]

24.   Ma, C.X.; White, J.C.; Dhankher, O.P.; Xing, B. Metal-based nanotoxicity and detoxification pathways in higher plants. *Environ. Sci. Technol.* **2015**, *49*, 7109–7122. [CrossRef] [PubMed]

25.   Tripathi, D.K.; Shweta; Singh, S.; Singh, S.; Pandey, R.; Singh, V.P.; Sharma, N.C.; Prasad, S.M.; Dubey, N.K.; Chauhan, D.K. An overview on manufactured nanoparticles in plants: Uptake, translocation, accumulation and phytotoxicity. *Plant Physiol. Biochem.* **2017**, *110*, 2–12. [CrossRef] [PubMed]

26.   Moustaka, J.; Moustakas, M. Photoprotective mechanism of the non-target organism Arabidopsis thaliana to paraquat exposure. *Pest. Biochem. Physiol.* **2014**, *111*, 1–6. [CrossRef] [PubMed]

27.   Moustaka, J.; Ouzounidou, G.; Bayçu, G.; Moustakas, M. Aluminum resistance in wheat involves maintenance of leaf Ca^{2+} and Mg^{2+} content, decreased lipid peroxidation and Al accumulation, and low photosystem II excitation pressure. *BioMetals* **2016**, *29*, 611–623. [CrossRef] [PubMed]

28. Foyer, C.H. Reactive oxygen species, oxidative signaling and the regulation of photosynthesis. *Environ. Exp. Bot.* **2018**, *154*, 134–142. [CrossRef]

29. Moustaka, J.; Panteris, E.; Adamakis, I.D.S.; Tanou, G.; Giannakoula, A.; Eleftheriou, E.P.; Moustakas, M. High anthocyanin accumulation in poinsettia leaves is accompanied by thylakoid membrane unstacking, acting as a photoprotective mechanism, to prevent ROS formation. *Environ. Exp. Bot.* **2018**, *154*, 44–55. [CrossRef]

30. Malea, P.; Zikidou, C. Temporal variation in biomass partitioning of the seagrass *Cymodocea nodosa* at the Gulf of Thessaloniki, Greece. *J. Biol. Res.-Thessalon.* **2011**, *5*, 75–90.

31. Malea, P.; Kevrekidis, T. Trace element (Al, As, B, Ba, Cr, Mo, Ni, Se, Sr, Tl, U and V) distribution and seasonality in compartments of the seagrass Cymodocea nodosa. *Sci. Total Environ.* **2013**, *463–464*, 611–623. [CrossRef] [PubMed]

32. Malea, P.; Adamakis, I.-D.; Kevrekidis, T. Microtubule integrity and cell viability under metal (Cu, Ni and Cr) stress in the seagrass *Cymodocea nodosa*. *Chemosphere* **2013**, *93*, 1035–1042. [CrossRef] [PubMed]

33. Malea, P.; Adamakis, I.-D.; Kevrekidis, T. Effects of lead uptake on microtubule cytoskeleton organization and cell viability in the seagrass *Cymodocea nodoca*. *Ecotoxicol. Environ. Saf.* **2014**, *104*, 175–181. [CrossRef] [PubMed]

34. Moustakas, M.; Malea, P.; Zafeirakoglou, A.; Sperdouli, I. Photochemical changes and oxidative damage in the aquatic macrophyte *Cymodocea nodosa* exposed to paraquat-induced oxidative stress. *Pest. Biochem. Physiol.* **2016**, *126*, 28–34. [CrossRef] [PubMed]

35. Batley, G.E.; Kirby, J.K.; McLaughlin, M.J. Fate and risks of nanomaterials in aquatic and terrestrial environments. *Acc. Chem. Res.* **2013**, *46*, 854–862. [CrossRef]

36. Perreault, F.; Oukarroum, A.; Melegari, S.P.; Matias, W.G.; Popovic, R. Polymer coating of copper oxide nanoparticles increases nanoparticles uptake and toxicity in the green alga *Chlamydomonas reinhardtii*. *Chemosphere* **2012**, *87*, 1388–1394. [CrossRef]

37. Dallas, P.; Georgakilas, V.; Niarchos, D.; Komninou, P.; Kehagias, T.; Petridis, D. Synthesis, characterization and thermal properties of polymer/magnetite nanocomposites. *Nanotechnology* **2006**, *17*, 2046–2053. [CrossRef]

38. Malea, P.; Adamakis, I.-D.; Kevrekidis, T. Kinetics of cadmium accumulation and its effects on microtubule integrity and cell viability in the seagrass *Cymodocea nodoca*. *Aquat. Toxicol.* **2013**, *144–145*, 257–264. [CrossRef]

39. Malea, P.; Kevrekidis, T.; Chatzipanagiotou, K.-R.; Mogias, A. Cadmium uptake kinetics in parts of the seagrass *Cymodocea nodosa* at high exposure concentrations. *J. Biol. Res.-Thessalon.* **2018**, *25*, 5. [CrossRef]

40. Fernández, J.A.; Vázquez, M.D.; López, J.; Carvalleira, A. Modeling the extra and intracellular uptake and discharge of heavy metals in *Fontinalis antipyretica* transplanted along a heavy metal and pH contamination gradient. *Environ. Pollut.* **2006**, *139*, 21–31. [CrossRef]

41. Diaz, S.; Villares, R.; Carballeira, A. Uptake kinetics of As, Hg, Sb, and Se in the aquatic moss *Fontinalis antipyretica* Hedw. *Water Air Soil Pollut.* **2012**, *223*, 3409–3423. [CrossRef]

42. Bayçu, G.; Gevrek-Kürüm, N.; Moustaka, J.; Csatári, I.; Rognes, S.E.; Moustakas, M. Cadmium zinc accumulation and photosystem II responses of *Noccaea caerulescens* to Cd and Zn exposure. *Environ. Sci. Pollut. Res.* **2017**, *24*, 2840–2850. [CrossRef] [PubMed]

43. Sandalio, L.M.; Rodríguez-Serrano, M.; Romero-Puertas, M.C.; Del Río, L.A. Imaging of reactive oxygen species and nitric oxide in vivo in plant tissues. *Method. Enzymol.* **2008**, *440*, 397–409.

44. Moustaka, J.; Tanou, G.; Adamakis, I.D.; Eleftheriou, E.P.; Moustakas, M. Leaf age dependent photoprotective and antioxidative mechanisms to paraquat-induced oxidative stress in *Arabidopsis thaliana*. *Int. J. Mol. Sci.* **2015**, *16*, 13989–14006. [CrossRef] [PubMed]

45. Villares, R.; Puente, X.; Carballeira, A. Seasonal variarion and background levels of heavy metals in two green seaweeds. *Environ. Pollut.* **2002**, *119*, 79–90. [CrossRef]

46. Miller, R.J.; Lenihan, H.S.; Muller, E.B.; Tseng, N.; Hanna, S.K.; Keller, A.A. Impacts of metal oxide nanoparticles on marine phytoplankton. *Environ. Sci. Technol.* **2010**, *44*, 7329–7334. [CrossRef] [PubMed]

47. Matranga, V.; Corsi, I. Toxic effects of engineered nanoparticles in the marine environment: Model organisms and molecular approaches. *Mar. Environ. Res.* **2012**, *76*, 32–40. [CrossRef] [PubMed]

48. Wong, S.W.Y.; Leung, P.T.; Djurišić, A.B.; Leung, K.M.Y. Toxicities of nano zinc oxide to five marine organisms: Influences of aggregate site and ion solubility. *Anal. Bioanal. Chem.* **2010**, *396*, 609–618. [CrossRef]

49. Klaine, S.J.; Alvarez, P.J.J.; Batley, G.E.; Fernandes, T.F.; Handy, R.D.; Lyon, D.Y.; Mahendra, S.; Mclaughlin, M.J.; Lead, J.R. Nanomaterials in the environment: Behavior, fate, bioavailability, and effects: A critical review. *Environ. Toxicol. Chem.* **2008**, *27*, 1825–1851. [CrossRef]

50. Zlobin, I.E.; Ivanov, Y.V.; Kartashov, A.V.; Sarvin, B.A.; Stavrianidi, A.N.; Kreslavski, V.D.; Kuznetsov, V.V. Impact of weak water deficit on growth, photosynthetic primary processes and storage processes in pine and spruce seedlings. *Photosynth. Res.* **2019**, *139*, 307–323. [CrossRef]

51. Hatami, M.; Kariman, K.; Ghorbanpour, M. Engineered nanomaterial-mediated changes in the metabolism of terrestrial plants. *Sci. Total Environ.* **2016**, *571*, 275–291. [CrossRef]

52. Apel, K.; Hirt, H. Reactive oxygen species: Metabolism, oxidative stress and signal transduction. *Annu. Rev. Plant Biol.* **2004**, *55*, 373–399. [CrossRef] [PubMed]

53. Gawroński, P.; Witoń, D.; Vashutina, K.; Bederska, M.; Betliński, B.; Rusaczonek, A.; Karpiński, S. Mitogen-activated protein kinase 4 is a salicylic acid-independent regulator of growth but not of photosynthesis in *Arabidopsis*. *Mol. Plant* **2014**, *7*, 1151–1166. [CrossRef] [PubMed]

54. Sperdouli, I.; Moustakas, M. Differential blockage of photosynthetic electron flow in young and mature leaves of *Arabidopsis thaliana* by exogenous proline. *Photosynthetica* **2015**, *53*, 471–477. [CrossRef]

55. Borisova-Mubarakshina, M.M.; Vetoshkina, D.V.; Ivanov, B.N. Antioxidant and signaling functions of the plastoquinone pool in higher plants. *Physiol. Plant.* **2019**, *166*, 181–198. [CrossRef] [PubMed]

56. Wilson, K.E.; Ivanov, A.G.; Öquist, G.; Grodzinski, B.; Sarhan, F.; Huner, N.P.A. Energy balance, organellar redox status, and acclimation to environmental stress. *Can. J. Bot.* **2006**, *84*, 1355–1370. [CrossRef]

57. Sperdouli, I.; Moustakas, M. A better energy allocation of absorbed light in photosystem II and less photooxidative damage contribute to acclimation of *Arabidopsis thaliana* young leaves to water deficit. *J. Plant Physiol.* **2014**, *171*, 587–593. [CrossRef]

58. Agathokleous, E.; Mouzaki-Paxinou, A.C.; Saitanis, C.J.; Paoletti, E.; Manning, W.J. The first toxicological study of the antiozonant and research tool ethylene diurea (EDU) using a *Lemna minor* L. bioassay: Hints to its mode of action. *Environ. Pollut.* **2016**, *213*, 996–1006. [CrossRef]

59. Sperdouli, I.; Moustakas, M. Leaf developmental stage modulates metabolite accumulation and photosynthesis contributing to acclimation of *Arabidopsis thaliana* to water deficit. *J. Plant Res.* **2014**, *127*, 481–489. [CrossRef]

60. Mittler, R.; Vanderauwera, S.; Suzuki, N.; Miller, G.; Tognetti, V.B.; Vandepoele, K.; Gollery, M.; Shulaev, V.; Van Breusegem, F. ROS signaling: The new wave? *Trends Plant Sci.* **2011**, *16*, 300–309. [CrossRef]

61. Foyer, C.H.; Shigeoka, S. Understanding oxidative stress and antioxidant functions to enhance photosynthesis. *Plant Physiol.* **2011**, *155*, 93–100. [CrossRef] [PubMed]

62. Fiedot-Tobola, M.; Ciesielska, M.; Maliszewska, I.; Rac-Rumijowska, O.; Suchorska-Wozniak, P.; Teterycz, H.; Bryjak, M. Deposition of zinc oxide on different polymer textiles and their antibacterial properties. *Materials* **2018**, *11*, 707. [CrossRef] [PubMed]

63. Gavrilenko, E.A.; Goncharova, D.A.; Lapin, I.N.; Nemoykina, A.L; Svetlichnyi, V.A.; Aljulaih, A.A.; Mintcheva, N.; Kulinich, S.A. Comparative study of physicochemical and antibacterial properties of ZnO nanoparticles prepared by laser ablation of Zn target in water and air. *Materials* **2019**, *12*, 186. [CrossRef] [PubMed]

64. Singh, R.; Singh, S.; Parihar, P.; Mishra, R.K.; Tripathi, D.K.; Singh, V.P.; Chauhan, D.K.; Prasad, S.M. Reactive oxygen species (ROS): Beneficial companions of plants' developmental processes. *Front. Plant Sci.* **2016**, *7*, 1299. [CrossRef] [PubMed]

65. Agathokleous, E. Environmental hormesis, a fundamental non-monotonic biological phenomenon with implications in ecotoxicology and environmental safety. *Ecotoxicol. Environ. Saf.* **2018**, *148*, 1042–1053. [CrossRef]

66. Agathokleous, E.; Kitao, M.; Calabrese, E.J. Hormesis: A compelling platform for sophisticated plant science. *Trends Plant Sci.* **2019**, *24*, 318–326. [CrossRef] [PubMed]

67. Agathokleous, E.; Kitao, M.; Harayama, H. On the nonmonotonic, hormetic photoprotective response of plants to stress. *Dose-Response* **2019**, *17*, 1559325819838420. [CrossRef] [PubMed]

68. Kitao, M.; Tobita, H.; Kitaoka, S.; Harayama, H.; Yazaki, K.; Komatsu, M.; Agathokleous, E.; Koike, T. Light energy partitioning under various environmental stresses combined with elevated CO_2 in three deciduous broadleaf tree species in Japan. *Climate* **2019**, *7*, 79. [CrossRef]

69. Agathokleous, E.; Kitao, M.; Calabrese, E.J. Environmental hormesis and its fundamental biological basis: Rewriting the history of toxicology. *Environ. Res.* **2018**, *165*, 274–278. [CrossRef] [PubMed]

Article

High Efficiency Mercury Sorption by Dead Biomass of *Lysinibacillus sphaericus*—New Insights into the Treatment of Contaminated Water

J. David Vega-Páez [1], Ricardo E. Rivas [2] and Jenny Dussán-Garzón [1,*]

[1] Microbiological Research Center (CIMIC), Department of Biological Sciences, Universidad de Los Andes, Bogotá 111711, Colombia; jd.vega1754@uniandes.edu.co

[2] Department of Chemistry, Universidad de Los Andes, Bogotá 111711, Colombia; re.rivas@uniandes.edu.co

* Correspondence: jdussan@uniandes.edu.co; Tel.: +57-1-3394949

Received: 28 March 2019; Accepted: 12 April 2019; Published: 19 April 2019

Abstract: Mercury (Hg) is a toxic metal frequently used in illegal and artisanal extraction of gold and silver which makes it a cause of environmental poisoning. Since biosorption of other heavy metals has been reported for several *Lysinibacillus sphaericus* strains, this study investigates Hg removal. Three *L. sphaericus* strains previously reported as metal tolerant (CBAM5, Ot4b31, and III(3)7) were assessed with mercury chloride ($HgCl_2$). Bacteria were characterized by scanning electron microscopy coupled with energy dispersive spectroscopy (EDS-SEM). Sorption was evaluated in live and dead bacterial biomass by free and immobilized cells assays. Hg quantification was achieved through spectrophotometry at 508 nm by reaction of Hg supernatants with dithizone prepared in Triton X-114 and by graphite furnace atomic absorption spectroscopy (GF-AAS). Bacteria grew up to 60 ppm of $HgCl_2$. Non-immobilized dead cell mixture of strains III(3)7 and Ot4b31 showed a maximum sorption efficiency of 28.4 µg Hg/mg bacteria during the first 5 min of contact with $HgCl_2$, removing over 95% of Hg. This process was escalated in a semi-batch bubbling fluidized bed reactor (BFB) using rice husk as the immobilization matrix leading to a similar level of efficiency. EDS-SEM analysis showed that all strains can adsorb Hg as particles of nanometric scale that can be related to the presence of S-layer metal binding proteins as shown in previous studies. These results suggest that *L. sphaericus* could be used as a novel biological method of mercury removal from polluted wastewater.

Keywords: mercury; biosorption; dead cells; *Lysinibacillus sphaericus*; dithizone; GF-AAS; EDS-SEM

---

1. Introduction

Mercury (Hg) is a highly toxic metal widely dispersed and because of its frequent use in several industries, it is a serious cause of poisoning due to bioaccumulation [1]. Symptoms of Hg poisoning are usually nonspecific and include fatigue, anxiety, depression, paresthesia, weight loss, memory loss, difficulty concentrating, and fetus malformation [2–4], but most importantly, these are usually manifested after months or years of continuous exposure to low doses of the metal [4,5].

The cycles followed by Hg are linked to liquid sources [6,7] and, according to their interaction with other compounds and organisms, they can change their oxidation state [6,8,9]. However, Hg methylation is one of the most important processes, since organic Hg compounds are the most toxic for humans [10–12]. Although methylmercury slowly undergoes demethylation processes [13] it accumulates in the brain, liver, kidneys, placenta and brain of the fetus, peripheral nerves, and bone marrow [14] causing damage.

In 2013, most countries worldwide signed the Minamata Convention on Mercury [15]. However, several places have reported Hg contamination events through these years [16–19] and Colombia's high rates in artisanal gold mining using Hg [20–22] are proportionally related to reports of clinical cases [23,24]. In fact, in 2012, it was reported that the population of Segovia in Antioquia, Colombia, had the highest per capita Hg contamination in the world, due to artisanal gold extraction [22] and other cases of people with high levels of Hg have been widely documented [25,26].

There are various physicochemical methods to treat water contaminated with Hg and research is currently focused on the development of nanomaterials to detect and remove mercury [27–29]. Different types of reactors are often escalated and coupled to classical techniques such as filtration, chemical coagulation, or sedimentation which can remove up to 80% of inorganic Hg and 40% of organic Hg [30], but all those techniques are often considered expensive because of the materials used in the design of the membranes or the reactants used in metal complexation. Despite problems in the escalation process [31], bubbling fluidizing bed reactors (BFB) are often a good alternative in biological treatments of high volumes of water samples contaminated with metals [32,33].

Biological treatment of organic Hg has been tested mainly using bacteria with alkylmercury lyase (MerB) which can transform organic Hg into the inorganic form [34]. The structure of this enzyme has been extensively studied as has its kinetics [35]. However, most of these microorganisms present risks in pathogenicity for humans—such as *Staphylococcus aureus*, *Shigella flexneri*, *Escherichia coli*, and *Pseudomonas sp.* [36]—so their use is not recommended in bioremediation.

Microorganisms have several types of interactions with Hg due to specific biochemical characteristics of the species such as the availability of sulfur-rich functional groups on their surface, or specific proteins involved in transport and/or oxidation-reduction processes of Hg species [34–38]. These types of interactions are usually classified as tolerance and resistance, where "tolerant" interactions are those in which the microorganism is able to capture (adsorb) the metal but not to metabolize it, and the "resistant" interactions can transform the metal from one species to another, often by redox reactions at intracellular level after being absorbed [39,40].

Tolerance of some bacilli strains to Hg was reported [41,42] and later it was identified that *Lysinibacillus sphaericus*—an aerobic gram positive spore-forming bacterium, nonpathogenic in humans—is able to adsorb metals [43] such as iron (Fe), cadmium (Cd), arsenic (As) [44], chromium (Cr), and lead (Pb) [45]. This demonstrates its tolerance and resistance to toxic metals by a mechanism which is known as crossed regulation [46]. Also, it was recently proven that *L. sphaericus* can adsorb gold (Au) and probably even synthesize nanoparticles [47].

Furthermore, dead cells of different *L. sphaericus* strains have also been proven as efficient metal accumulators by passive process [48] and the presence of S-layer proteins in metals binding has been widely discussed for the genus [49–51] even in sporulated forms. The advantages of dead over live cells in biological treatment of pollutants include a lack of risk to the environment and no need to be preserved since binding mechanisms are non-metabolic [52–54]. Thus, the objective of this work was to establish the Hg removal capability of three different dead *L. sphaericus* strains to assess its possible use in the treatment of water contaminated with this metal.

2. Materials and Methods

The three Colombian *L. sphaericus* strains from the CIMIC culture collection used in this study are shown in Table 1. All the strains were grown in nutrient agar for 24 h at 30 °C. A colony of each strain was incubated in 20 mL of nutrient broth for 24 h at 30 °C before the assessment of Hg minimum inhibitory concentration (MIC).

Table 1. *Lysinibacillus sphaericus* strains used.

Strain	Origin	Ref.Seq Assembly Accession Number	Metals Removal References
III(3)7	Soil sample from oak forest	GCF_001598075.1	[44,46,55,56]
Ot4b31	Larvae, Beetles	GCF_000392615.1	[47,48,57]
CBAM5	Subsurface soil sample of petroleum exploration area	GCF_000568835.1	[47,55,58]

2.1. Mercury Removal by Free Cells

A selective pressure with Hg was performed to identify minimum inhibitory concentration (MIC) on all *L. sphaericus* strains using nutrient broth (NB) and minimal salt medium supplemented with sodium acetate and yeast extract (MSM). Bacteria growth was evaluated in $HgCl_2$ solutions from 5 mg/mL to 80 mg/L. Growth was determined with an OD600 of at least 0.1 in less than 10 days.

Live and dead *L. sphaericus* cells were tested to differentiate between absorption and adsorption mechanisms. Strains were used alone and mixed. 100 µL of each strain growth at the highest $HgCl_2$ concentration were taken and grown in an overnight culture (ON) of 100 mL NB at 30 °C under agitation at 150 rpm to late exponential growth (OD600 between 0.5 and 0.8 monitored by a UV–vis BioMate 3 spectrometer, Thermo Scientific, Roskilde, Denmark). Cells were then centrifuged at 11,500 rpm for 30 min at 4 °C and pellets were washed three times with deionized water. Dead bacteria were obtained by treating pellets with 10% formalin in phosphate-buffered saline (PBS) for an hour [59] and dead pellets samples were cultured in nutrient agar to verify unviability.

Live and dead pellets were dissolved in 40 mL of $HgCl_2$. Bioassays were incubated under agitation at 30 °C and 150 rpm. 10 mL aliquots were taken at 5, 10, 15 and 60 min and centrifuged for 2 min. Supernatants were also filtered through 0.22 µm pore and quantified through UV–vis spectrophotometry. Removal assays were made in triplicate with two biological replicates.

2.2. Mercury Removal by Immobilized Dead Cells

The most efficient treatment found among non-immobilized cells was used in an escalation process. Dead cells were immobilized in sterile rice husk (RH) and a filter filled with RH was designed to be used in a semi-batch bubbling fluidized bed reactor (BFB) (Figure S1). BFB was supported by a secure vessel in case of foam excess production during bubbling process. The set point was established as Hg concentration: 60 ppm. Operative volume was 8 L of $HgCl_2$ loaded at time 0 working at 30 °C. Treatment was performed over 5 days with a residence time of 24 h and 50% recycle. Aliquots were taken each day and quantified through GF-AAS.

2.3. Mercury Quantification in Free Cells

Mercury quantification in supernatants of non-immobilized cell treatments was performed by producing a mercury dithizonate complex in Triton X-114 (Hg-Dz), which could be read by absorbance measures using a UV–vis spectrometer (BioMate 3S Spectrophotometer, Thermo Fisher Scientific, Waltham, MA, USA). Preliminarily, dithizone solutions (Dz) were prepared by dissolving 5 mg in different solvents (methanol, ethanol, butanol, toluene, carbon tetrachloride, chloroform, dichloromethane, and micellar medium of Triton X-114) to identify optimum polarity medium whose absorption spectrum of dithizone would not interfere with Hg-Dz complex at the supernatant pH. Reaction was read at 508 nm after 30 s. This colorimetric method was validated by inductively coupled plasma optical emission spectrometry (ICP-OES, iCAP 6500, Thermo Scientific, Waltham, MA, USA) under routine conditions for mercury analysis. All working standard solutions were prepared using Hg Panreac standard (lote: R3861502) and measurements were made in triplicate.

2.4. Mercury Quantification in Immobilized Cells

Aliquots from the escalated process were measured by atomic absorption spectrometry with electrothermal atomization because yellow color in the samples after treatment with RH could generate interference in the Hg-Dz complex spectrum. High-resolution continuum source atomic absorption spectrometer (RH-CSAAS, ContrAA 800, commercially available from Analytik Jena, Jena, Germany) in graphite furnace mode was used (GF-AAS). To avoid mercury volatilization during heating steps, 1% palladium (m/v) (Merck) dilution was used as a modifier.

2.5. EDS-SEM Analysis

10 µL of diluted bacteria samples from the free cells assays were taken and allowed to dry on a coin covered with sterile aluminum foil. This was then analyzed at the Uniandes Microscopy Center by SEM observation and metal semi quantification with EDS using a Tescan LYRA 3 Scanning electron microscope (TESCAN, Brno, Czech Republic). Qualitative results were compared with a control: *Escherichia coli* K12 C600.

RH (Rice Husk) was also observed by drying samples at 30 °C for 6 h and applying a gold sputter coating to inhibit charging, reduce thermal damage, and improve the secondary electron signal required for topographic examination of bacteria and HR in SEM.

2.6. Statistical Analysis

Minitab® v.17 software was used for the statistical analysis of data. Analysis of Variances (ANOVA) was used to determine whether there were significant statistical differences between treatments. Analysis of Means (ANOM) graphs were obtained to show the effects of variables (strains: III(3)7, Ot4b31, CBAM5, mixtures of two or three strains; and state: live or dead) in mercury removal among treatments.

3. Results

3.1. Minimum Inhibitory Growth Concentration (MIC)

Minimum inhibitory concentration is defined as the minimum concentration of an antimicrobial agent, to which a microorganism does not show evident growth [60]. Table 2 shows the results of the selection pressure and highlights the MIC for the strains evaluated.

L. sphaericus highest growth concentration with $HgCl_2$ was at 60 mg/L achieved in NB for all strains. These bacteria were stored as metal tolerant strains and used in further assays.

Table 2. MIC of $HgCl_2$ on *L. sphaericus* strains.

L. sphaericus Strain	Liquid Culture Media (MSM/NB)					
	5 ppm	10 ppm	20 ppm	40 ppm	60 ppm	80 ppm
III(3)7	+/+	+/+	−/+	−/+	−/+	−/−
Ot4b31	+/+	−/+	−/+	−/+	−/+	−/−
CBAM5	+/+	+/+	−/+	−/+	−/+	−/−

− No Growth; + Growth

3.2. Mercury Removal by Non-Immobilized L. sphaericus Cells

Hg spectrophotometric quantification through Dz was made and the detection limit of the method was in the range of 2 ppm to 4 ppm for calibration curves between 2 ppm and 10 ppm, and bacteria wet weight was determined to calculate Hg removal efficiency (Figures 1 and S2). Spectrophotometric quantification showed no significant differences with ICP-OES (Figure 2).

Dz reaction with Hg forms an orange color complex which can be read by spectrophotometry in the range of 488–510 nm [61]. Excess dithizone was necessary to acquire a broad range of linearity in calibration curves for Hg determination by achieving a quantitative reaction throughout the linear working range (Figure S3). Dz color must be emerald green in order to get absorption peaks that do not interfere with Hg-Dz. A pH of 5–6 was optimal for Hg-Dz formation in excess of dithizone since, at that pH, there is no background interference (Figure S4).

Dead bacteria showed higher Hg removal efficiency than live bacteria in all treatments except for strain III(3)7. Mixtures of two strains showed higher efficiency results compared to each strain. Dead *L. sphaericus* III(3)7 + Ot4b31 cells showed a maximum removal of 32.85 µg/mg bacteria after an hour and 28.4 µg/mg bacteria in less than 5 min (more than 95% of Hg). A mixture of three strains showed lower efficiency behavior than the rest of treatments (Figure 3).

Efficiency differences in Hg removal among strains, can be associated with S-layer protein quantity in these bacteria and by synergic effects between living cells as will be discussed later.

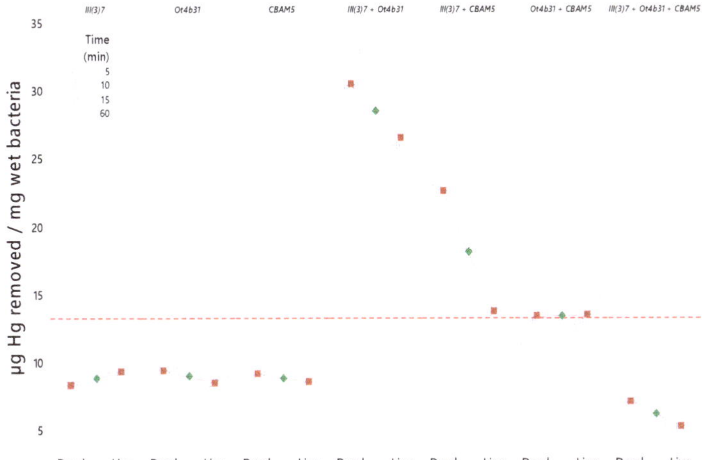

Figure 1. Efficiency in mercury removal by *L. sphaericus* (strains alone and mixtures) over time.

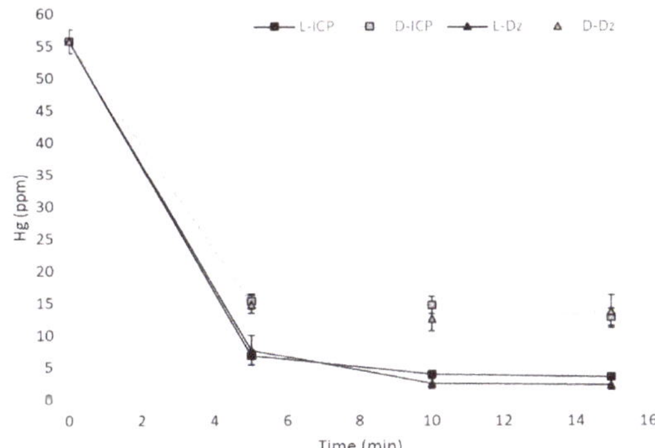

Figure 2. Comparison of Hg quantification with ICP-OES. Data of *L. sphaericus* III(3)7 treatment. L: Live; D: Dead.

3.3. Mercury Removal by Immobilized Mixture of Dead L. sphaericus Strains III(3)7 and Ot4b31 on RH

RH with and without bacteria was used in a BFB reactor to test Hg removal in a proportion of 5-times more Hg than bacteria. AAS has a high-resolution monochromator which identifies possible spectral interferences caused by the sample matrix at the wavelength of the most sensitive working atomic line for Hg, 253.652 nm. Signal detection showed no background interference (Figure 3: top left).

Mercury removal efficiency with RH and bacteria was 2.64 times higher than RH alone (control) as shown in Figure 3 and Table 3. However, the effect of the RH matrix was eliminated so the net effect of dead mixture of *L. sphaericus* III(3)7 and Ot4b31 was calculated as being 18% less efficient than the non-escalated process (Table 3).

Table 3. Mercury removal efficiency comparison of free cells (after 1 h) vs. escalated process.

Treatment	Mercury Removal Efficiency (µg Hg/mg bacteria)
Non-escalated mixture	32.85
Mixture with RH	71.76
Mixture without RH effect (normalized)	27.04

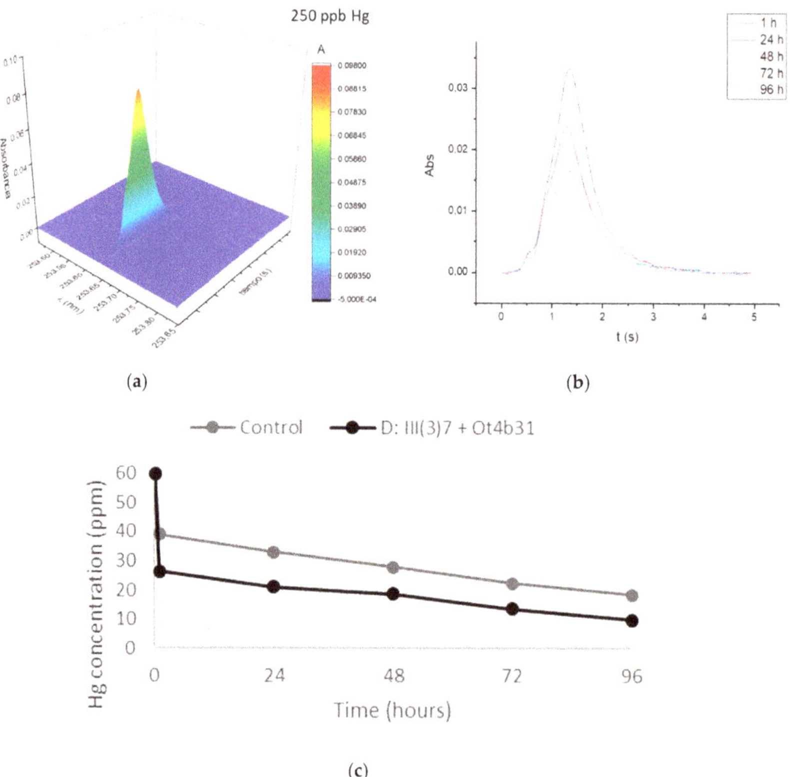

(a)

(b)

(c)

Figure 3. GF-AAS quantification of Hg aliquots of escalated process. (**a**) Hg signal of standard solution diluted 1/240. (**b**) Time-resolved absorbance signals for samples. (**c**) Hg concentration over time in escalated process with (black) and without bacteria (grey).

3.4. EDS-SEM

The analysis of energy dispersion spectroscopy coupled with the scanning electron microscopy (EDS-SEM), shows the cells on a micrometric scale, detecting the most abundant elements in the generated image. Metals can be easily detected by high-intensity reflecting electrons irradiated on the sample.

EDS-SEM shows that mercury is accumulated as spherical particles on *L. sphaericus* (Figures 4 and S5) while this attachment behavior is not observed on surface of *E. coli* (Figure S6). There is also a possibility that Hg could be absorbed by bacteria, but a dense peptidoglycan cell wall prevents the beam from being reflected. Dot pattern found in *L. sphaericus* III(3)7 after an hour of contact with Hg was also observed in strains Ot4b31 and CBAM5 and no differences between living and dead cells were observed (Figure 5).

RH was also observed after Hg treatment to verify cell attachment to RH surface and Hg sorption in cells. Hg dot pattern found in non-immobilized cells was also found on the immobilized ones. RH alone showed large Hg particles attached to its surface (Figure S7).

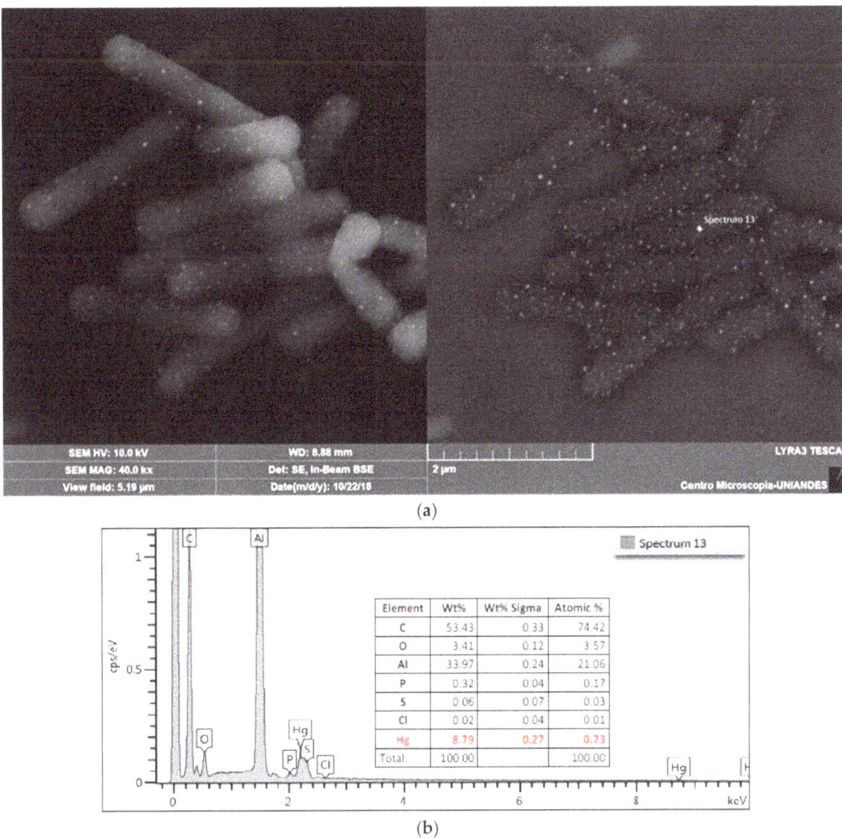

(a)

(b)

Figure 4. EDS-SEM analysis of *L. sphaericus* III(3)7 live cells after 1 h in contact with $HgCl_2$. (**a**) Image taken at 10 kV by detection of secondary electrons and backscattered electrons; (**b**) Data from energy dispersive spectroscopy of punctual area "Spectrum 13".

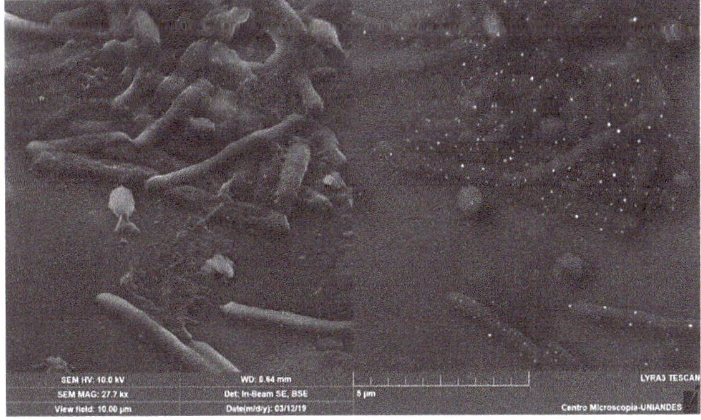

Figure 5. EDS-SEM of dead *L. sphaericus* III(3)7 and Ot4b31 immobilized in RH after in contact with $HgCl_2$.

3.5. Statistical Analysis

ANOM analysis showed a confidence interval type of approach that allowed us to determine which, if any, of the levels had a significantly different mean from the overall average of all the group means combined. Figure 6 shows that both the strain and the treatment factors are significant in this study since mercury removal is clearly improved using dead cells and mixtures of *L. sphaericus* III(3)7 + Ot4b31 or III(3)7 + CBAM5 strains.

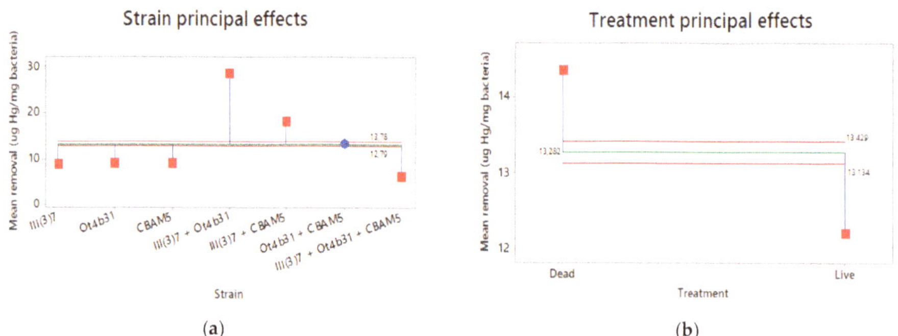

Figure 6. ANOM graphs of principal effects on mercury removal by *L. sphaericus*. (**a**) Effect of bacteria strain (alone or mixed); (**b**) Effect of bacteria treatment (dead or live).

4. Discussion

Colombia is a country where illegal and artisanal goldmining has led to high rates of Hg contamination throughout over 60% of the country [62]. Even though Colombia signed the Minamata Convention on Mercury in 2013, it was only last year that the government set the law to reduce Hg reduction over the next 5 years [63]. However, Hg pollution continues to be a problem, especially for mining communities where water treatment is difficult to achieve due to their political and economic contexts. In fact, no treatment is currently implemented in any of the affected communities.

Biological treatment of Hg has been investigated using microorganisms that are easily found in polluted sediments, but sorption mechanisms must be further studied in order to establish whether Hg particles are transformed into more toxic or difficult to handle species or not. Hg^0 can be rapidly volatilized by reductases codified by merA genes or sulfate-reducing mechanisms [64,65]. Methylated species are the most dangerous species and biological systems tends to methylate Hg when hgcAB genes are expressed [66]. Therefore, Hg removal by tolerant microorganisms capable of adsorbing the heavy metal is highly desirable but transforming it by resistance mechanisms is not.

Biosorption can be defined as the property of biological materials to accumulate heavy metals from aqueous polluted solutions through physicochemical pathways of uptake or by binding and concentration of heavy metals from even very dilute amounts [67,68]. Since active transport of metals requires energy and is not performed by dead biomass, bioaccumulation by absorption mechanisms, which leads to potential changes in mercury species characteristic from resistant bacteria, cannot happen.

$HgCl_2$ solution is mainly composed by irregular Hg salt crystals, so spheres morphology is achieved by an unknown mechanism on the surface of *L. sphaericus* cells. Particle sizes ranged between 70 nm and 120 nm so possible formation of nanoparticles could be occurring if there was an active Hg reductase protein being expressed in the cells. However, no merA genes have been annotated in the genomes of all three strains, nor have merB or hgcAB genes that codify to alkyl mercury lyase and mercury methylase, respectively [43].

Gómez and Dussan [43] proposed a metabolic model overview based on annotated *L. sphaericus* toxic lineage genomes where there are some specific transport proteins for metals such as As/Sb, Fe, and Cr among others. However, Hg complexation with sulfhydryl rich molecules—such as cysteine from proteins—has been shown to facilitate Hg entering the cells [69], and that could be a reason for the size of the Hg particles found in EDS-SEM results.

However, chemical interactions of metals in biological models are often unspecific since most of them can be transported by different channels. MerT transporter has not been identified in *L. sphaericus* strains so there could be a lack of *mer* operon that would lead to Hg specific transformation. Hg reduction could be occurring also by presence of quinones and/or other molecules that could react spontaneously with Hg [70,71]. That would also explain the size of Hg particles found both in living and dead cells in EDS-SEM images.

Results of Hg removal by living and dead bacteria are consistent with previous studies with other metals where it is a fact that dead bacteria tend to be more efficient than living bacteria [42,52,72]. This could be due to active transport mechanisms on living bacteria that constantly uptake and desorb Hg by unspecific channels. Moreover, it is necessary to highlight that Hg is not accumulated in cytoplasm since resistance genes from the *mer* operon are absent.

Higher efficiency of mixed strains both in live and dead treatments is probably related to the amount of S-layer proteins in each strain since III(3)7 has 13 full copies, Ot4b31 has 32 copies and CBAM5 has 21 copies but most of them are truncated [56–58]. These S-layer proteins are usually immersed in an exopolysaccharide matrix as shown by Francois and collaborators [72] that could be making high complexity structures by molecular interactions.

Also, it is important to notice that CBAM5 is the only strain isolated from a heavily polluted environment, which may have specific genes to degrade carbohydrates instead of specific metals tolerance like Hg. This is unusual in petroleum exploitation areas [73] and all evolutive pressures of this strain could be a reason for low affinity when working with other strains, even with unspecific interactions. The negative synergy of this strain in Hg removal requires further research.

Dead bacteria present several advantages because of their lack of metabolic susceptibility to changes from the environment by undesired horizontal gene transfer from other microorganisms [74], and also because they can be more easily accepted for people than "live" bacteria since all microorganisms tend to be more related to pathogens and illness [75], especially among Hg affected rural communities due to their low access to education [76]. Dead bacteria have proven to be efficient in mercury removal, but Colombian strains tested in this study against Hg concentrations as high as 60 ppm showed a promising inexpensive method to provide water treatment in communities that are still receiving mercury pollution from mining activities.

A pilot BFB reactor proved that an agricultural residue like RH is a good matrix to immobilize dead bacteria and to improve Hg removal efficiency from polluted water. This matrix could later be easily disposed of since solid wastes are easier to handle and treated water could be less harmful to communities. Further studies are needed to implement an in-field removal process by adjusting parameters to ensure a maximum concentration of 1 ppb in treated water.

5. Conclusions

L. sphaericus is a well-known bacterium capable of metal adsorption and absorption. Colombian strains III(3)7, Ot4b31, and CBAM5 have been widely tested with metals such Cd, Fe, Pb, As, Co, Cr, Zn and, more recently, with Au. However, this is the first Hg wide study using these strains with an important result in Colombia's context of water pollution due to illegal and artisanal mining. Both dead and live cells can be used in Hg removal but dead cells of mixed strains III(3)7 and Ot4b31 are the most efficient.

Also, immobilization in RH matrix—often an agricultural residue—provided a good yield of removal in a BFB pilot reactor. These results support the propose of a novel clean method to purify heavily polluted water at low cost as an alternative to expensive traditional methods. Moreover, this could be the first approach to solving the current issue of Hg pollution in Colombia's rivers. Studies conducted in the field or with water samples from affected zones are necessary to assess bacterial behavior in non-synthetic mercury polluted samples since more elements and molecules could improve or interfere in the high efficiency of the proposed method.

Supplementary Materials: The following are available online at http://www.mdpi.com/1996-1944/12/8/1296/s1, Figure S1. Diagram of BFB reactor designed for escalated treatment of Hg. V-101: Hg loading vessel. V-102: Secure vessel. P-101 and P-102: Air pumps for bubbling mixing in RH bed. R-101: Packed reactor with a filter filled with RH. Figure S2. Efficiency in mercury removal. (a) Single strains efficiency. (b) Mixed strains efficiency. D: Dead; L: Live. Figure S3. Calibration curves for dithizone. (a) Reaction in excess of mercury (0 ppm, 8 ppm and 10 ppm). (b) Reaction in excess of dithizone (0 ppm, 8 ppm). (c) Calibration curves for different proportions of Dz-Hg. Figure S4. Stability of Dz in Triton X114 at different pH. From left to right: pH 3,5,6,7,8. Figure S5. Detail of a *L. sphaericus* cell attached to RH surface. Figure S6. EDS-SEM of E. coli K12 C600 after 1 h in contact with $HgCl_2$. Figure S7. RH with a large Hg particle attached to its surface.

Author Contributions: Conceptualization, J.D.V.-P. and J.D.-G.; Formal analysis, J.D.V.; Investigation, J.D.V.-P. and J.D.-G.; Methodology, J.D.V.-P., R.E.R., and J.D.-G.; Project administration, J.D.-G.; Supervision, J.D.-G.; Writing—original draft, J.D.V.-P.; Writing—review & editing, J.D.V.-P., R.E.R., and J.D.-G.

Funding: This research was funded by the Research Fund at the School of Science at Universidad de los Andes call 2018-1 for the Financing of Projects of Investigation and Participation in Academic Events Category Master Students, project code INV-2017-25-1053, and the Microbiological Research Center (CIMIC). APC was funded by School of Science at Universidad de los Andes, and the Microbiological Research Center (CIMIC).

Acknowledgments: We would like to acknowledge Monica Beatriz López Santos for all her support at EDS-SEM analysis at Microscopy Center in Universidad de Los Andes. We would also like to thank Juan Felipe Correal Bohorquez for his kind assistance during removal bioassays. We also thank Tiziana Laudato for the English grammar corrections provided.

Conflicts of Interest: The authors declare no conflict of interest.

References

1. Marrugo-Negrete, J.; Pinedo-Hernández, J.; Díez, S. Geochemistry of mercury in tropical swamps impacted by gold mining. *Chemosphere* **2015**, *134*, 44–51. [CrossRef]

2. Ha, E.; Basu, N.; Bose-O'Reilly, S.; Dórea, J.G.; McSorley, E.; Sakamoto, M.; Chan, H.M. Current progress on understanding the impact of mercury on human health. *Environ. Res.* **2017**, *152*, 419–433. [CrossRef]

3. UNEP. Global Mercury Assessment 2013 Sources, Emissions, Releases and Environmental Transport. *UN Environment.* 2013. Available online: http://www.unenvironment.org/es/node/15600 (accessed on 21 September 2018).

4. Gupta, R.C.; Milatovic, D.; Lall, R.; Srivastava, A. Mercury. In *Veterinary Toxicology*; Elsevier: Amsterdam, The Netherlands, 2018; pp. 455–462.

5. Branco, V.; Caito, S.; Farina, M.; da Rocha, J.T.; Aschner, M.; Carvalho, C. Biomarkers of mercury toxicity: Past, present, and future trends. *J. Toxicol. Environ. Health Part B* **2017**, *20*, 119–154. [CrossRef] [PubMed]

6. Fitzgerald, W.F.; Lamborg, C.H. Geochemistry of Mercury in the Environment. In *Treatise on Geochemistry*; Elsevier: Amsterdam, The Netherlands, 2014; pp. 91–129.

7. Richard, J.-H.; Bischoff, C.; Ahrens, C.G.M.; Biester, H. Mercury (II) reduction and co-precipitation of metallic mercury on hydrous ferric oxide in contaminated groundwater. *Sci. Total Environ.* **2016**, *539*, 36–44. [CrossRef] [PubMed]

8. Kotnik, J.; Horvat, M.; Jereb, V. Modelling of mercury geochemical cycle in Lake Velenje, Slovenia. *Environ. Model. Softw.* **2002**, *17*, 593–611. [CrossRef]

9. Watras, C.J.; Morrison, K.A.; Rubsam, J.L.; Rodger, B. Atmospheric mercury cycles in northern Wisconsin. *Atmos. Environ.* **2009**, *43*, 4070–4077. [CrossRef]

10. Wells, E.M.; Herbstman, J.B.; Lin, Y.H.; Hibbeln, J.R.; Halden, R.U.; Witter, F.R.; Goldman, L.R. Methyl mercury, but not inorganic mercury, associated with higher blood pressure during pregnancy. *Environ. Res.* **2017**, *154*, 247–252. [CrossRef] [PubMed]

11. Rudd, J.W.M.; Bodaly, R.A.; Fisher, N.S.; Kelly, C.A.; Kopec, D.; Whipple, C. Fifty years after its discharge, methylation of legacy mercury trapped in the Penobscot Estuary sustains high mercury in biota. *Sci. Total Environ.* **2018**, *642*, 1340–1352. [CrossRef] [PubMed]

12. Buchanan, S.; Anglen, J.; Turyk, M. Methyl mercury exposure in populations at risk: Analysis of NHANES 2011–2012. *Environ. Res.* **2015**, *140*, 56–64. [CrossRef] [PubMed]

13. Suda, I.; Takahashi, H. Degradation of methyl and ethyl mercury into inorganic mercury by other reactive oxygen species besides hydroxyl radical. *Arch. Toxicol.* **1992**, *66*, 34–39. [CrossRef]

14. Nordberg, G.; Fowler, B.; Nordberg, M. (Eds.) *Handbook on the Toxicology of Metals*, 4th ed.; Academic Press: Cambridge, MA, USA, 2014.

15. UNEP. Parties and Signatories. *Minamata Convention on Mercury*, 2018. Available online: http://www.mercuryconvention.org/Countries/Parties/tabid/3428/language/en-US/Default.aspx (accessed on 28 December 2018).

16. Gao, Z.; Ying, X.; Yan, J.; Wang, J.; Cai, S.; Yan, C. Acute mercury vapor poisoning in a 3-month-old infant: A case report. *Clin. Chim. Acta* **2017**, *465*, 119–122. [CrossRef] [PubMed]

17. Ori, M.R.; Larsen, J.B.; Shirazi, F.M. Mercury Poisoning in a Toddler from Home Contamination due to Skin-Lightening Cream. *J. Pediatr.* **2018**, *196*, 314–317. [CrossRef] [PubMed]

18. Eqani, S.A.; Bhowmik, A.K.; Qamar, S.; Shah, S.T.; Sohail, M.; Mulla, S.I.; Fasola, M.; Shen, H. Mercury contamination in deposited dust and its bioaccumulation patterns throughout Pakistan. *Sci. Total Environ.* **2016**, *569–570*, 585–593. [CrossRef] [PubMed]

19. Budnik, L.T.; Casteleyn, L. Mercury pollution in modern times and its socio-medical consequences. *Sci. Total Environ.* **2019**, *654*, 720–734. [CrossRef]

20. Betancur-Corredor, B.; Loaiza-Usuga, J.C.; Denich, M.; Borgemeister, C. Gold mining as a potential driver of development in Colombia: Challenges and opportunities. *J. Clean. Prod.* **2018**, *199*, 538–553. [CrossRef]

21. Pinedo-Hernández, J.; Marrugo-Negrete, J.; Díez, S. Speciation and bioavailability of mercury in sediments impacted by gold mining in Colombia. *Chemosphere* **2015**, *119*, 1289–1295. [CrossRef] [PubMed]

22. Webster, P.C. Not all that glitters: Mercury poisoning in Colombia. *Lancet* **2012**, *379*, 1379–1380. [CrossRef]

23. Palacios-Torres, Y.; Caballero-Gallardo, K.; Olivero-Verbel, J. Mercury pollution by gold mining in a global biodiversity hotspot, the Choco biogeographic region, Colombia. *Chemosphere* **2018**, *193*, 421–430. [CrossRef] [PubMed]

24. Narváez, D.M.; Groot, H.; Diaz, S.M.; Palma, R.M.; Muñoz, N.; Cros, M.P.; Hernández-Vargas, H. Oxidative stress and repetitive element methylation changes in artisanal gold miners occupationally exposed to mercury. *Heliyon* **2017**, *3*, e00400. [CrossRef]

25. Salazar-Camacho, C.; Salas-Moreno, M.; Marrugo-Madrid, S.; Marrugo-Negrete, J.; Díez, S. Dietary human exposure to mercury in two artisanal small-scale gold mining communities of northwestern Colombia. *Environ. Int.* **2017**, *107*, 47–54. [CrossRef]

26. Gutiérrez-Mosquera, H.; Sujitha, S.B.; Jonathan, M.P.; Sarkar, S.K.; Medina-Mosquera, F.; Ayala-Mosquera, H.; Morales-Mira, G.; Arreola-Mendoza, L. Mercury levels in human population from a mining district in Western Colombia. *J. Environ. Sci.* **2018**, *68*, 83–90. [CrossRef]

27. Zolnikov, T.R.; Ortiz, D.R. A systematic review on the management and treatment of mercury in artisanal gold mining. *Sci. Total Environ.* **2018**, *633*, 816–824. [CrossRef]

28. Vikrant, K.; Kim, K.-H. Nanomaterials for the adsorptive treatment of Hg(II) ions from water. *Chem. Eng. J.* **2019**, *358*, 264–282. [CrossRef]

29. Wang, H.; Huang, W.; Tang, L.; Chen, Y.; Zhang, Y.; Wu, M.; Song, Y.; Wen, S. Electrospun nanofibrous mercury filter: Efficient concentration and determination of trace mercury in water with high sensitivity and tunable dynamic range. *Anal. Chim. Acta* **2017**, *982*, 96–103. [CrossRef] [PubMed]

30. World Health Organization. *Mercury in Drinking-Water. Background Document for Development of WHO Guidelines for Drinking-Water Quality*; World Health Organization: Geneva, Switzerland, 2005.

31. Rüdisüli, M.; Schildhauer, T.J.; Biollaz, S.M.A.; van Ommen, J.R. Scale-up of bubbling fluidized bed reactors—A review. *Powder Technol.* **2012**, *217*, 21–38. [CrossRef]

32. Mack, C.; Burgess, J.; Duncan, J. Membrane bioreactors for metal recovery from wastewater: A review. *Water SA* **2004**, *30*, 521–532. [CrossRef]

33. Anderson, E.L. *Analysis of Various Bioreactor Configurations for Heavy Metal Removal Using the Fungus Penicillium ochro-chloron*; Worcester Polytechnic Institute: Worcester, MA, USA, 1999.

34. Lafrance-Vanasse, J.; Lefebvre, M.; di Lello, P.; Sygusch, J.; Omichinski, J.G. Crystal Structures of the Organomercurial Lyase MerB in Its Free and Mercury-bound Forms: Insights into the mechanism of methylmercury degradation. *J. Biol. Chem.* **2009**, *284*, 938–944. [CrossRef] [PubMed]

35. Moore, M.J.; Distefano, M.D.; Zydowsky, L.D.; Cummings, R.T.; Walsh, C.T. Organomercurial lyase and mercuric ion reductase: Nature's mercury detoxification catalysts. *Acc. Chem. Res.* **1990**, *23*, 301–308. [CrossRef]

36. Hassen, A.; Saidi, N.; Cherif, M.; Boudabous, A. Effects of heavy metals on *Pseudomonas aeruginosa* and *Bacillus thuringiensis*. *Bioresour. Technol.* **1998**, *65*, 73–82. [CrossRef]

37. Al Rmalli, S.W.; Dahmani, A.A.; Abuein, M.M.; Gleza, A.A. Biosorption of mercury from aqueous solutions by powdered leaves of castor tree *Ricinus communis* L. *J. Hazard. Mater.* **2008**, *152*, 955–959. [CrossRef]

38. Amin, F.; Talpur, F.N.; Balouch, A.; Chandio, Z.A.; Surhio, M.A.; Afridi, H.I. Biosorption of mercury(II) from aqueous solution by fungal biomass *Pleurotus eryngii*: Isotherm, kinetic, and thermodynamic studies: Remediation/Treatment. *Environ. Prog. Sustain. Energy* **2016**, *35*, 1274–1282. [CrossRef]

39. Yu, J.G.; Yue, B.Y.; Wu, X.W.; Liu, Q.; Jiao, F.P.; Jiang, X.Y.; Chen, X.Q. Removal of mercury by adsorption: A review. *Environ. Sci. Pollut. Res.* **2016**, *23*, 5056–5076. [CrossRef]

40. Castillo González, C.; Dussán Garzón, J. Caracterización Molecular de la Resistencia a Mercurio Mercúrico de Bacterias Nativas Colombianas. Ph.D. Dissertation, University of Los Andes, Bogota, Colombia, 2007.

41. Eom, Y.; Won, J.H.; Ryu, J.-Y.; Lee, T.G. Biosorption of mercury(II) ions from aqueous solution by garlic *Allium sativum* L. powder. *Korean J. Chem. Eng.* **2011**, *28*, 1439–1443. [CrossRef]

42. Takahashi, Y.; Châtellier, X.; Hattori, K.H.; Kato, K.; Fortin, D. Adsorption of rare earth elements onto bacterial cell walls and its implication for REE sorption onto natural microbial mats. *Chem. Geol.* **2005**, *219*, 53–67. [CrossRef]

43. Gómez-Garzón, C.; Hernández-Santana, A.; Dussán, J. A genome-scale metabolic reconstruction of *Lysinibacillus sphaericus* unveils unexploited biotechnological potentials. *PLoS ONE* **2017**, *12*, e0179666. [CrossRef]

44. Vargas, J.E.; Dussan, J. Adsorption of Toxic Metals and Control of Mosquitos-borne Disease by *Lysinibacillus sphaericus*: Dual Benefits for Health and Environment. *Biomed. Environ. Sci. BES* **2016**, *29*, 187–196.

45. Shaw, D.R.; Dussan, J. Transcriptional analysis and molecular dynamics simulations reveal the mechanism of toxic metals removal and efflux pumps in *Lysinibacillus sphaericus* OT4b.31. *Int. Biodeterior. Biodegrad.* **2018**, *127*, 46–61. [CrossRef]

46. Shaw, D.R.; Dussan, J. Mathematical Modelling of Toxic Metal Uptake and Efflux Pump in Metal-Resistant Bacterium *Bacillus cereus* Isolated From Heavy Crude Oil. *Water Air Soil Pollut.* **2015**, *226*, 112. [CrossRef]

47. Bustos, M.; Ibarra, H.; Dussán, J. The Golden Activity of *Lysinibacillus sphaericus*: New Insights on Gold Accumulation and Possible Nanoparticles Biosynthesis. *Materials* **2018**, *11*, 1587. [CrossRef]

48. Velásquez, L.; Dussan, J. Biosorption and bioaccumulation of heavy metals on dead and living biomass of *Bacillus sphaericus*. *J. Hazard. Mater.* **2009**, *167*, 713–716. [CrossRef]

49. Merroun, M.L.; Raff, J.; Rossberg, A.; Hennig, C.; Reich, T.; Selenska-Pobell, S. Complexation of Uranium by Cells and S-Layer Sheets of *Bacillus sphaericus* JG-A12. *Appl. Environ. Microbiol.* **2005**, *71*, 5532–5543. [CrossRef]

50. Ilk, N.; Egelseer, E.M.; Sleytr, U.B. S-layer fusion proteins—Construction principles and applications. *Curr. Opin. Biotechnol.* **2011**, *22*, 824–831. [CrossRef] [PubMed]

51. Allievi, M.C. Metal Biosorption by Surface-Layer Proteins from Bacillus Species. *J. Microbiol. Biotechnol.* **2011**, *21*, 147–153. [CrossRef] [PubMed]

52. Javanbakht, V.; Alavi, S.A.; Zilouei, H. Mechanisms of heavy metal removal using microorganisms as biosorbent. *Water Sci. Technol.* **2014**, *69*, 1775–1787. [CrossRef]

53. Gupta, P.; Diwan, B. Bacterial Exopolysaccharide mediated heavy metal removal: A Review on biosynthesis, mechanism and remediation strategies. *Biotechnol. Rep.* **2017**, *13*, 58–71. [CrossRef]

54. Andleeb, S.; Batool, I.; Ali, S.; Akhtar, K.; Ali, N.M. Accumulation of Heavy Metals by Living and Dead Bacteria as Biosorbents: Isolated from Waste Soil. *Biol. Sci. PJSIR* **2017**, *60*, 102–111.

55. Lozano, L.C.; Dussán, J. Metal tolerance and larvicidal activity of *Lysinibacillus sphaericus*. *World J. Microbiol. Biotechnol.* **2013**, *29*, 1383–1389. [CrossRef] [PubMed]

56. Rey, A.; Silva-Quintero, L.; Dussán, J. Complete genome sequencing and comparative genomic analysis of functionally diverse Lysinibacillus sphaericus III(3)7. *Genom. Data* **2016**, *9*, 78–86. [CrossRef] [PubMed]

57. Peña-Montenegro, T.D.; Dussán, J. Genome sequence and description of the heavy metal tolerant bacterium *Lysinibacillus sphaericus* strain OT4b.31. *Stand. Genom. Sci.* **2013**, *9*, 42–56. [CrossRef]

58. Peña-Montenegro, T.D.; Lozano, L.; Dussán, J. Genome sequence and description of the mosquitocidal and heavy metal tolerant strain *Lysinibacillus sphaericus* CBAM5. *Stand. Genom. Sci.* **2015**, *10*, 2. [CrossRef]

59. Chao, Y.; Zhan, T.G. Optimization of fixation methods for observation of bacterial cell morphology and surface ultrastructures by atomic force microscopy. *Appl. Microbiol. Biotechnol.* **2011**, *92*, 381–392. [CrossRef]

60. Andrews, J.M. Determination of minimum inhibitory concentrations. *J. Antimicrob. Chemother.* **2001**, *48*, 5–16. [CrossRef]

61. Orozco, J.M.G.; Cañizares Macías, M.D.P. Determinación de mercurio en formulaciones farmacéuticas utilizando un sistema de flujo continuo y ditizona en medio micelar. *Rev. Mex. Cienc. Farm.* **2010**, *41*, 37–43.

62. Suárez, L.G.; Aristizabal, J.D. Mercury and gold mining in Colombia: A failed state. *Univ. Sci.* **2013**, *18*, 33–49.

63. WWF. Congreso Aprobó Vinculación de Colombia al Convenio de Minamata. Available online: http://www.wwf.org.co/?uNewsID=325172 (accessed on 8 April 2019).

64. Sotero-Martins, A.; de Jesus, M.S.; Lacerda, M.; Moreira, J.C.; Filgueiras, A.L.L.; Barrocas, P.R.G. A conservative region of the mercuric reductase gene (mera) as a molecular marker of bacterial mercury resistance. *Braz. J. Microbiol.* **2008**, *39*, 307–310. [CrossRef]

65. Pedrero, Z.; Bridou, R.; Mounicou, S.; Guyoneaud, R.; Monperrus, M.; Amouroux, D. Transformation, localization, and biomolecular binding of Hg species at subcellular level in methylating and nonmethylating sulfate-reducing bacteria. *Environ. Sci. Technol.* **2012**, *46*, 11744–11751. [CrossRef] [PubMed]

66. Ndu, U.; Christensen, G.A.; Rivera, N.A.; Gionfriddo, C.M.; Deshusses, M.A.; Elias, D.A.; Hsu-Kim, H. Quantification of Mercury Bioavailability for Methylation Using Diffusive Gradient in Thin-Film Samplers. *Environ. Sci. Technol.* **2018**, *52*, 8521–8529. [CrossRef]

67. Volesky, B. *Sorption and Biosorption*; BV Sorbex: Montreal, QC, Canada, 2003.

68. Fourest, E.; Roux, J.-C. Heavy metal biosorption by fungal mycelial by-products: Mechanisms and influence of pH. *Appl. Microbiol. Biotechnol.* **1992**, *37*, 399–403. [CrossRef]

69. Schaefer, J.K.; Rocks, S.S.; Zheng, W.; Liang, L.; Gu, B.; Morel, F.M.M. Active transport, substrate specificity, and methylation of Hg(II) in anaerobic bacteria. *Proc. Natl. Acad. Sci. USA* **2011**, *108*, 8714–8719. [CrossRef] [PubMed]

70. Gilmour, C.C.; Podar, M.; Bullock, A.L.; Graham, A.M.; Brown, S.D.; Somenahally, A.C.; Johs, A.; Hurt, R.A., Jr.; Bailey, K.L.; Elias, D.A. Mercury Methylation by Novel Microorganisms from New Environments. *Environ. Sci. Technol.* **2013**, *47*, 11810–11820. [CrossRef] [PubMed]

71. Zheng, W.; Liang, L.; Gu, B. Mercury Reduction and Oxidation by Reduced Natural Organic Matter in Anoxic Environments. *Environ. Sci. Technol.* **2012**, *46*, 292–299. [CrossRef]

72. François, F.; Lombard, C.; Guigner, J.M.; Soreau, P.; Brian-Jaisson, F.; Martino, G.; Vandervennet, M.; Garcia, D.; Molinier, A.L.; Pignol, D.; et al. Isolation and Characterization of Environmental Bacteria Capable of Extracellular Biosorption of Mercury. *Appl. Environ. Microbiol.* **2012**, *78*, 1097–1106. [CrossRef] [PubMed]

73. Nadkarni, R.A. *Modern Instrumental Methods of Elemental Analysis of Petroleum Products and Lubricants*, ASTM International: West Conshohocken, PA, USA, 1991.

74. Lal, D.; Lal, R. Evolution of mercuric reductase (merA) gene: A case of horizontal gene transfer. *Mikrobiologiia* **2010**, *79*, 524–531. [CrossRef] [PubMed]

75. Venkova, T.; Yeo, C.C.; Espinosa, M. Editorial: The Good, the Bad, and the Ugly: Multiple Roles of Bacteria in Human Life. *Front. Microbiol.* **2018**, *9*, 1702. [CrossRef] [PubMed]

76. Ashford, M. Could Humans Live without Bacteria? *Live Science*. Available online: https://www.livescience.com/32761-good-bacteria-boost-immune-system.html (accessed on 8 April 2019).

Article

Chlorophyll Fluorescence Imaging Analysis for Elucidating the Mechanism of Photosystem II Acclimation to Cadmium Exposure in the Hyperaccumulating Plant *Noccaea caerulescens*

Gülriz Bayçu [1], Julietta Moustaka [2], Nurbir Gevrek [1] and Michael Moustakas [1,3,*]

[1] Division of Botany, Department of Biology, Faculty of Science, Istanbul University, 34134 Istanbul, Turkey; gulrizb@istanbul.edu.tr (G.B.); ngevrek@gmail.com (N.G.)

[2] Department of Plant and Environmental Sciences, University of Copenhagen, Thorvaldsensvej 40, DK-1871 Frederiksberg C, Denmark; moustaka@plen.ku.dk

[3] Department of Botany, Aristotle University of Thessaloniki, 54124 Thessaloniki, Greece

[*] Correspondence: moustak@bio.auth.gr; Tel.: +30-2310-99-8335

Received: 11 November 2018; Accepted: 13 December 2018; Published: 18 December 2018

Abstract: We provide new data on the mechanism of *Noccaea caerulescens* acclimation to Cd exposure by elucidating the process of photosystem II (PSII) acclimation by chlorophyll fluorescence imaging analysis. Seeds from the metallophyte *N. caerulescens* were grown in hydroponic culture for 12 weeks before exposure to 40 and 120 µM Cd for 3 and 4 days. At the beginning of exposure to 40 µM Cd, we observed a spatial leaf heterogeneity of decreased PSII photochemistry, that later recovered completely. This acclimation was achieved possibly through the reduced plastoquinone (PQ) pool signaling. Exposure to 120 µM Cd under the growth light did not affect PSII photochemistry, while under high light due to a photoprotective mechanism (regulated heat dissipation for protection) that down-regulated PSII quantum yield, the quantum yield of non-regulated energy loss in PSII (Φ_{NO}) decreased even more than control values. Thus, *N. caerulescens* plants exposed to 120 µM Cd for 4 days exhibited lower reactive oxygen species (ROS) production as singlet oxygen ($^1O_2$). The response of *N. caerulescens* to Cd exposure fits the 'Threshold for Tolerance Model', with a lag time of 4 d and a threshold concentration of 40 µM Cd required for the induction of the acclimation mechanism.

Keywords: Cd toxicity; detoxification mechanism; photochemical quenching; photosynthetic heterogeneity; photoprotective mechanism; phytoremediation; plastoquinone pool; redox state; spatiotemporal variation

---

1. Introduction

Cadmium is a non-essential heavy metal that can occur in the environment in high concentrations as a consequence of numerous human activities, thus becoming toxic to all organisms [1–5]. Plants have developed several exclusive and effective mechanisms for Cd detoxification and tolerance, including control of Cd influx and acceleration of Cd efflux, Cd chelation and sequestration, Cd remobilization, and scavenging of Cd-induced reactive oxygen species [5–10].

Hyperaccumulators are plant species that vigorously take up heavy metals, translocate them into the above-ground parts and isolate them into a risk-free state [4,11]. These plants can accumulate several percent of heavy metals in their dry mass [4]. Hyperaccumulators also have to stock the absorbed heavy metal in a manner that is not detrimental to vital enzymes and especially photosynthesis [12,13]. Hyperaccumulators can be used for phytoremediation and also for phytomining [4,14–16]. Phytoremediation is a cost-effective and environmentally-friendly technology that uses plants to remove the toxic metals from soils; it has been widely used in practice [14,17].

Noccaea caerulescens is known as a zinc–cadmium–nickel hyperaccumulator because it can accumulate these metals at extremely high concentrations in its aboveground tissues [18], and has been proposed as an ideal species for examining metal tolerance and hyperaccumulation [19]. It has recently gained a lot of attention due to its potential use in phytoremediation and phytomining [20,21]. Certain ecotypes of *N. caerulescens* can store as much as 14,000 μg Cd g^{-1} dry biomass without showing toxicity signs [22–24]. Cadmium concentrations in the leaves above 0.01% dry biomass are considered extraordinary and are the limit level for Cd hyperaccumulation [24,25].

Photosynthesis has been shown to be very sensitive to Cd either directly or indirectly [4,26–31]. A Cd-induced decrease in photosynthetic efficiency may result from disturbances in the electron transport [32,33], enzymatic activities involved in CO_2 fixation [34,35], or from stomatal closure [36,37]. Photosystem II (PSII) is extremely sensitive to Cd that exerts multiple effects on both donor (it inhibits oxygen evolution) and acceptor sites (it inhibits electron transfer from quinone A, QA to quinone B, QB) [28,33,38,39]. The less susceptible component of the photosynthetic apparatus to Cd is thought to be PSI [28,40].

We investigated photosynthetic acclimation to Cd toxicity using the hyperaccumulator *Noccaea caerulescens*. Our previous study indicated that despite the substantial high toxicity levels of Zn and Cd in *N. caerulescens* aboveground tissues, the photochemical energy use at PSII did not differ compared to controls [13]. However, the underlying mechanism of photosynthetic acclimation has not been elucidated. In the present study, in order to investigate the mechanism of *N. caerulescens* acclimation to Cd exposure and to clarify the process of photosynthetic acclimation, we treated in hydroponic culture *N. caerulescens* plants with 40 and 120 μM Cd for 3 and 4 days.

2. Materials and Methods

2.1. Seed Collection and Experimental Design

Seeds of *Noccaea caerulescens* F.K. Mey collected from a former Copper Mine area at Røros (Norway) were cultivated hydroponically in an environmental growth chamber as described previously [13].

After growth for 12 weeks, some plants were exposed to 40 or 120 μM Cd (supplied as $3CdSO_4.8H_2O$) for 3 and 4 days while others were left to control growth conditions.

2.2. Chlorophyll Fluorescence Imaging Analysis

Chlorophyll fluorescence measurements were carried out with an Imaging-PAM Chlorophyll Fluorometer (Walz, Effeltrich, Germany) in dark-adapted leaves (15 min) of *N. caerulescens* plants, grown at 0 (control), 40 or 120 μM Cd for 3 and 4 days, as described previously [13,41]. Five leaves were measured from five different plants with eight areas of interest in each leaf. Two light intensities were selected for chlorophyll fluorescence measurements, a low light intensity that was similar to the growth light (300 μmol photons m^{-2} s^{-1}, GL) and a high light intensity (1000 μmol photons m^{-2} s^{-1}, HL, more than three times that of the growth light). The measured and calculated chlorophyll fluorescence parameters with their definitions are given in Table 1.

Representative results of the measured chlorophyll fluorescence parameters are also displayed as color-coded images.

2.3. Statistical Analyses

All measurements that are expressed as mean $\pm$ SD were analyzed by student *t*-test ($p < 0.05$). Five leaves from five different plants were analyzed in each treatment. In all graphs, the error bars are standard deviations, while columns with the same letter are not statistically different at $p < 0.05$.

Table 1. Definitions of all measured and calculated chlorophyll fluorescence parameters.

Chlorophyll Fluo-Rescence Parameter	Definition	Calculation
F_o	Minimum chlorophyll *a* fluorescence in the dark-adapted leaf (PSII centers open)	Obtained by applying measuring photon irradiance of 1.2 µmol photons m^{-2} s^{-1}
F_m	Maximum chlorophyll *a* fluorescence in the dark-adapted leaf (PSII centers closed)	Obtained with a saturating pulse (SP) of 6000 µmol photons m^{-2} s^{-1}
F_s	Steady-state photosynthesis	Measured after 5 min illumination time before switching off the actinic light (AL) of 300 µmol photons m^{-2} s^{-1} or 1000 µmol photons m^{-2} s^{-1}
$F_o{}'$	Minimum chlorophyll *a* fluorescence in the light-adapted leaf	It was computed by the Imaging Win software (Heinz Walz GmbH, Effeltrich, Germany) as Fo′ = Fo/(Fv/Fm + Fo/Fm′) [42]
$F_m{}'$	Maximum chlorophyll *a* fluorescence in the light-adapted leaf	Measured with saturating pulses (SPs) every 20 s for 5 min after application of the actinic light (AL) of 300 µmol photons m^{-2} s^{-1} or 1000 µmol photons m^{-2} s^{-1}
F_v/F_m	The maximum quantum efficiency of PSII photochemistry	Calculated as $(F_m - F_o)/F_m$
Φ_{PSII}	The effective quantum yield of photochemical energy conversion in PSII	Calculated as $(F_m' - F_s)/F_m'$
q_P	The redox state of QA	Calculated as $(F_m' - F_s)/(F_m' - F_o')$
NPQ	The non-photochemical quenching that reflects heat dissipation of excitation energy	Calculated as $(F_m - F_m')/F_m'$
ETR	The relative PSII electron transport rate	Calculated as Φ_{PSII} x Photosynthetic Photon Flux Density $\times$ 0.5 $\times$ 0.84
Φ_{NPQ}	The quantum yield of regulated non-photochemical energy loss in PSII, that is the quantum yield for dissipation by down regulation in PSII	Calculated as $F_s/F_m' - F_s/F_m$
Φ_{NO}	The quantum yield of non-regulated energy loss in PSII	Calculated as F_s/F_m
$1 - q_P$	The fraction of closed PSII reaction centers	Calculated as $1 - q_P$

3. Results

3.1. Changes in the Maximum Quantum Efficiency of PSII Photochemistry after Cd Exposure

At the beginning of exposure to 40 µM Cd, the maximum quantum efficiency of PSII photochemistry (F_v/F_m) in *N. caerulescens* decreased significantly but increased to control values at 120 µM Cd (Figure 1).

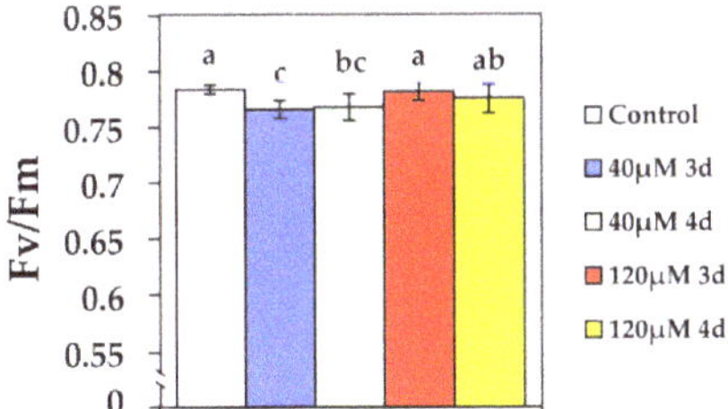

Figure 1. Changes in the maximum quantum efficiency of PSII (F_v/F_m) in *N. caerulescens* plants grown at 0 (control), 40 or 120 µM Cd^{2+} for 3 and 4 days.

3.2. Changes in the Allocation of Absorbed Light Energy in PSII after Cd Exposure

The quantum yield of photochemical energy conversion in PSII (Φ_{PSII}), at both growth light (GL) and high light (HL) intensity decreased significantly compared to the control, after 3 d at 40 µM Cd, while it improved during the 4 d (Figure 2). However, Φ_{PSII} increased to control values after 3 d at 120 µM Cd at GL and stabilized to control values after 4 days of exposure (Figure 2a). High light (HL) exposure to 120 µM Cd resulted in decreased Φ_{PSII} compared to controls (Figure 2b).

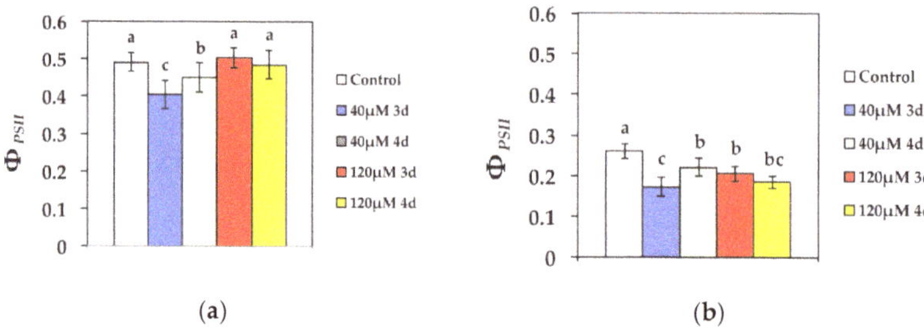

Figure 2. Changes in the quantum efficiency of PSII photochemistry (Φ_{PSII}) in *N. caerulescens* measured (**a**) at 300 µmol photons m^{-2} s^{-1} or (**b**) 1000 µmol photons m^{-2} s^{-1}. *N. caerulescens* plants were grown at 0 (control), 40, or 120 µM Cd^{2+} for 3 and 4 days.

The quantum yield of regulated non-photochemical energy loss in PSII (Φ_{NPQ}) decreased significantly compared to the control after 3 d at 40 µM Cd at GL and increased to control values during the 4 d (Figure 3a). Exposure to 120 µM Cd resulted in decreased Φ_{NPQ} at GL compared to controls during the 4 d (Figure 3a). At HL, Φ_{NPQ} remained unchanged at 40 µM Cd, but increased significantly at 120 µM Cd (Figure 3b).

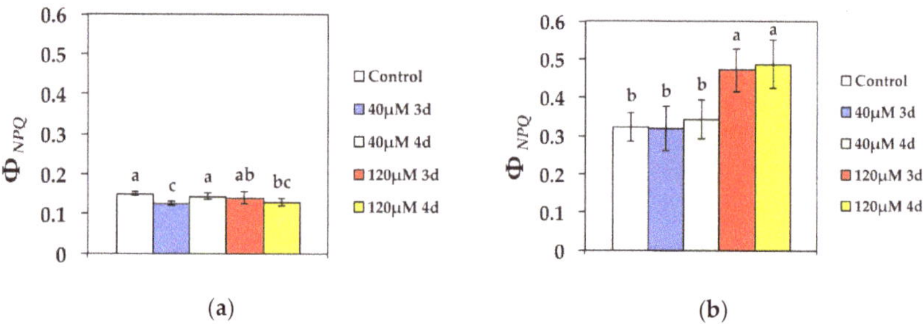

Figure 3. Changes in the quantum yield for dissipation by down regulation in PSII (regulated heat dissipation, a loss process serving for protection) (Φ_{NPQ}) measured at (**a**) 300 µmol photons m^{-2} s^{-1} or (**b**) 1000 µmol photons m^{-2} s^{-1}. *N. caerulescens* plants were grown at 0 (control), 40, or 120 µM Cd^{2+} for 3 and 4 days.

The quantum yield of non-regulated energy loss in PSII (Φ_{NO}), a loss process due to PSII inactivity, at both GL and HL intensity, increased significantly compared to the control after 3 d exposure to 40 µM Cd, while during the 4 d it decreased compared to 3 d (Figure 4). After exposure to 120 µM Cd for 3 d at GL, Φ_{NO} retained the same values compared to the controls, but increased during the 4 d (Figure 4a). However, Φ_{NO} decreased more than the control values at 120 µM Cd at HL (Figure 4b).

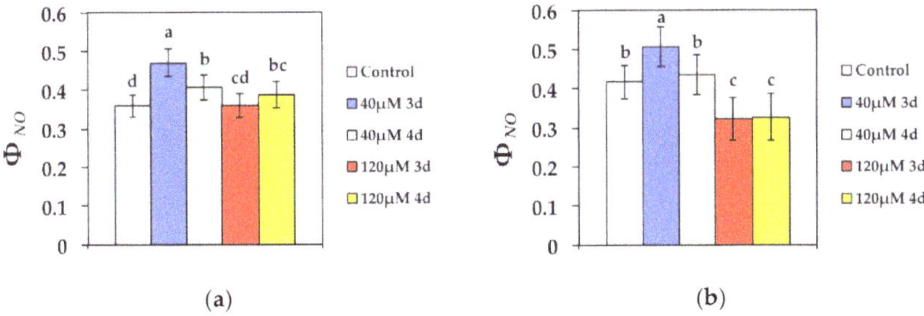

(a)   (b)

Figure 4. Changes in the quantum yield of non-regulated energy dissipated in PSII (non-regulated heat dissipation, a loss process due to PSII inactivity) (Φ_{NO}) measured at (**a**) 300 μmol photons m^{-2} s^{-1} or (**b**) 1000 μmol photons m^{-2} s^{-1}. *N. caerulescens* plants were grown at 0 (control), 40, or 120 μM Cd^{2+} for 3 and 4 days.

3.3. Non-Photochemical Quenching and Electron Transport Rate in Response to Cd

Non-photochemical quenching (NPQ) that reflects heat dissipation of excitation energy, decreased significantly compared to the control after 3 d at 40 μM Cd at GL, while it improved during the 4 d (Figure 5a). Exposure to 120 μM Cd resulted in decreased NPQ at GL compared to controls during the 4 d (Figure 5a). At HL, NPQ decreased significantly compared to the control after 3 d exposure to 40 μM Cd, and increased to control values during the 4 d, while after exposure to 120 μM Cd increased significantly compared to the controls (Figure 5b).

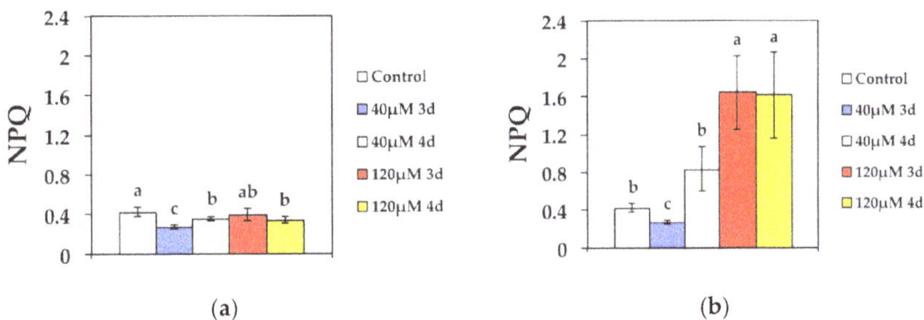

(a)   (b)

Figure 5. Changes in non-photochemical fluorescence quenching (NPQ) measured at (**a**) 300 μmol photons m^{-2} s^{-1} or (**b**) 1000 μmol photons m^{-2} s^{-1}. *N. caerulescens* plants were grown at 0 (control), 40 or 120 μM Cd^{2+} for 3 and 4 days.

The electron transport rate (ETR), at both GL and HL intensity, decreased significantly compared to the control after 3 d at 40 μM Cd, while it improved during the 4 d (Figure 6). However, ETR increased to control values after 3 d exposure to 120 μM Cd at GL and stabilized to control values after 4 d exposure (Figure 6a). High light exposure to 120 μM Cd resulted in decreased ETR compared to the controls (Figure 6b).

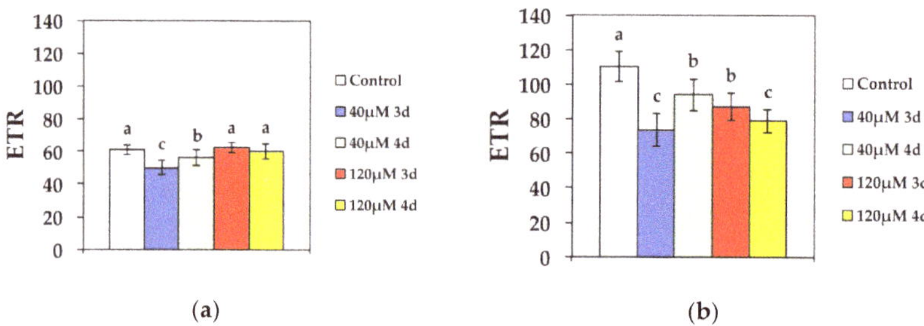

(a)   (b)

Figure 6. Changes in the relative PSII electron transport rate (ETR) measured at (**a**) 300 µmol photons m^{-2} s^{-1} or (**b**) 1000 µmol photons m^{-2} s^{-1}. *N. caerulescens* plants were grown at 0 (control), 40 or 120 µM Cd^{2+} for 3 and 4 days.

3.4. Changes in the Redox State of PSII after Cd Exposure

The redox state of QA (q_P) that is a measure of the fraction of open PSII reaction centers, at both GL and HL intensity, decreased significantly compared to the control after 3 d at 40 µM Cd, while it improved during the 4 d (Figure 7). However, q_P increased to control values after 3 d exposure to 120 µM Cd at GL and stabilized to control values after 4 d exposure (Figure 7a). High light exposure to 120 µM Cd resulted in a more reduced redox state of QA compared to controls, i.e., a lower fraction of open PSII reaction centers (Figure 7b).

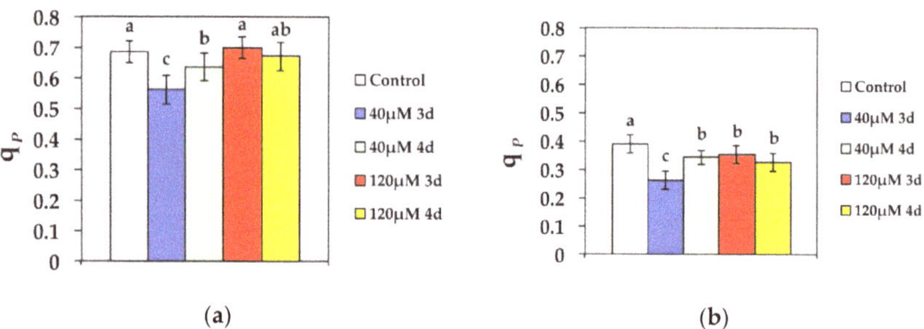

(a)   (b)

Figure 7. Changes in the photochemical fluorescence quenching, that is the relative reduction state of Q$_A$, reflecting the fraction of open PSII reaction centers (q_P) measured at (**a**) 300 µmol photons m^{-2} s^{-1} or (**b**) 1000 µmol photons m^{-2} s^{-1}. *N. caerulescens* plants were grown at 0 (control), 40, or 120 µM Cd^{2+} for 3 and 4 days.

3.5. Spatiotemporal Variation of PSII Responses to Cd Exposure

The major veins (mid-vein, first- and second-order veins) in *N. caerulescens* leaves grown under control growth conditions at both GL and HL defined areas with a lower fraction of open PSII reaction centers or a more reduced redox state of QA, while mesophyll cells expressed larger spatial heterogeneity with a larger fraction of open PSII reaction centers or a more oxidized redox state (Figures 8e and 9d).

The maximum quantum efficiency of PSII photochemistry (F_v/F_m) show the smallest spatial heterogeneity even though it decreased significantly at 40 µM Cd and increased to control values at 120 µM Cd (Figure 8a). The quantum yield of photochemical energy conversion in PSII (Φ_{PSII}) decreased significantly after 3 d at 40 µM Cd at GL, while it improved during the 4 d, showing a high spatiotemporal leaf heterogeneity (Figure 8b). Among the chlorophyll fluorescence parameters with high spatiotemporal heterogeneity observed at GL, were the images of the quantum yield of non-regulated energy dissipated in PSII (non-regulated heat dissipation, a loss process due to

PSII inactivity) (Φ_{NO}) (Figure 8d) and the images of the redox state of the PQ pool (q_P) (Figure 8e). The most severely affected leaf area after 3 d at 40 µM Cd, was the left and right leaf side, while the central area was less affected (Figure 8d,e). At the left and right leaf side after 3 d exposure to 40 µM Cd, the quantum yield of non-regulated energy loss in PSII (Φ_{NO}) increased; thus, these areas exhibited increased singlet oxygen ($^1O_2$) production (Figure 8d), and also presented the lower q_P values (Figure 8e). However, in the left and right leaf side after 4 d exposure to Cd, Φ_{NO} decreased (Figure 8d) and the redox state of the PQ pool increased (q_P) (Figure 8e). At exposure to 120 µM Cd at GL, leaf spatial heterogeneity decreased, and both Φ_{NO} (Figure 8d) and q_P (Figure 8e) stabilized to control values.

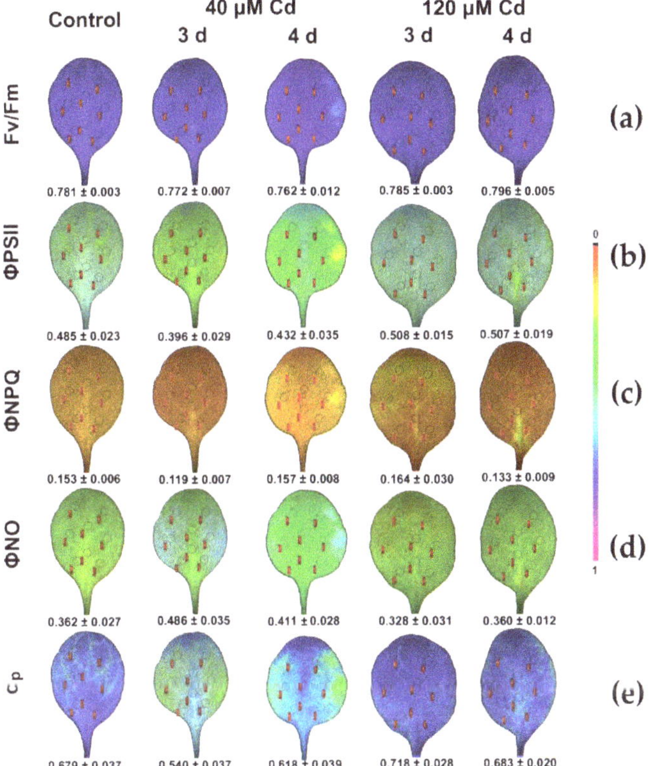

Figure 8. Representative chlorophyll fluorescence images of the maximum quantum efficiency (F_v/F_m) of PSII after 15 min dark adaptation (**a**) and after 5 min illumination at 300 µmol photons m^{-2} s^{-1} actinic light; of the actual (effective) quantum yield of PSII photochemistry (Φ_{PSII}) (**b**), the quantum yield for dissipation by downregulation in PSII (Φ_{NPQ}) (**c**), the quantum yield of non-regulated energy loss in PSII (Φ_{NO}) (**d**), and the relative reduction state of Q_A, reflecting the fraction of open PSII reaction centers (q_P) (**e**). *N. caerulescens* plants were grown at 0 (control), 40 or 120 µM Cd^{2+} for 3 and 4 days. The colour code depicted at the right side of the images ranges from black (pixel values 0.0) to purple (1.0). The eight areas of interest are shown in each image. The average value of each photosynthetic parameter of the leaf is presented in the figure.

Exposure of *N. caerulescens* to HL increased the spatiotemporal leaf heterogeneity (Figure 9) and the plants suffered more from Cd toxicity during the 3 d of exposure to 40 µM Cd, but they recovered during the 4 d. However, exposure to 120 µM Cd at HL revealed mild effects. This was realized by an increase in Φ_{NPQ} (Figure 9b) that down-regulated PSII quantum yield (Φ_{PSII}) (Figure 9a) and decreased the quantum yield of non-regulated energy loss in PSII (Φ_{NO}) (Figure 9c).

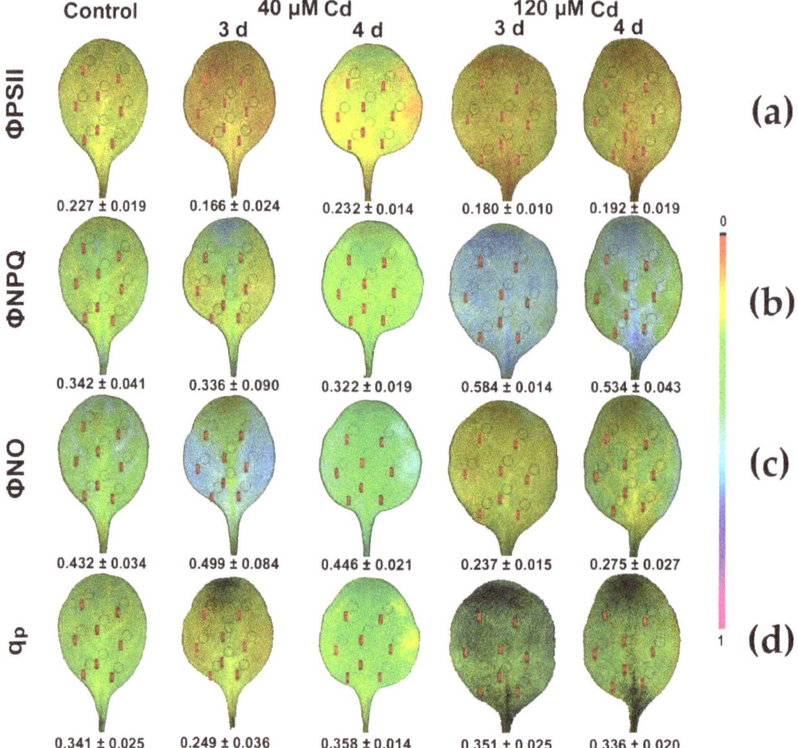

Figure 9. Representative chlorophyll fluorescence images after 5 min illumination at 1000 µmol photons m^{-2} s^{-1} actinic light; of the actual (effective) quantum yield of PSII photochemistry (Φ_{PSII}) (a), the quantum yield for dissipation by downregulation in PSII (Φ_{NPQ}) (b), the quantum yield of non-regulated energy loss in PSII (Φ_{NO}) (c), and the relative reduction state of Q_A, reflecting the fraction of open PSII reaction centers (q_P) (d) *N. caerulescens* plants were grown at 0 (control), 40 or 120 µM Cd^{2+} for 3 and 4 days. The colour code depicted at the right side of the images ranges from black (pixel values 0.0) to purple (1.0). The eight areas of interest are shown in each image. The average value of each photosynthetic parameter of the leaf is presented in the figure.

4. Discussion

The type of damage on PSII that has frequently been identified as the main target of Cd toxicity on photosynthesis strongly depends on light conditions [4,43–46]. At GL, the damage of the PSII function is mainly due to the impairment that results from the replacement by Cd^{2+} of the Mg^{2+} ion in the chlorophyll molecules of the light-harvesting complex II, while in HL it is mainly from direct damage to the PSII reaction center [4,44–46].

N. caerulescens leaves grown under control growth conditions at both GL and HL show a spatial heterogeneity in PSII functionality (Figures 8 and 9). This spatial heterogeneity may be attributed to 'patchy stomatal behavior', in which stomata in adjacent regions exhibit significantly different mean apertures from each other, resulting in significantly different stomatal conductance (g_s) [47,48]. Stomatal conductance decreases when the stomata close; this is used as an indicator of the extent of stomatal opening [49,50]. It is assumed that spatial variation in the quantum efficiency of PSII photochemistry (Φ_{PSII}) arises from local differences in internal CO_2 concentrations, which in turn result from changes in stomatal conductance due to patchy stomatal behavior [51]. A body of evidence suggests that patterns of Φ_{PSII} can be used to calculate stomatal conductance [51–55].

At the beginning of exposure to 40 µM, Cd Φ_{PSII} decreased significantly at the left and right leaf sides (Figures 8b and 9a), with a simultaneous decrease in Φ_{NPQ} (Figures 8c and 9b) resulting in an increase of the quantum yield of non-regulated non-photochemical energy loss (Φ_{NO}) (Figures 8d and 9c). The increase in Φ_{NO} indicates that photochemical energy conversion and photoprotective regulatory mechanism were insufficient, pointing to serious problems of the plant to cope with the absorbed light energy [56,57]. Φ_{NO} consists of chlorophyll fluorescence internal conversions and intersystem crossing, which indicate the formation of singlet oxygen ($^1O_2$) via the triplet state of chlorophyll (^{3}chl *) [13,58,59]. After 3 d exposure to 40 µM Cd, *N. caerulescens* leaves exhibited increased $^1O_2$ production at the left and right leaf sides, since Φ_{NO} increased significantly at those areas. Thus, although Cd^{2+} is a redox-inert element, it produces reactive oxygen species [28]. The simultaneous reduced PQ pool that was observed mainly at the left and right leaf sides mediated stomatal closure probably through the generation of mesophyll chloroplastic hydrogen peroxide (H_2O_2) [60]. The stomatal closure at these areas implies decreased transpiration rates that slow down Cd supply.

During the 4 d exposure to 40 µM Cd, Φ_{PSII} increased at the left and right leaf sides (Figures 8b and 9a), with a simultaneous increase in Φ_{NPQ} (Figures 8c and 9b) resulting in a decrease of Φ_{NO} (Figures 8d and 9c) compared to 3 d exposure. This response is attributed to both the possible Cd detoxification mechanism achieved by vacuolar sequestration, that seems to be the main mechanism for Cd detoxification [61–63], and to the reduced plastoquinone (PQ) pool that mediated stomatal closure and decreased Cd supply at the affected leaf area, leading to the acclimation of *N. caerulescens* to Cd exposure. Under exposure to 120 µM Cd at HL, the quantum yield of non-regulated energy loss in PSII (Φ_{NO}) decreased even more than control values, and thus exhibited lower singlet oxygen ($^1O_2$) production. This was due to the photoprotective mechanism that can divert absorbed light to other processes such as thermal dissipation, preventing the photosynthetic apparatus from oxidative damage [64–70].

The observed spatial heterogeneity in the quantum yield of linear electron transport (Φ_{PSII}) in *N. caerulescens* leaves exposed to 40 µM Cd for 3 d (Figures 8b and 9a) is in accordance to elemental imaging using laser ablation inductively-coupled plasma mass spectrometry, performed on whole leaves of the hyperaccumulator *N. caerulescens* that revealed differences in the supply of Cd over the whole leaf area, suggesting a heterogeneous distribution across the leaf [71]. Useful information can be obtained by combining chlorophyll fluorescence images, followed by laser ablation inductively-coupled plasma mass spectrometry on whole leaves of the hyperaccumulator *N. caerulescens* exposed to Cd.

It seems that spatiotemporal variations in the redox state of the PQ pool related to stomatal conductance, an indicator of the extent of stomatal opening [50], are interconnected to the heterogeneous distribution of Cd over the entire leaf area [71]. Thus, the spatial heterogeneity in the redox state of the PQ pool throughout the whole leaf area (Figures 8e and 9d) reveals a spatial supply of Cd across the leaf. Recently, Cd^{2+} root influx has been shown to exhibit spatiotemporal patterns [72]. A heterogeneous distribution of a reduced PQ pool gives rise to a spatial distribution of H_2O_2 accumulation [73]. Still, reactive oxygen species ($O_2{}^-$, H_2O_2) production corresponds to spatial accumulation metal patterns [74].

In our work, the response of *N. caerulescens* to Cd exposure fits the 'Threshold for Tolerance Model', with a lag time or/and a threshold concentration required for the induction of a tolerance mechanism [75–78]. Concurrent to this model, mild stress or short exposure times can produce significant effects on plants, while moderate stress or longer exposure times have less or no effect [79]. In accordance with this model, 40 µM Cd and 3d exposure time caused significant effects on PSII functioning, while 120 µM Cd or 4d exposure time have less or no effect. A lag-time of 4d exposure to 40 µM Cd was required for *N. caerulescens* to activate stress-coping mechanisms.

5. Conclusions

Acclimation to Cd exposure was achieved through the possible Cd detoxification mechanism done by vacuolar sequestration and the reduced plastoquinone (PQ) pool signaling that mediated stomatal closure and decreased Cd supply at the affected leaf area. The response of *N. caerulescens* to Cd exposure fits the 'Threshold for Tolerance Model', with a lag time of 4 d and a threshold concentration of 40 µM Cd required for the induction of the acclimation mechanism through the reduced PQ pool that mediated stomatal closure probably by the generation of mesophyll chloroplastic hydrogen peroxide (H_2O_2) [60], which acts as a fast acclimation signaling molecule [73,80], as well as activates the Cd detoxification mechanism through vacuolar sequestration [61–63]. The mode of Cd damage on PSII strongly depends on the irradiance conditions [4,43–46]. Chlorophyll fluorescence imaging analysis is a non-invasive tool to assess the physiological status of plants and detect the impacts of environmental stress [81–83], permitting also the visualization of the spatiotemporal variations in PSII efficiency [76]. As it was shown in our experiments, it is also capable of elucidating the mechanism of photosystem II acclimation to Cd exposure.

Author Contributions: G.B. and M.M. conceived and designed the experiments and analyzed the data; J.M. and N.G. performed the experiments and analyzed the data; M.M. wrote the paper; All authors review and approved the manuscript.

Funding: This research was funded by the Istanbul University Scientific Research Projects Thesis-2015-44682 and BEK-2016-21903.

Conflicts of Interest: The authors declare no conflict of interest. The funders had no role in the design of the study; in the collection, analyses, or interpretation of data; in the writing of the manuscript, and in the decision to publish the results.

References

1. Lagerwerff, J.V.; Specht, A.W. Contamination of roadside soil and vegetation with cadmium, nickel, lead, and zinc. *Environ. Sci. Technol.* **1970**, *4*, 583–586. [CrossRef]
2. Buchauer, M.J. Contamination of soil and vegetation near a zinc smelter by zinc, cadmium, copper, and lead. *Environ. Sci. Technol.* **1973**, *7*, 131–135. [CrossRef]
3. McBride, M.B.; Richards, B.K.; Steenhuis, T.; Russo, J.J.; Sauvé, S. Mobility and solubility of toxic metals and nutrients in soil fifteen years after sewage sludge application. *Soil Sci.* **1997**, *162*, 487–500. [CrossRef]
4. Küpper, H.; Parameswaran, A.; Leitenmaier, B.; Trtílek, M.; Šerlík, I. Cadmium induced inhibition of photosynthesis and long-term acclimation to cadmium stress in the hyperaccumulator *Thlaspi caerulescens*. *New Phytol.* **2007**, *175*, 655–674. [CrossRef] [PubMed]
5. Yao, X.; Cai, Y.; Yu, D.; Liang, G. bHLH104 confers tolerance to cadmium stress in *Arabidopsis thaliana*. *J. Integr. Plant Biol.* **2018**, *60*, 691–702. [CrossRef] [PubMed]
6. Hall, J.L. Cellular mechanisms for heavy metal detoxification and tolerance. *J. Exp. Bot.* **2002**, *53*, 1–11. [CrossRef] [PubMed]
7. Kim, D.Y.; Bovet, L.; Maeshima, M.; Martinoia, E.; Lee, Y. The ABC transporter AtPDR8 is a cadmium extrusion pump conferring heavy metal resistance. *Plant J.* **2007**, *50*, 207–218. [CrossRef] [PubMed]
8. Dalcorso, G.; Farinati, S.; Furini, A. Regulatory networks of cadmium stress in plants. *Plant Signal. Behav.* **2010**, *5*, 663–667. [CrossRef] [PubMed]
9. Lin, Y.F.; Aarts, M.G. The molecular mechanism of zinc and cadmium stress response in plants. *Cell Mol. Life Sci.* **2012**, *69*, 3187–3206. [CrossRef] [PubMed]
10. Shi, Y.Z.; Zhu, X.F.; Wan, J.X.; Li, G.X.; Zheng, S.J. Glucose alleviates cadmium toxicity by increasing cadmium fixation in root cell wall and sequestration into vacuole in Arabidopsis. *J. Integr. Plant Biol.* **2015**, *57*, 830–837. [CrossRef] [PubMed]
11. Brooks, R.R.; Lee, J.; Reeves, R.D.; Jaffre, T. Detection of nickeliferous rocks by analysis of herbarium species of indicator plants. *J. Geochem. Explor.* **1977**, *7*, 49–57. [CrossRef]
12. Leitenmaier, B.; Küpper, H. Cadmium uptake and sequestration kinetics in individual leaf cell protoplasts of the Cd/Zn hyperaccumulator *Thlaspi caerulescens*. *Plant Cell Environ.* **2011**, *34*, 208–219. [CrossRef] [PubMed]

13. Bayçu, G.; Gevrek-Kürüm, N.; Moustaka, J.; Csatári, I.; Rognes, S.E.; Moustakas, M. Cadmium-zinc accumulation and photosystem II responses of *Noccaea caerulescens* to Cd and Zn exposure. *Environ. Sci. Pollut. Res.* **2017**, *24*, 2840–2850. [CrossRef] [PubMed]

14. Baker, A.J.M.; McGrath, S.P.; Sidoli, C.M.D.; Reeves, R.D. The possibility of situ heavy metal decontamination of polluted soils using crops of metal-accumulating plants. *Resour. Conserv. Recycl.* **1994**, *11*, 41–49. [CrossRef]

15. McGrath, S.P.; Zhao, F.J. Phytoextraction of metals and metalloids from contaminated soils. *Curr. Opin. Biotechnol.* **2003**, *14*, 277–282. [CrossRef]

16. Chaney, R.L.; Angle, J.S.; McIntosh, M.S.; Reeves, R.D.; Li, Y.M.; Brewer, E.P.; Chen, K.Y.; Roseberg, R.J.; Perner, H.; Synkowski, E.C.; et al. Using hyperaccumulator plants to phytoextract Soil Ni and Cd. *Z. Naturforsch. C* **2005**, *60*, 190–198. [PubMed]

17. Zhang, X.D.; Wang, Y.; Li, H.B.; Yang, Z.M. Isolation and identification of rapeseed (*Brassica napus*) cultivars for potential higher and lower Cd accumulation. *J. Plant Nutr. Soil Sci.* **2018**, *181*, 479–487. [CrossRef]

18. Escarré, J.; Lefèbvre, C.; Gruber, W.; Leblanc, M.; Lepart, J.; Rivière, Y.; Delay, B. Zinc and cadmium hyperaccumulation by *Thlaspi caerulescens* from metalliferous and nonmetalliferous sites in the Mediterranean area: Implications for phytoextraction. *New Phytol.* **2000**, *145*, 429–437.

19. Assunção, A.G.L.; Bookum, W.M.; Nelissen, H.J.M.; Vooijs, R.; Schat, H.; Ernst, W.H.O. Differential metal-specific tolerance and accumulation patterns among *Thlaspi caerulescens* populations originating from different soil types. *New Phytol.* **2003**, *159*, 411–419. [CrossRef]

20. Escarré, J.; Lefèbvre, C.; Frérot, H.; Mahieu, S.; Noret, N. Metal concentration and metal mass of metallicolous, non metallicolous and serpentine *Noccaea caerulescens* populations, cultivated in different growth media. *Plant Soil* **2013**, *370*, 197–221. [CrossRef]

21. Mousset, M.; David, P.; Petit, C.; Pouzadoux, J.; Hatt, C.; Flaven, E.; Ronce, O.; Mignot, A. Lower selfing rates in metallicolous populations than in non-metallicolous populations of the pseudometallophyte *Noccaea caerulescens* (Brassicaceae) in Southern France. *Ann. Bot.* **2016**, *117*, 507–519. [CrossRef] [PubMed]

22. Lasat, M.M.; Pence, N.S.; Garvin, D.F.; Ebbs, S.D.; Kochian, L.V. Molecular physiology of zinc transport in the Zn hyperaccumulator *Thlaspi caerulescens*. *J. Exp. Bot.* **2000**, *51*, 71–79. [CrossRef] [PubMed]

23. Lombi, E.; Zhao, F.J.; Dunham, S.J.; McGrath, S.P. Cadmium accumulation in populations of *Thlaspi caerulescens* and *Thlaspi goesingense*. *New Phytol.* **2000**, *145*, 11–20. [CrossRef]

24. Assunção, A.G.L.; Schat, H.; Aarts, M.G.M. *Thlaspi caerulescens*, an attractive model species to study heavy metal hyperaccumulation in plants. *New Phytol.* **2003**, *159*, 351–360. [CrossRef]

25. Baker, A.J.M.; McGrath, S.P.; Reeves, D.R.; Smith, J.A.C. Metal hyperaccumulator plants: A review of the ecology and physiology of a biological resource for phytoremediation of metal-polluted soils. In *Phytoremediation of Contaminated Soils and Water*; Terry, N., Banuelos, G., Eds.; CRC Press: Boca Raton, FL, USA, 2000; pp. 171–188.

26. Greger, M.; Ögren, E. Direct and indirect effects of Cd^{2+} on photosynthesis in sugar beet (*Beta vulgaris*). *Physiol. Plant.* **1991**, *83*, 129–135. [CrossRef]

27. Ouzounidou, G.; Moustakas, M.; Eleftheriou, E.P. Physiological and ultrastructural effects of cadmium on wheat (*Triticum aestivum* L.) leaves. *Arch. Environ. Contam. Toxicol.* **1997**, *32*, 154–160. [CrossRef] [PubMed]

28. Ekmekçi, Y.; Tanyolaç, D.; Ayhan, B. Effects of cadmium on antioxidant enzyme and photosynthetic activities in leaves of two maize cultivars. *J. Plant Physiol.* **2008**, *165*, 600–611. [CrossRef] [PubMed]

29. Liu, C.; Guo, J.; Cui, Y.; Lü, T.; Zhang, X.; Shi, G. Effects of cadmium and salicylic acid on growth, spectral reflectance and photosynthesis of castor bean seedlings. *Plant Soil* **2011**, *344*, 131–141. [CrossRef]

30. Parmar, P.; Kumari, N.; Sharma, V. Structural and functional alterations in photosynthetic apparatus of plants under cadmium stress. *Bot. Stud.* **2013**, *54*, 45. [CrossRef] [PubMed]

31. Xue, Z.C.; Gao, H.Y.; Zhang, L.T. Effects of cadmium on growth, photosynthetic rate, and chlorophyll content in leaves of soybean seedlings. *Biol. Plant.* **2013**, *57*, 587–590. [CrossRef]

32. Baszynski, T.; Wajda, L.; Krol, M.; Wolinska, D.; Krupa, Z.; Tukendorf, A. Photosynthetic activities of cadmium-treated tomato plants. *Physiol. Plant.* **1980**, *48*, 365–370. [CrossRef]

33. Sigfridsson, K.G.V.; Bernát, G.; Mamedov, F.; Styring, S. Molecular interference of Cd^{2+} with Photosystem II. *Biochim. Biophys. Acta* **2004**, *1659*, 19–31. [CrossRef] [PubMed]

34. Malik, D.; Sheoran, I.S.; Singh, R. Carbon metabolism in leaves of cadmium treated wheat seedlings. *Plant Physiol. Biochem.* **1992**, *30*, 223–229.

35. Krantev, A.; Yordanova, R.; Janda, T.; Szalai, G.; Popova, L. Treatment with salicylic acid decreases the effect of cadmium on photosynthesis in maize plants. *J. Plant Physiol.* **2008**, *165*, 920–931. [CrossRef] [PubMed]

36. Poschenrieder, C.; Gunse, B.; Barcelo, J. Influence of cadmium on water relations, stomatal resistance, and abscisic acid content in expanding bean leaves. *Plant Physiol.* **1989**, *90*, 1365–1371. [CrossRef] [PubMed]

37. Shi, G.R.; Cai, Q.S. Photosynthetic and anatomic responses of peanut leaves to cadmium stress. *Photosynthetica* **2008**, *46*, 627–630. [CrossRef]

38. Krupa, Z.; Moniak, M. The stage of leaf maturity implicates the response of the photosynthetic apparatus to cadmium toxicity. *Plant Sci.* **1998**, *138*, 149–156. [CrossRef]

39. Pagliano, C.; Raviolo, M.; Dalla Vecchia, F.; Gabbrielli, R.; Gonnelli, C.; Rascio, N.; Barbato, R.; La Rocca, N. Evidence for PSII-donor-side damage and photoinhibition induced by cadmium treatment on rice (*Oryza sativa* L.). *J. Photochem. Photobiol. B Biol.* **2006**, *84*, 70–78. [CrossRef] [PubMed]

40. Chugh, L.K.; Sawhney, S.K. Photosynthetic activities of *Pisum sativum* seedlings grown in presence of cadmium. *Plant Physiol. Biochem.* **1999**, *37*, 297–303. [CrossRef]

41. Moustaka, J.; Panteris, E.; Adamakis, I.D.S.; Tanou, G.; Giannakoula, A.; Eleftheriou, E.P.; Moustakas, M. High anthocyanin accumulation in poinsettia leaves is accompanied by thylakoid membrane unstacking, acting as a photoprotective mechanism, to prevent ROS formation. *Environ. Exp. Bot.* **2018**, *154*, 44–55. [CrossRef]

42. Oxborough, K.; Baker, N.R. Resolving chlorophyll a fluorescence images of photosynthetic efficiency into photochemical and non-photochemical components-calculation of qP and Fv-/Fm- without measuring Fo-. *Photosynth. Res.* **1997**, *54*, 135–142. [CrossRef]

43. Cedeno-Maldonado, A.; Swader, J.A.; Heath, R.L. The cupric ion as an inhibitor of photosynthetic electron transport in isolated chloroplasts. *Plant Physiol.* **1972**, *50*, 698–701. [CrossRef] [PubMed]

44. Küpper, H.; Küpper, F.; Spiller, M. Environmental relevance of heavy metal substituted chlorophylls using the example of water plants. *J. Exp. Bot.* **1996**, *47*, 259–266. [CrossRef]

45. Küpper, H.; Küpper, F.; Spiller, M. In situ detection of heavy metal substituted chlorophylls in water plants. *Photosynth. Res.* **1998**, *58*, 125–133. [CrossRef]

46. Küpper, H.; Setlík, I.; Spiller, M.; Küpper, F.C.; Prásil, O. Heavy metal-induced inhibition of photosynthesis: Targets of in vivo heavy metal chlorophyll formation. *J. Phycol.* **2002**, *38*, 429–441. [CrossRef]

47. Weyers, J.D.B.; Lawson, T. Heterogeneity in stomatal characteristics. *Adv. Bot. Res.* **1997**, *26*, 317–352.

48. Lawson, T.; Morison, J. Visualising patterns of CO_2 diffusion in leaves. *New Phytol.* **2006**, *169*, 637–640. [CrossRef] [PubMed]

49. Omasa, K.; Hashimoto, Y.; Aiga, I. Observation of stomatal movements of intact plants using an image instrumentation system with a light microscope. *Plant Cell Physiol.* **1983**, *24*, 281–288. [CrossRef]

50. Jones, H.G. Use of thermography for quantitative studies of spatial and temporal variation of stomatal conductance over leaf surfaces. *Plant Cell Environ.* **1999**, *22*, 1043–1055. [CrossRef]

51. Lawson, T.; Weyers, J. Spatial and temporal variation in gas exchange over the lower surface of *Phaseolus vulgaris* L. primary leaves. *J. Exp. Bot.* **1999**, *50*, 1381–1391. [CrossRef]

52. Eckstein, J.; Beyschlag, W.; Mott, K.A.; Ryel, R.J. Changes in photon flux can induce stomatal patchiness. *Plant Cell Environ.* **1996**, *12*, 559–568. [CrossRef]

53. Genty, B.; Meyer, S. Quantitative mapping of leaf photosynthesis using chlorophyll fluorescence imaging. *Aust. J. Plant Physiol.* **1994**, *22*, 277–284. [CrossRef]

54. Mott, K.A. Effects of patchy stomatal closure on gas exchange measurements following abscisic acid treatment. *Plant Cell Environ.* **1995**, *18*, 1291–1300. [CrossRef]

55. Omasa, K.; Takayama, K. Simultaneous measurement of stomatal conductance, non-photochemical quenching, and photochemical yield of photosystem II in intact leaves by thermal and chlorophyll fluorescence imaging. *Plant Cell Physiol.* **2003**, *44*, 1290–1300. [CrossRef] [PubMed]

56. Calatayud, A.; Roca, D.; Martínez, P.F. Spatial–temporal variations in rose leaves under water stress conditions studied by chlorophyll fluorescence imaging. *Plant Physiol. Biochem.* **2006**, *44*, 564–573. [CrossRef] [PubMed]

57. Moustakas, M.; Malea, P.; Zafeirakoglou, A.; Sperdouli, I. Photochemical changes and oxidative damage in the aquatic macrophyte *Cymodocea nodosa* exposed to paraquat-induced oxidative stress. *Pest. Biochem. Physiol.* **2016**, *126*, 28–34. [CrossRef] [PubMed]

58. Apel, K.; Hirt, H. Reactive oxygen species: Metabolism, oxidative stress, and signal transduction. *Annu. Rev. Plant Biol.* **2004**, *55*, 373–399. [CrossRef] [PubMed]

59. Gawroński, P.; Witoń, D.; Vashutina, K.; Bederska, M.; Betliński, B.; Rusaczonek, A.; Karpiński, S. Mitogen-activated protein kinase 4 is a salicylic acid-independent regulator of growth but not of photosynthesis in Arabidopsis. *Mol. Plant* **2014**, *7*, 1151–1166. [CrossRef] [PubMed]

60. Wang, W.H.; He, E.M.; Chen, J.; Guo, Y.; Chen, J.; Liu, X.; Zheng, H.L. The reduced state of the plastoquinone pool is required for chloroplast-mediated stomatal closure in response to calcium stimulation. *Plant J.* **2016**, *86*, 132–144. [CrossRef] [PubMed]

61. Küpper, H.; Mijovilovich, A.; Meyer-klaucke, W.; Kroneck, P.M.H. Tissue- and age-dependent differences in the complexation of cadmium and zinc in the cadmium/zinc hyperaccumulator *Thlaspi caerulescens* (Ganges Ecotype) revealed by X-ray absorption spectroscopy. *Plant Physiol.* **2004**, *134*, 748–757.

62. Ma, J.F.; Ueno, D.; Zhao, F.J.; McGrath, S.P. Subcellular localization of Cd and Zn in the leaves of a Cd-hyperaccumulating ecotype of *Thlaspi caerulescens*. *Planta* **2005**, *220*, 731–736. [CrossRef] [PubMed]

63. Ebbs, S.D.; Zambrano, M.C.; Spiller, S.M.; Newville, M. Cadmium sorption, influx, and efflux at the mesophyll layer of leaves from ecotypes of the Zn/Cd hyperaccumulator *Thlaspi caerulescens*. *New Phytol.* **2009**, *181*, 626–636. [CrossRef] [PubMed]

64. Demmig-Adams, B.; Adams, W.W.A., III. Photoprotection in an ecological context: The remarkable complexity of thermal energy dissipation. *New Phytol.* **2006**, *172*, 11–21. [CrossRef] [PubMed]

65. Moustaka, J.; Moustakas, M. Photoprotective mechanism of the non-target organism *Arabidopsis thaliana* to paraquat exposure. *Pest. Biochem. Physiol.* **2014**, *111*, 1–6. [CrossRef] [PubMed]

66. Agathokleous, E. Environmental hormesis, a fundamental non-monotonic biological phenomenon with implications in ecotoxicology and environmental safety. *Ecotoxicol. Environ. Saf.* **2018**, *148*, 1042–1053. [CrossRef]

67. Georgieva, K.; Doncheva, S.; Mihailova, G.; Petkova, S. Response of sun- and shade-adapted plants of *Haberlea rhodopensis* to desiccation. *Plant Growth Regul.* **2012**, *67*, 121–132. [CrossRef]

68. Jusovic, M.; Velitchkova, M.Y.; Misheva, S.P.; Börner, A.; Apostolova, E.L.; Dobrikova, A.G. Photosynthetic responses of a wheat mutant (*Rht-B1c*) with altered DELLA proteins to salt stress. *J. Plant Growth Regul.* **2018**, *37*, 645–656. [CrossRef]

69. Moustaka, J.; Ouzounidou, G.; Sperdouli, I.; Moustakas, M. Photosystem II is more sensitive than photosystem I to Al^{3+} induced phytotoxicity. *Materials* **2018**, *11*, 1772. [CrossRef] [PubMed]

70. Moustaka, J.; Ouzounidou, G.; Bayçu, G.; Moustakas, M. Aluminum resistance in wheat involves maintenance of leaf Ca^{2+} and Mg^{2+} content, decreased lipid peroxidation and Al accumulation, and low photosystem II excitation pressure. *BioMetals* **2016**, *29*, 611–623. [CrossRef] [PubMed]

71. Callahan, D.L.; Hare, D.J.; Bishop, D.P.; Doble, P.A.; Roessner, U. Elemental imaging of leaves from the metal hyperaccumulating plant *Noccaea caerulescens* shows different spatial distribution of Ni, Zn and Cd. *RSC Adv.* **2016**, *6*, 2337–2344. [CrossRef]

72. Chen, X.; Ouyang, Y.; Fan, Y.; Qiu, B.; Zhang, G.; Zeng, F. The pathway of transmembrane cadmium influx via calcium-permeable channels and its spatial characteristics along rice root. *J. Exp. Bot.* **2018**, *69*, 5279–5291. [CrossRef] [PubMed]

73. Antonoglou, O.; Moustaka, J.; Adamakis, I.D.; Sperdouli, I.; Pantazaki, A.; Moustakas, M.; Dendrinou-Samara, C. Nanobrass CuZn nanoparticles as foliar spray non phytotoxic fungicides. *ACS Appl. Mater. Interfaces* **2018**, *10*, 4450–4461. [CrossRef] [PubMed]

74. Hanć, A.; Małecka, A.; Kutrowska, A.; Bagniewska-Zadworna, A.; Tomaszewska, B.; Barałkiewicz, D. Direct analysis of elemental biodistribution in pea seedlings by LA-ICP-MS, EDX and confocal microscopy: Imaging and quantification. *Microchem. J.* **2016**, *128*, 305–311. [CrossRef]

75. Barceló, J.; Poschenrieder, C. Fast root growth responses, root exudates, and internal detoxification as clues to the mechanisms of aluminium toxicity and resistance: A review. *Environ. Exp. Bot.* **2002**, *48*, 75–92. [CrossRef]

76. Sperdouli, I.; Moustakas, M. Spatio-temporal heterogeneity in *Arabidopsis thaliana* leaves under drought stress. *Plant Biol.* **2012**, *14*, 118–128. [CrossRef] [PubMed]

77. Sperdouli, I.; Moustakas, M. Leaf developmental stage modulates metabolite accumulation and photosynthesis contributing to acclimation of *Arabidopsis thaliana* to water deficit. *J. Plant Res.* **2014**, *127*, 481–489. [CrossRef] [PubMed]

78. Moustakas, M.; Malea, P.; Haritonidou, K.; Sperdouli, I. Copper bioaccumulation, photosystem II functioning and oxidative stress in the seagrass *Cymodocea nodosa* exposed to copper oxide nanoparticles. *Environ. Sci. Pollut. Res.* **2017**, *24*, 16007–16018. [CrossRef] [PubMed]

79. Lichtenthaler, H.K. The stress concept in plants: An introduction. *Ann. N.Y. Acad. Sci.* **1998**, *851*, 187–198. [CrossRef] [PubMed]

80. Wilson, K.E.; Ivanov, A.G.; Öquist, G.; Grodzinski, B.; Sarhan, F.; Huner, N.P.A. Energy balance, organellar redox status, and acclimation to environmental stress. *Can. J. Bot.* **2006**, *84*, 1355–1370. [CrossRef]

81. Guidi, L.; Calatayud, A. Non-invasive tools to estimate stress-induced changes in photosynthetic performance in plants inhabiting Mediterranean areas. *Environ. Exp. Bot.* **2014**, *103*, 42–52. [CrossRef]

82. Gorbe, E.; Calatayud, A. Applications of chlorophyll fluorescence imaging technique in horticultural research: A review. *Sci. Hortic.* **2012**, *138*, 24–35. [CrossRef]

83. Moustaka, J.; Tanou, G.; Adamakis, I.D.; Eleftheriou, E.P.; Moustakas, M. Leaf age dependent photoprotective and antioxidative mechanisms to paraquat-induced oxidative stress in *Arabidopsis thaliana*. *Int. J. Mol. Sci.* **2015**, *16*, 13989–14006. [CrossRef] [PubMed]

Article

Photosystem II Is More Sensitive than Photosystem I to Al^{3+} Induced Phytotoxicity

Julietta Moustaka [1,2], **Georgia Ouzounidou** [3], **Ilektra Sperdouli** [1,4] **and Michael Moustakas** [1,5,*]

[1] Department of Botany, Aristotle University of Thessaloniki, GR-54124 Thessaloniki, Greece; ilektras@bio.auth.gr
[2] Department of Plant and Environmental Sciences, University of Copenhagen, Thorvaldsensvej 40, DK-1871 Frederiksberg C, Denmark; moustaka@plen.ku.dk
[3] Institute of Food Technology, Hellenic Agricultural Organization—Demeter, 1 S. Venizelou Str., GR-14123 Lycovrissi, Greece; geouz@yahoo.gr
[4] Institute of Plant Breeding and Genetic Resources, Hellenic Agricultural Organisation–Demeter, Thermi, GR-57001 Thessaloniki, Greece
[5] Division of Botany, Department of Biology, Faculty of Science, Istanbul University, 34134 Istanbul, Turkey
[*] Correspondence: moustak@bio.auth.gr; Tel.: +30-2310-99-8335

Received: 14 August 2018; Accepted: 17 September 2018; Published: 19 September 2018

Abstract: Aluminium (Al) the most abundant metal in the earth's crust is toxic in acid soils (pH < 5.5) mainly in the ionic form of Al^{3+} species. The ability of crops to overcome Al toxicity varies among crop species and cultivars. Here, we report for a first time the simultaneous responses of photosystem II (PSII) and photosystem I (PSI) to Al^{3+} phytotoxicity. The responses of PSII and PSI in the durum wheat (*Triticum turgidum* L. cv. 'Appulo E') and the triticale (X Triticosecale Witmark cv. 'Dada') were evaluated by chlorophyll fluorescence quenching analysis and reflection spectroscopy respectively, under control ($-$Al, pH 6.5) and 148 µM Al ($+$Al, pH 4.5) conditions. During control growth conditions the high activity of PSII in 'Appulo E' led to a rather higher electron flow to PSI, which induced a higher PSI excitation pressure in 'Appulo E' than in 'Dada' that presented a lower PSII activity. However, under 148 µM Al the triticale 'Dada' presented a lower PSII and PSI excitation pressure than 'Appulo E'. In conclusion, both photosystems of 'Dada' displayed a superior performance than 'Appulo E' under Al exposure, while in both cultivars PSII was more affected than PSI from Al^{3+} phytotoxicity.

Keywords: aluminium; chlorophyll fluorescence; durum wheat; excitation pressure; non-photochemical quenching; photosynthesis; photoprotection; photoinhibition; reactive oxygen species; triticale

---

1. Introduction

Aluminium (Al) is considered as the most abundant metal in the earth's crust, comprising approximately 7% of the soil [1,2]. Although Al is nontoxic as a metal, with very low solubility in the neutral pH range (6.0–8.0), its solubility increases and becomes toxic to all living cells under acidic or alkaline pH conditions where is present mainly in the ionic form of Al^{3+} species (at pH < 5.5) or as aluminate $Al[OH]_4^-$ (at pH > 8.5) [3,4]. Aluminium toxicity is limiting crop production on acid soils through inhibition of root elongation, which occurs within hours of exposure to Al^{3+}, disturbance of nutrient uptake and other metabolic functions, affecting also the process of photosynthesis [5–20]. Cereals differ significantly in their response to Al toxicity and genetic variation has been found between species as well as between cultivars [21–23], revealing distinct Al-tolerance mechanisms [2,16,18,24,25].

In the form of the trivalent cation Al^{3+}, that is toxic to most plants at relatively low concentrations, it is the main limiting factor in the world's arable non-irrigated crop production to over 40% [26,27]. With the world's population forecasted to reach nine billion by 2050, cereal production needs to increase by 50% by 2030 [28]. Consequently, increasing cereal yields is now one of the top priorities for agricultural research [28]. Since plant production is driven by photosynthesis, studying Al^{3+} toxicity effects on photosynthesis has the potential to increase cereal yields by understanding the factors that influence negatively the molecular mechanism of absorbed light energy utilization.

Photosynthesis is the process by which organisms convert the absorbed solar energy into chemical energy via photosystem II (PSII) and photosystem I (PSI). Light reactions of photosynthesis are driven by the cooperation of PSII and PSI that work coordinately to transfer photosynthetic electrons efficiently and are located in the photosynthetic membranes of chloroplasts, the thylakoids [29,30]. Chloroplasts exhibit stacked and unstacked thylakoid membranes, designated as grana and stroma thylakoids, respectively [29,31]. The two photosystems, PSI and PSII, are laterally and functionally separated mainly in stroma (non-appressed) and grana (appressed) thylakoid membranes, respectively [29,30], that allows the regulation of the excitation energy distribution between the two photosystems [32]. Photosystem II and PSI are working in connection in the linear electron transport catalyzing the transfer of electrons from H_2O to $NADP^+$ through the formation of strong reductants, and of a proton gradient that is used to drive ATP synthesis [33,34].

Aluminium toxicity has been shown to reduce the photochemical efficiency of PSII in several plant species [35–40] by causing inhibition of electron transfer between the first stable electron acceptor of PSII, quinone A (Q_A) and the quinone B (Q_B) [15], and closing PSII reaction centers (RCs) [6,37,38,41,42]. Thus, Al-toxicity increases the percentage of closed PSII RCs and reduces the rate of photosynthesis with subsequent reduced growth and development [15,42,43]). However, Al-resistant cultivars keep a larger fraction of PSII RCs in an open configuration [43]. Al^{3+} concentrations resulted in a reduction of the energy transfer from light harvesting complex (LHCI) to RCs of PSI, followed by an impairment of PSI RCs and electron transfer of PSI [44].

Absorption of more light than what can be used to drive photosynthesis, causes photodamage to the photosynthetic apparatus and the light-processing structures, primarily PSII, resulting in a decrease in the photosynthetic activity causing reduced plant growth and productivity [45,46]. Among the photoprotective mechanisms that plants have developed to counteract the effects of excessive harmful energy is the dissipation of excessive energy as heat and the scavenging of reactive oxygen species (ROS) by enzymatic and non-enzymatic antioxidant molecules [47–50]. Dissipation of the excess light energy as heat in the antenna or PSII RCs is believed to be the main mechanism that plants use to deal with excessive light energy and this process is called non-photochemical quenching (NPQ) [47,51–53].

Chlorophyll fluorescence measurements and in particular measurements of PSII excitation pressure, that is the redox state of the plastoquinone (PQ) pool, have been proposed as a sensitive bio-indicator to measure Al effects on plants [43]. Chlorophyll fluorescence quenching analysis has been extensively applied as a probe of photosynthesis research and has been successfully used to assess the changes in the function of PSII under different environmental conditions [54–56]. In addition, the measurement of the fraction of closed and open reaction centers of PSI can be evaluated by reflection spectroscopy [57,58].

A number of studies as already mentioned have examined the functioning of PSII under Al toxicity [6,35–38,41–43], but the functioning of PSI under Al toxicity, as far as we know, was investigated only once [44]. Nevertheless, due to differences in the experimental conditions, it is problematic to acquire comprehensive information regarding the proportional resistance to Al toxicity of the two photosystems if they have not been examined concurrently. To the best of our knowledge, a simultaneous comparative study of the two photosystems to Al toxicity has not been addressed. Here, we report the concurrent responses of PSII and PSI to Al^{3+} toxicity, in the durum wheat (*Triticum turgidum* L. cv. 'Appulo E') and the triticale (X Triticosecale Witmark cv. 'Dada'). Triticale is considered as more Al-tolerant species than durum wheat (*Triticum turgidum*); this difference is largely attributed to its superior ability to grow better under acidic conditions [25].

2. Materials and Methods

2.1. Plant Material and Growth Conditions

Durum wheat (*Triticum turgidum* L. cv. 'Appulo E') and the triticale (X Triticosecale Witmark cv. 'Dada') were used to compare the tolerance of the two photosystems to Al toxicity. Seeds obtained from the Institute of Plant Breeding and Genetic Resources, Thermi, Greece, germinated at $22 \pm 1\ ^\circ\text{C}$ for 2 d in the dark. The germinated seeds were transferred in a growth chamber and mounted on nylon-mesh floats on plastic vessels filled with nutrient solution at pH 6.5 [23]. The seedlings were grown in hydroponic culture at controlled environmental conditions as described previously [43].

2.2. Al Treatment

Al was supplied at 148 µM as $\text{KAl(SO}_4)_2 12\text{H}_2\text{O}$ for 14 days. Al-containing pots (nutrient solution plus 148 µM Al) were acidified initially to pH 4.5 with 1N HCl [25], while growth solutions of control plants (nutrient solution only) were adjusted to pH 6.5. According to the GEOCHEM-EZ speciation programme [59] the free Al^{3+} activities were calculated to be 16.8 µM [43].

2.3. Lipid Peroxidation Measurements

The level of lipid peroxidation of controls and 14-days Al^{3+} treated plants was measured as malondialdehyde (MDA) content, as described previously [60], according to the method of Heath and Packer [61]. The concentration of MDA was calculated from the difference of the absorbance at 532 and 600 nm and expressed as nmol (MDA) g^{-1} fresh weight.

2.4. Measurements of Chlorophyll a Fluorescence

Chlorophyll a fluorescence was measured in dark-adapted (20 min) leaf samples, using a pulse amplitude modulation fluorometer (PAM, Walz, Effeltrich, Germany), as described before [35,43]. First, minimal chlorophyll *a* fluorescence (F_o) was measured by application of a weak modulated light beam (L_1) followed by a saturating light pulse (L_2) to measure maximal chlorophyll *a* fluorescence (F_m) in the dark adapted (20 min) samples. Then, by application of the actinic light (L_A) and saturating light pulses, maximum chlorophyll *a* fluorescence in the light (F_m') was measured, while to assess steady-state photosynthesis (F_s) values, the actinic light (L_A) alone was applied. Minimum chlorophyll *a* fluorescence in the light (F_o') was measured immediately after turning off the actinic light (L_A) (Figure 1). The calculated chlorophyll fluorescence parameters with their definitions are given in Table 1.

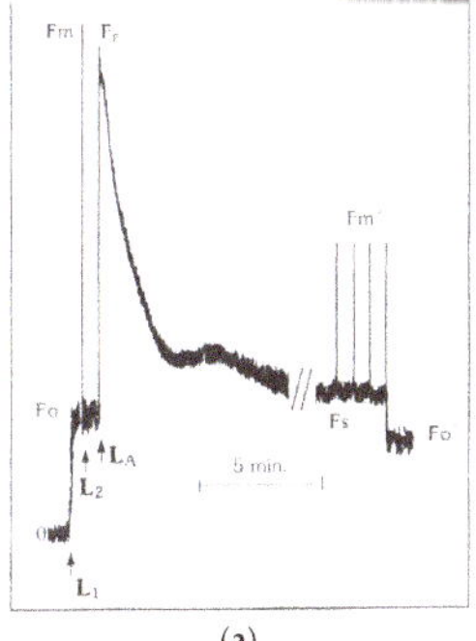

(a)

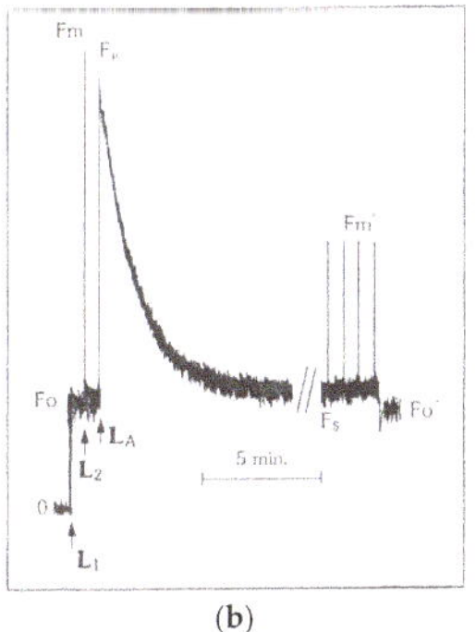

(b)

Figure 1. Typical modulated fluorescence signals obtained by the triticale (X Triticosecale Witmark cv. 'Dada') after 20 min dark adaptation; (**a**) control leaves from plants in the nutrient solution at pH 6.5; and (**b**) leaves from plants in the nutrient solution plus 148 µM Al at pH 4.5; L_1, arrow denotes onset of a weak modulated light beam; L_2, arrow denotes onset of a saturating light pulse (approximately 8000 µmol photons $\text{m}^{-2}\ \text{s}^{-1}$); L_A, arrow denotes continuous actinic light.

Table 1. Definitions of the calculated chlorophyll fluorescence parameters with their calculation formula.

Chlorophyll Fluorescence Parameter	Definition	Calculation
F_v/F_m	The maximum quantum efficiency of photosystem II (PSII) photochemistry	Calculated as $(F_m - F_o)/F_m$
$F_v{}'/F_m{}'$	The PSII maximum efficiency is an estimate of the maximum efficiency of PSII photochemistry at a given PPFD (photosynthetic photon flux density)	Calculated as $(F_m{}' - F_o{}')/F_m{}'$
Φ_{PSII}	The effective quantum yield of photochemical energy conversion in PSII estimating the efficiency at which light absorbed by PSII is used for photochemistry, that means is used for reduction of the primary acceptor of PSII quinone A (QA)	Calculated as $(F_m{}' - F_s)/F_m{}'$
q_P	The photochemical quenching is a measure of the fraction of open PSII reaction centers, that is the redox state of QA	Calculated as $(F_m{}' - F_s)/(F_m{}' - F_o{}')$
NPQ	The non-photochemical quenching that reflects heat dissipation of excitation energy	Calculated as $(F_m - F_m{}')/F_m{}'$
ETR	The relative PSII electron transport rate	Calculated as $\Phi_{PSII} \times$ PPFD $\times$ c $\times$ abs, where c is 0.5 since the absorbed light energy is assumed to be equally distributed between PSII and PSI, and abs is the total light absorption of the leaf taken as 0.84.
Φ_{NPQ}	The quantum yield of regulated non-photochemical energy loss in PSII, that is the quantum yield for dissipation by down regulation in PSII	Calculated as $F_s/F_m{}' - F_s/F_m$
Φ_{NO}	The quantum yield of non-regulated energy loss in PSII, a loss process due to PSII inactivity	Calculated as F_s/F_m
$1 - q_P$	Excitation pressure of PSII, or the fraction of closed PSII reaction centers	Calculated as $1 - q_P$

2.5. Measurements of Leaf Absorbance Changes at 820 nm

A Hansatech $P_{700}+$ measuring system was employed to monitor light-induced changes in leaf absorbance at around 820 nm according to Havaux et al. [57], as described before [58]. The fraction of closed PSI reaction centers (B_1) was calculated as: $B_1 = \Delta S/(\Delta S)max = (Rfr - R')/(Rfr - R)$.

2.6. Statistical Analysis

Data are presented as the mean $\pm$ SD. Statistical analysis was performed using the Student's *t*-test. Differences were considered statistically significant at $p < 0.05$.

3. Results

3.1. Allocation of the Absorbed Light Energy in PSII under Normal Growth and Al^{3+} Exposure

Under control growth conditions at pH 6.5, the durum wheat 'Appulo E' presented higher effective quantum yield of photochemical energy conversion in PSII (Φ_{PSII}) (Figure 2a) and lower quantum yield of regulated non-photochemical energy loss in PSII (Φ_{NPQ}) (Figure 2b), with no difference in the quantum yield of non-regulated energy loss in PSII (Φ_{NO}) (Figure 3a), compared with triticale 'Dada'. Under 148 µM Al at pH 4.5 the triticale 'Dada' had higher Φ_{PSII} and lower Φ_{NPQ} (Figure 2), but higher Φ_{NO} (that did not differ from control conditions) (Figure 3a), than the durum wheat 'Appulo E'. However, 'Appulo E' due to the efficient photoprotective mechanism, that is the quantum yield for dissipation by down regulation in PSII, possessed lower Φ_{NO} even though from control conditions (Figure 3a).

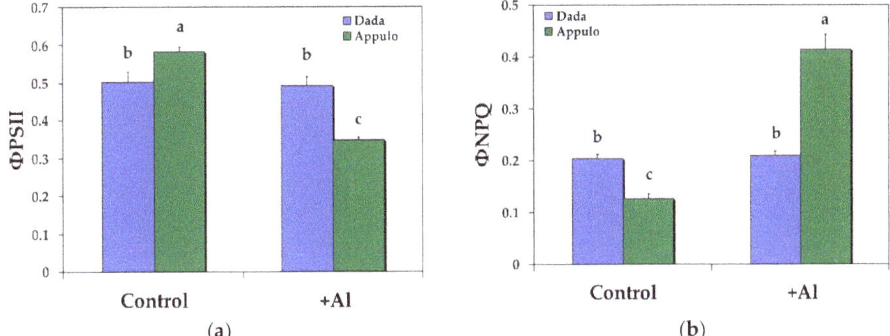

(a)   (b)

Figure 2. Changes in the balance between light capture and energy use in the triticale 'Dada' and the durum wheat 'Appulo E'; (**a**) the quantum efficiency of photosystem II (PSII) photochemistry (photochemical utilization) (Φ_{PSII}); and (**b**) the quantum yield for dissipation by down regulation in PSII (regulated heat dissipation, a loss process serving for protection) (Φ_{NPQ}); under normal growth conditions (control) and under Al^{3+} exposure (+Al). Error bars on columns are standard deviations based on five leaves from five plants. Columns with different letters are statistically different ($p < 0.05$).

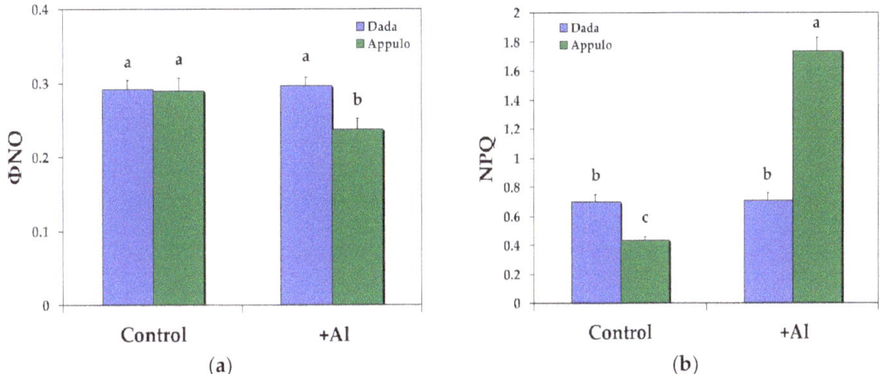

(a)   (b)

Figure 3. Changes in the quantum yield of non-regulated energy dissipated in PSII (non-regulated heat dissipation, a loss process due to PSII Inactivity) (Φ_{NO}) (**a**), and changes in non photochemical fluorescence quenching (NPQ) (**b**); in the triticale 'Dada' and the durum wheat 'Appulo E', under normal growth conditions (control) and under Al^{3+} exposure (+Al). Error bars on columns are standard deviations based on five leaves from five plants. Columns with different letters are statistically different ($p < 0.05$).

3.2. Non-Photochemical Quenching under Normal Growth and Al^{3+} Exposure

The triticale 'Dada' had higher non-photochemical fluorescence quenching (NPQ) under control growth conditions (pH 6.5) than the durum wheat 'Appulo E' but under 148 µM Al at pH 4.5 it was the reverse (Figure 3b).

3.3. Electron Transport Rate and the Redox State of PSII under Normal Growth and Al^{3+} Exposure

Under control growth conditions the durum wheat 'Appulo E' presented higher electron transport rate (ETR) (Figure 4a) and a more oxidized redox state of PSII (q_P) (Figure 4b), than the triticale 'Dada'. Under Al exposure the triticale 'Dada' had higher ETR than the durum wheat 'Appulo E' (Figure 4a), but the same redox state of PSII (q_P) (Figure 4b) with durum wheat 'Appulo E'.

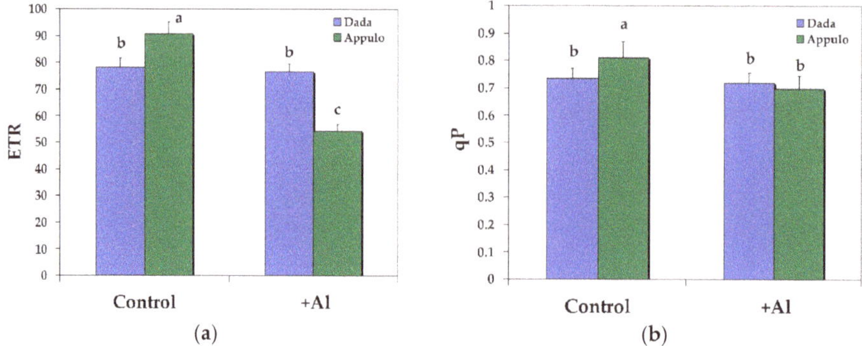

(a)   (b)

Figure 4. Changes in the relative PSII electron transport rate (ETR) (**a**); and changes in the photochemical fluorescence quenching, that is the relative reduction state of Q_A, reflecting the fraction of open PSII reaction centers (q_P) (**b**); in the triticale 'Dada' and the durum wheat 'Appulo E', under normal growth conditions (control) and under Al^{3+} exposure (+Al). Error bars on columns are standard deviations based on five leaves from five plants. Columns with different letters are statistically different ($p < 0.05$).

3.4. The Maximum PSII Quantum Efficiency (F_v/F_m) and PSII Maximum Efficiency in Light (F_v'/F_m') under Normal Growth and Al^{3+} Exposure

The maximum quantum efficiency of PSII (F_v/F_m) under normal growth conditions was higher in the durum wheat 'Appulo E', but under 148 µM Al it was higher in the triticale 'Dada' (Figure 5a). The maximum efficiency of PSII in the light (F_v'/F_m') was similar under control growth conditions (Figure 5b), but under 148 µM Al it was higher in the triticale 'Dada' suggesting a higher quantum yield of the open, functional reaction centers, than in the durum wheat 'Appulo E' (Figure 5b).

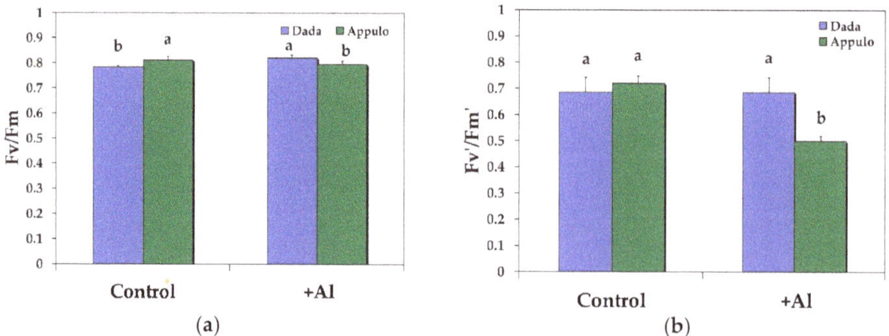

(a)   (b)

Figure 5. The maximum quantum efficiency of PSII (F_v/F_m) (**a**); and the maximum efficiency of PSII in the light (F_v'/F_m') (**b**); in the triticale 'Dada' and the durum wheat 'Appulo E', under normal growth conditions (control) and under Al^{3+} exposure (+Al). Error bars on columns are standard deviations based on five leaves from five plants. Columns with different letters are statistically different ($p < 0.05$).

3.5. Oxidative Damage under Normal Growth and Al^{3+} Exposure

Under Al exposure the level of lipid peroxidation measured as malondialdehyde (MDA) content and expressed as nmol (MDA) g^{-1} fresh weight increased compared with control growth conditions, but it was the same in both the triticale 'Dada' and the durum wheat 'Appulo E', while under normal growth conditions it was higher in 'Dada' (Figure 6).

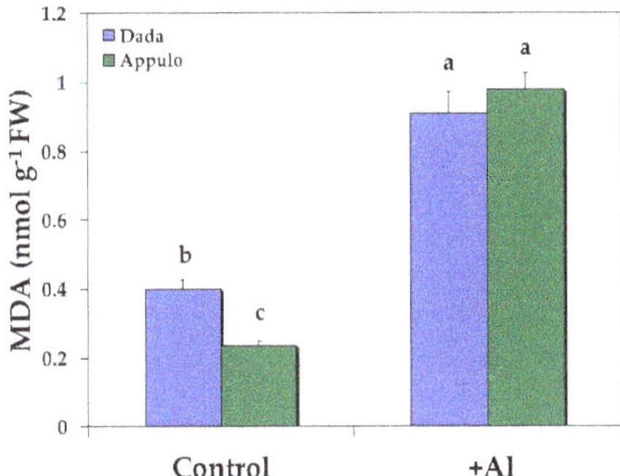

Figure 6. Changes in the level of lipid peroxidation measured as malondialdehyde (MDA) content and expressed as nmol (MDA) g^{-1} fresh weight in the triticale 'Dada' and the durum wheat 'Appulo E', under normal growth conditions (control) and under Al^{3+} exposure (+Al). Error bars on columns are standard deviations based on five leaves from five plants. Columns with different letters are statistically different ($p < 0.05$).

3.6. Excitation Pressure in PSI and PSII under Normal Growth and Al^{3+} Exposure

The fraction of closed PSI reaction centers (B$_1$) or PSI excitation pressure under both control growth conditions and Al exposure was higher in 'Appulo E' (Table 2), while the fraction of closed PSII reaction centers (PSII excitation pressure) under control growth conditions was higher in the triticale 'Dada', but under Al exposure was higher in 'Appulo E' (Table 2). Under 148 µM Al, PSII excitation pressure in both triticale 'Dada' and durum wheat 'Appulo E' was higher than PSI excitation pressure (Table 2).

Table 2. PSI and PSII excitation pressure in the triticale 'Dada' and the durum wheat 'Appulo E', under normal growth conditions, under Al^{3+} exposure, and the percentage change.

Chlorophyll Fluorescence Parameter	Control Growth	+148 µM Al	Change %
B1 (excitation pressure in PSI) 'Dada'	0.259	0.265	+2.3
B1 (excitation pressure in PSI) 'Appulo E'	0.280	0.339	+21.0
1-qp (excitation pressure in PSII) 'Dada'	0.266	0.281	+5.6
1-qp (excitation pressure in PSII) 'Appulo E'	0.188	0.303	+61.2

4. Discussion

In a hydroponic solution as summarized by Famoso et al. [2], Al may be found either (a) as free Al^{3+}, that actively inhibits root growth; (b) precipitated with other elements and essentially non-toxic to plant growth; (c) different hydroxyl Al monomers also non-toxic to roots [62]; or (d) complexed with other elements in an equilibrium between its active and inactive states. Thus, the degree of Al toxicity to plants is primarily related to the activity of free Al^{3+} in solution [63]. In our experiment, according to the GEOCHEM-EZ speciation program [59], the free Al^{3+} activities in the nutrient solutions were calculated to be 16.8 µM.

The significant lower quantum efficiency of PSII photochemistry in 'Dada' (Φ_{PSII}) under control growth conditions (Figure 2a) was compensated by a significant higher regulated heat dissipation, a loss process serving for protection (Φ_{NPQ}) (Figure 2b), that was sufficient enough to retain the same quantum yield of non-regulated energy dissipated in PSII (Φ_{NO}) in both 'Dada' and 'Appulo E' (Figure 3a). Under Al exposure we observed a reverse situation, with the significant higher photoprotective heat dissipation (Φ_{NPQ}) in 'Appulo E' (Figure 2b) not only to compensate the significant lower quantum efficiency of PSII photochemistry (Φ_{PSII}) (Figure 2a), but even more, to lower the quantum yield of non-regulated energy dissipated in PSII (Φ_{NO}) compared to 'Dada' (Figure 3a).

The most vulnerable component of the photosynthetic machinery to abiotic stresses is considered to be PSII [64]. However, despite the fact that PSI was shown to be more resistant to mild water deficit than PSII, it was heavily damaged by prolonged water deficit [64]. PSI is impaired when electron flow from PSII to PSI exceeds the capability of PSI electron carriers to manage the electrons [65,66]. Proton gradient (ΔpH)-dependent slow-down of electron transfer from PSII to PSI protects PSI from excess electrons [66]. As occurred in our experiment, PSI in Appulo E was more inhibited under control growth conditions than in Dada. During control growth conditions the high activity of PSII in Appulo E led to a rather higher electron flow to PSI, causing probably the formation of ROS within PSI complex [67,68], which induced a higher PSI excitation pressure in Appulo E than in Dada (Table 2) that presented a lower PSII photochemistry (Figure 2a) and lower PSI excitation pressure (Table 2). This higher PSI photoinhibition in Appulo E than in Dada under control growth conditions was alleviated by the absence of PSII photoinhibition in Appulo E as indicated by the F_v/F_m value (Figure 5). The absence of the photoprotective mechanism of NPQ in Appulo E under control growth conditions (Figure 3b), that slows-down the electron transfer from PSII to PSI, could not protect PSI from excess electrons. However, this absence of the photoprotective mechanism of NPQ in Appulo E (Figure 3b) did not cause any problem on the fraction of open PSII reaction centers (Figure 4b). Thus, under control growth conditions Appulo E shows lower PSII excitation pressure and a better PSII function, despite a higher PSI photoinhibition. Hence, regardless of the ROS formation within PSI complex in Appulo E, the level of lipid peroxidation, measured by MDA accumulation that reflects ROS formation and corresponds to oxidative damage, was shown to be less than in Dada, under control growth conditions (Figure 6). Current evidence suggests that ROS production can serve as the signal that triggers the expression of genes that may serve to alleviate electron pressure on the reducing side of PSI [69].

After Al exposure, the electron flow from PSII to PSI in 'Appulo E' was suppressed, but excitation pressure was increased in both photosystems, although more in PSII. This slow-down of electron transfer from PSII to PSI in 'Appulo E' protects PSI from excess electrons. A proper regulation of ETR is crucial in the protection of PSI against photoinhibition [70]. However, PSI photoinhibition may represent a kind of protective mechanism against over-reduction of PSI acceptor side, diminishing creation of huge amount of ROS and avoiding extensive cell injury [71,72]. The controlled photoinhibition of PSII in 'Appulo E', under Al exposure, as indicated by the F_v/F_m value (Figure 5a), was also able to protect PSI from permanent photodamage [66].

The high excitation pressure in PSII (1 − qp) under Al exposure, observed in 'Appulo E' (Table 2), indicates an imbalance between energy supply and demand [73]. However, the significant increase of NPQ processes in PSII (Figure 3b), that reflects the dissipation of excess excitation energy in the form of harmless heat [47,51,52,74], seems that protected 'Appulo E' plants under Al exposure from the destructive effects of ROS. It appears that under Al exposure, NPQ increase in PSII was sufficient enough to protect 'Appulo E' plants from ROS production since the quantum yield of non-regulated non-photochemical energy loss (Φ_{NO}) decreased significantly (Figure 3a), thus exhibited lower singlet oxygen ($^1O_2$) production. The quantum yield of non-regulated non-photochemical energy loss (Φ_{NO}) consists of chlorophyll fluorescence internal conversions and intersystem crossing, which leads to the formation of $^1O_2$ via the triplet state of chlorophyll (^{3}chl*) [75–77]. The increased NPQ in 'Appulo E' under Al exposure (Figure 3b) was also capable to keep the same fraction of open reaction centers as in Dada (Figure 4b) and also the same level of lipid peroxidation (Figure 6), thus the same degree of oxidative damage. The photoprotective mechanism of NPQ can divert absorbed light to other processes, such as thermal dissipation, preventing the photosynthetic apparatus from oxidative damage [47,48,78–81].

5. Conclusions

In conclusion, we confirmed that the triticale cv. 'Dada' was more tolerant to Al phytotoxicity than durum wheat 'Appulo E', as reflected by the better PSII functionality under Al acidic conditions. However, under normal growth conditions ($-$Al, pH 6.5), durum wheat 'Appulo E' displayed a better PSII functionality. Yet, under Al exposure, PSII was more affected than PSI from Al^{3+} phytotoxicity in both cultivars.

Author Contributions: G.O. and M.M. conceived and designed the experiments and analyzed the data; J.M. and I.S. performed the experiments and analyzed the data; M.M. wrote the paper; all authors review and approved the manuscript.

Funding: This research received no external funding.

Conflicts of Interest: The authors declare no conflict of interest.

References

1.   Wolt, J. Applications to Environmental Science and Agriculture. In *Soil Solution Chemistry*; John Wiley & Sons: New York, NY, USA, 1994.

2.   Famoso, A.N.; Clark, R.T.; Shaff, J.E.; Craft, E.; McCouch, S.R.; Kochian, L.V. Development of a novel aluminum tolerance phenotyping platform used for comparisons of cereal aluminum tolerance and investigations into rice aluminum tolerance mechanisms. *Plant Physiol.* **2010**, *153*, 1678–1691. [CrossRef] [PubMed]

3.   Kinraide, T.B. Identity of the rhizotoxic aluminium species. *Plant Soil* **1991**, *134*, 167–178. [CrossRef]

4.   Eleftheriou, E.P.; Moustakas, M.; Fragiskos, N. Aluminate-induced changes in morphology and ultrastructure of Thinopyrum roots. *J. Exp. Bot.* **1993**, *44*, 427–436. [CrossRef]

5.   Rengel, Z. Role of calcium in aluminium toxicity. *New Phytol.* **1992**, *121*, 499–513. [CrossRef]

6.   Moustakas, M.; Ouzounidou, G.; Lannoye, R. Aluminum effects on photosynthesis and elemental uptake in an aluminum-tolerant and non-tolerant wheat cultivar. *J. Plant Nutr.* **1995**, *18*, 669–683. [CrossRef]

7.   Barceló, J.; Poschenrieder, C. Fast root growth responses, root exudates, and internal detoxification as clues to the mechanisms of aluminium toxicity and resistance: A review. *Environ. Exp. Bot.* **2002**, *48*, 75–92. [CrossRef]

8.   Tamás, L.; Simonovicová, M.; Huttová, J.; Mistrík, I. Aluminium stimulated hydrogen peroxide production of germinating barley seeds. *Environ. Exp. Bot.* **2004**, *51*, 281–288. [CrossRef]

9.   Doncheva, S.; Amenos, M.; Poschenrieder, C.; Barcelo, J. Root cell patterning: A primary target for aluminium toxicity in maize. *J. Exp. Bot.* **2005**, *56*, 1213–1220. [CrossRef] [PubMed]

10.   Poschenrieder, C.; Gunsé, B.; Corrales, I.; Barceló, J. A glance into aluminum toxicity and resistance. *Sci. Total Environ.* **2008**, *40*, 356–368. [CrossRef] [PubMed]

11.   Chen, L.S.; Qi, Y.P.; Jiang, H.X.; Yang, L.T.; Yang, G.H. Photosynthesis and photoprotective systems of plants in response to aluminum toxicity. *Afr. J. Biotechnol.* **2010**, *9*, 9237–9247.

12.   Giannakoula, A.; Moustakas, M.; Mylona, P.; Papadakis, I.; Yupsanis, T. Aluminium tolerance in maize is correlated with increased levels of mineral nutrients, carbohydrates and proline and decreased levels of lipid peroxidation and Al accumulation. *J. Plant Physiol.* **2008**, *165*, 385–396. [CrossRef] [PubMed]

13.   Giannakoula, A.; Moustakas, M.; Syros, T.; Yupsanis, T. Aluminium stress induces up-regulation of an efficient antioxidant system in the Al-tolerant maize line but not in the Al-sensitive line. *Environ. Exp. Bot.* **2010**, *67*, 487–494. [CrossRef]

14.   Zelinová, V.; Halušková, L.; Huttová, J.; Illéš, P.; Mistrík, I.; Valentovičová, K.; Tamás, L. Short-term aluminium-induced changes in barley root tips. *Protoplasma* **2011**, *248*, 523–530. [CrossRef] [PubMed]

15.   Li, Z.; Xing, F.; Xing, D. Characterization of target site of aluminum phytotoxicity in photosynthetic electron transport by fluorescence techniques in tobacco leaves. *Plant Cell Physiol.* **2012**, *53*, 1295–1309. [CrossRef] [PubMed]

16.   Kochian, L.V.; Piñeros, M.A.; Jiping, L.; Magalhaes, J.V. Plant adaptation to acid soils: The molecular basis for crop aluminum resistance. *Ann. Rev. Plant Biol.* **2015**, *66*, 571–598. [CrossRef] [PubMed]

17. Zhu, H.; Wang, H.; Zhu, Y.; Zou, J.; Zhao, F.J.; Huang, C.F. Genome-wide transcriptomic and phylogenetic analyses reveal distinct aluminum-tolerance mechanisms in the aluminum-accumulating species buckwheat (*Fagopyrum tataricum*). *BMC Plant Biol.* **2015**, *15*, 16. [CrossRef] [PubMed]

18. Fan, W.; Xu, J.M.; Lou, H.Q.; Xiao, C.; Chen, W.W.; Yang, J.L. Physiological and molecular analysis of aluminium-induced organic acid anion secretion from grain amaranth (*Amaranthus hypochondriacus* L.) roots. *Int. J. Mol. Sci.* **2016**, *17*, 608. [CrossRef] [PubMed]

19. Li, H.; Yang, L.T.; Qi, Y.P.; Guo, P.; Lu, Y.B.; Chen, L.S. Aluminum toxicity-induced alterations of leaf proteome in two citrus species differing in aluminum tolerance. *Int. J. Mol. Sci.* **2016**, *17*, 1180. [CrossRef] [PubMed]

20. Singh, S.; Tripathi, D.K.; Singh, S.; Sharma, S.; Dubey, N.K.; Chauhan, D.K.; Vaculik, M. Toxicity of aluminium on various levels of plant cells and organism: A review. *Environ. Exp. Bot.* **2017**, *137*, 177–193. [CrossRef]

21. Foy, C.D. Plant adaptation to acid, aluminium-toxic soils. *Commun. Soil Sci. Plant Anal.* **1988**, *19*, 959–987. [CrossRef]

22. Garcia-Oliveira, A.L.; Martins-Lopes, P.; Tolrà, R.; Poschenrieder, C.; Guedes-Pinto, H.; Benito, C. Differential physiological responses of Portuguese bread wheat (*Triticum aestivum* L.) genotypes under aluminium stress. *Diversity* **2016**, *8*, 26. [CrossRef]

23. Moustakas, M.; Yupsanis, T.; Symeonidis, L.; Karataglis, S. Aluminum toxicity effects on durum wheat cultivars. *J. Plant Nutr.* **1992**, *15*, 627–638. [CrossRef]

24. Bona, L.; Wright, R.J.; Baligar, V.C.; Matuz, J. Screening wheat and other small grains for acid soil tolerance. *Landsc. Urban Plan.* **1993**, *27*, 175–178. [CrossRef]

25. Moustakas, M.; Ouzounidou, G.; Lannoye, R. Rapid screening for aluminum tolerance in cereals by use of the chlorophyll fluorescence test. *Plant Breed.* **1993**, *111*, 343–346. [CrossRef]

26. Von Uexküll, H.R.; Mutert, E. Global extent, development and economic impact of acid soils. *Plant Soil* **1995**, *171*, 1–15. [CrossRef]

27. Ma, J.F.; Chen, Z.C.; Shen, R.F. Molecular mechanisms of Al tolerance in gramineous plants. *Plant Soil* **2014**, *381*, 1–12. [CrossRef]

28. Allen, A.M.; Barker, G.L.A.; Berry, S.T.; Coghill, J.A.; Gwilliam, R.; Kirby, S.; Robinson, P.; Brenchley, R.C.; D'Amore, R.; McKenzie, N.; et al. Transcript-specific, single-nucleotide polymorphism discovery and linkage analysis in hexaploid bread wheat (*Triticum aestivum* L.). *Plant Biotechnol. J.* **2011**, *9*, 1086–1099. [CrossRef] [PubMed]

29. Anderson, J.M. Changing concepts about the distribution of photosystems I and II between grana-appressed and stroma-exposed thylakoid membranes. *Photosynth. Res.* **2002**, *73*, 157–164. [CrossRef] [PubMed]

30. Apostolova, E.L.; Dobrikova, A.G.; Ivanova, P.I.; Petkanchin, I.B.; Taneva, S.G. Relationship between the organization of the PSII supercomplex and the functions of the photosynthetic apparatus. *J. Photochem. Photobiol. B.* **2006**, *83*, 114–122. [CrossRef] [PubMed]

31. Mustárdy, L.; Garab, G. Granum revisited. A three-dimensional model-where things fall into place. *Trends Plant Sci.* **2003**, *8*, 117–122. [CrossRef]

32. Trissl, H.W.; Wilhelm, C. Why do thylakoid membranes from higher plants form grana stacks? *Trends Biochem. Sci.* **1993**, *18*, 415–419. [CrossRef]

33. Edwards, G.E.; Walker, D.A. *C3, C4: Mechanisms, and Cellular and Environmental Regulation, of Photosynthesis*; Blackwell Scientific: London/Oxford, UK, 1983.

34. Huang, W.; Zhang, S.B.; Xu, J.C.; Liu, T. Plasticity in roles of cyclic electron flow around photosystem I at contrasting temperatures in the chilling-sensitive plant *Calotropis gigantea*. *Environ. Exp. Bot.* **2017**, *141*, 145–153. [CrossRef]

35. Moustakas, M.; Ouzounidou, G.; Eleftheriou, E.P.; Lannoye, R. Indirect effects of aluminium stress on the function of the photosynthetic apparatus. *Plant Physiol. Biochem.* **1996**, *34*, 553–560.

36. Moustakas, M.; Eleftheriou, E.P.; Ouzounidou, G. Short-term effects of aluminium at alkaline pH on the structure and function of the photosynthetic apparatus. *Photosynthetica* **1997**, *34*, 169–177. [CrossRef]

37. Chen, L.S.; Qi, Y.P.; Liu, X.H. Effects of aluminum on light energy utilization and photoprotective systems in citrus leaves. *Ann. Bot.* **2005**, *96*, 35–41. [CrossRef] [PubMed]

38. Jiang, H.X.; Chen, L.S.; Zheng, J.G.; Han, S.; Tang, N.; Smith, B.R. Aluminum-induced effects on photosystem II photochemistry in citrus leaves assessed by the chlorophyll a fluorescence transient. *Tree Physiol.* **2008**, *28*, 1863–1871. [CrossRef] [PubMed]

39. Inostroza-Blancheteau, C.; Reyes-Díaz, M.; Aquea, F.; Nunes-Nesi, A.; Alberdi, M.; Arce-Johnson, P. Biochemical and molecular changes in response to aluminium-stress in highbush blueberry (*Vaccinium corymbosum* L.). *Plant Physiol. Biochem.* **2011**, *49*, 1005–1012. [CrossRef] [PubMed]

40. Hasni, I.; Hamdani, S.; Carpentier, R. Destabilization of the Oxygen Evolving Complex of Photosystem II by Al^{3+}. *Photochem. Photobiol.* **2013**, *89*, 1135–1142. [CrossRef] [PubMed]

41. Moustakas, M.; Ouzounidou, G. Increased non-photochemical quenching in leaves of aluminium-stressed wheat plants is due to Al^{3+}-induced elemental loss. *Plant Physiol. Biochem.* **1994**, *32*, 527–532.

42. Hasni, I.; Yaakoubi, H.; Hamdani, S.; Tajmir-Riahi, H.A.; Carpentier, R. Mechanism of interaction of Al^{3+} with the proteins composition of photosystem II. *PLoS ONE* **2015**, *10*, e0120876. [CrossRef] [PubMed]

43. Moustaka, J.; Ouzounidou, G.; Bayçu, G.; Moustakas, M. Aluminum resistance in wheat involves maintenance of leaf Ca^{2+} and Mg^{2+} content, decreased lipid peroxidation and Al accumulation, and low photosystem II excitation pressure. *BioMetals* **2016**, *29*, 611–623. [CrossRef] [PubMed]

44. Hasni, I.; Msilini, N.; Hamdani, S.; Tajmir-Riahi, H.A.; Carpentier, R. Characterization of the structural changes and photochemical activity of photosystem I under Al^{3+} effect. *J. Photochem. Photobiol. B* **2015**, *149*, 292–299. [CrossRef] [PubMed]

45. Demmig-Adams, B.; Stewart, J.J.; Adams, W.W., III. Multiple feedbacks between chloroplast and whole plant in the context of plant adaptation and acclimation to the environment. *Philos. Trans. R. Soc. B* **2014**, *369*, 20130244. [CrossRef] [PubMed]

46. Watanabe, C.K.A.; Yamori, W.; Takahashi, S.; Terashima, I.; Noguchi, K. Mitochondrial alternative pathway-associated photoprotection of photosystem II is related to the photorespiratory pathway. *Plant Cell Physiol.* **2016**, *57*, 1426–1431. [CrossRef] [PubMed]

47. Niyogi, K.K. Photoprotection revisited: Genetic and molecular approaches. *Annu. Rev. Plant Physiol. Plant Mol. Biol.* **1999**, *50*, 333–359. [CrossRef] [PubMed]

48. Demmig-Adams, B.; Adams, W.W., III. Photoprotection in an ecological context: The remarkable complexity of thermal energy dissipation. *New Phytol.* **2006**, *172*, 11–21. [CrossRef] [PubMed]

49. Takahashi, S.; Murata, N. How do environmental stresses accelerate photoinhibition? *Trends Plant Sci.* **2008**, *13*, 178–182. [CrossRef] [PubMed]

50. Moustaka, J.; Panteris, E.; Adamakis, I.D.S.; Tanou, G.; Giannakoula, A.; Eleftheriou, E.P.; Moustakas, M. High anthocyanin accumulation in poinsettia leaves is accompanied by thylakoid membrane unstacking, acting as a photoprotective mechanism, to prevent ROS formation. *Environ. Exp. Bot.* **2018**, *154*, 44–55. [CrossRef]

51. Muller, P.; Li, X.P.; Niyogi, K.K. Non-photochemical quenching. A response to excess light energy. *Plant Physiol.* **2001**, *125*, 1558–1566. [CrossRef] [PubMed]

52. Ruban, A.V. Non photochemical chlorophyll fluorescence quenching: Mechanism and effectiveness in protecting plants from photodamage. *Plant Physiol.* **2016**, *170*, 1903–1916. [CrossRef] [PubMed]

53. Sun, Y.; Geng, Q.; Du, Y.; Yang, X., Zhai, H. Induction of cyclic electron flow around photosystem I during heatstress in grape leaves. *Plant Sci.* **2017**, *256*, 65–71. [CrossRef] [PubMed]

54. Guidi, L.; Calatayud, A. Non-invasive tools to estimate stress-induced changes in photosynthetic performance in plants inhabiting Mediterranean areas. *Environ. Exp. Bot.* **2014**, *103*, 42–52. [CrossRef]

55. Gorbe, E.; Calatayud, A. Applications of chlorophyll fluorescence imaging technique in horticultural research: A review. *Sci. Hortic.* **2012**, *138*, 24–35. [CrossRef]

56. Kalaji, H.M.; Oukarroum, A.A.; Alexandrov, V.; Kouzmanova, M.; Brestic, M.; Zivcak, M.; Samborska, I.A.; Cetner, M.D.; Allakhverdiev, S.I. Identification of nutrient deficiency in maize and tomato plants by in vivo chlorophyll a fluorescence measurements. *Plant Physiol. Biochem.* **2014**, *81*, 16–25. [CrossRef] [PubMed]

57. Havaux, M.; Greppin, H.; Strasser, R.J. Functioning of photosystems I and II in pea leaves exposed to heat stress in the presence or absence of light. Analysis using in vivo fluorescence, absorbance, oxygen and photoacoustic measurements. *Planta* **1991**, *186*, 88–98. [CrossRef] [PubMed]

58. Ouzounidou, G.; Moustakas, M.; Strasser, R.J. Sites of action of copper in the photosynthetic apparatus of maize leaves. *Aust. J. Plant Physiol.* **1997**, *24*, 81–90. [CrossRef]

59. Shaff, J.E.; Schultz, B.A.; Craft, E.J.; Clark, R.T.; Kochian, L.V. GEOCHEM-EZ: A chemical speciation program with greater power and flexibility. *Plant Soil* **2010**, *330*, 207–214. [CrossRef]

60. Moustakas, M.; Sperdouli, I.; Kouna, T.; Antonopoulou, C.I.; Therios, I. Exogenous proline induces soluble sugar accumulation and alleviates drought stress effects on photosystem II functioning of *Arabidopsis thaliana* leaves. *Plant Growth Regul.* **2011**, *65*, 315–325. [CrossRef]

61. Heath, R.L.; Packer, L. Photoperoxidation in isolated chloroplasts. *Arch. Biochem. Biophys.* **1968**, *125*, 189–198. [CrossRef]

62. Parker, D.R.; Kinraide, T.B.; Zelazny, L.W. Aluminum speciation and phytotoxicity in dilute hydroxy-aluminum solutions. *Soil Sci. Soc. Am. J.* **1988**, *52*, 438–444. [CrossRef]

63. Kochian, L.V.; Hoekenga, O.A.; Piñeros, M.A. How do crop plants tolerate acid soils? Mechanisms of aluminum tolerance and phosphorous efficiency. *Ann. Rev. Plant Biol.* **2004**, *55*, 459–493. [CrossRef] [PubMed]

64. Zlobin, I.E.; Ivanov, Y.V.; Kartashov, A.V.; Sarvin, B.A.; Stavrianidi, A.N.; Kreslavski, V.D.; Kuznetsov, V.V. Impact of weak water deficit on growth, photosynthetic primary processes and storage processes in pine and spruce seedlings. *Photosynth. Res.* **2018**. [CrossRef] [PubMed]

65. Suorsa, M.; Jarvi, S.; Grieco, M.; Nurmi, M.; Pietrzykowska, M.; Rantala, M.; Kangasjarvi, S.; Paakkarinen, V.; Tikkanen, M.; Jansson, S.; et al. PROTON GRADIENT REGULATION5 is essential for proper acclimation of Arabidopsis photosystem I to naturally and artificially fluctuating light conditions. *Plant Cell* **2012**, *24*, 2934–2948. [CrossRef] [PubMed]

66. Tikkanen, M.; Mekala, N.R.; Aro, E.M. Photosystem II photoinhibition-repair cycle protects photosystem I from irreversible damage. *Biochim. Biophys. Acta* **2014**, *1837*, 210–215. [CrossRef] [PubMed]

67. Takagi, D.; Takumi, S.; Hashiguchi, M.; Sejima, T.; Miyake, C. Superoxide and singlet oxygen produced within the thylakoid membranes both cause photosystem I photoinhibition. *Plant Physiol.* **2016**, *171*, 1626–1634. [CrossRef] [PubMed]

68. Huang, W.; Yang, Y.J.; Hu, H.; Zhang, S.B. Moderate photoinhibition of photosystem II protects photosystem I from photodamage at chilling stress in tobacco leaves. *Front. Plant Sci.* **2016**, *7*, 182. [CrossRef] [PubMed]

69. Foyer, C.H. Reactive oxygen species, oxidative signaling and the regulation of photosynthesis. *Environ. Exp. Bot.* **2018**, *154*, 134–142. [CrossRef]

70. Brestic, M.; Zivcak, M.; Kunderlikova, K.; Sytar, O.; Shao, H.; Kalaji, H.M.; Allakhverdiev, S.I. Low PSI content limits the photoprotection of PSI and PSII in early growth stages of chlorophyll b-deficient wheat mutant lines. *Photosynth. Res.* **2015**, *125*, 151–166. [CrossRef] [PubMed]

71. Brestic, M.; Zivcak, M.; Kunderlikova, K.; Allakhverdiev, S.I. High temperature specifically affects the photoprotective responses of chlorophyll b-deficient wheat mutant lines. *Photosynth. Res.* **2016**, *130*, 251–266. [CrossRef] [PubMed]

72. Huang, W.; Yang, Y.J.; Hu, H.; Zhang, S.B. Responses of photosystem I compared with photosystem II to fluctuating light in the shade-establishing tropical tree species *Psychotria henryi*. *Front. Plant Sci.* **2016**, *7*, 1549. [CrossRef] [PubMed]

73. Takahashi, S.; Badger, M.R. Photoprotection in plants: A new light on photosystem II damage. *Trends Plant Sci.* **2011**, *16*, 53–60. [CrossRef] [PubMed]

74. Moustaka, J.; Tanou, G.; Adamakis, I.D.; Eleftheriou, E.P.; Moustakas, M. Leaf age dependent photoprotective and antioxidative mechanisms to paraquat-induced oxidative stress in *Arabidopsis thaliana*. *Int. J. Mol. Sci.* **2015**, *16*, 13989–14006. [CrossRef] [PubMed]

75. Apel, K.; Hirt, H. Reactive oxygen species: Metabolism, oxidative stress and signal transduction. *Ann. Rev. Plant Biol.* **2004**, *55*, 373–399. [CrossRef] [PubMed]

76. Gawroński, P.; Witoń, D.; Vashutina, K.; Bederska, M.; Betliński, B.; Rusaczonek, A.; Karpiński, S. Mitogen-activated protein kinase 4 is a salicylic acid-independent regulator of growth but not of photosynthesis in Arabidopsis. *Mol. Plant* **2014**, *7*, 1151–1166. [CrossRef] [PubMed]

77. Bayçu, G.; Gevrek-Kürüm, N.; Moustaka, J.; Csatári, I.; Rognes, S.E.; Moustakas, M. Cadmium-zinc accumulation and photosystem II responses of *Noccaea caerulescens* to Cd and Zn exposure. *Environ. Sci. Pollut. Res.* **2017**, *24*, 2840–2850. [CrossRef] [PubMed]

78. Moustaka, J.; Moustakas, M. Photoprotective mechanism of the non-target organism *Arabidopsis thaliana* to paraquat exposure. *Pest. Biochem. Physiol.* **2014**, *111*, 1–6. [CrossRef] [PubMed]

79. Agathokleous, E. Environmental hormesis, a fundamental non-monotonic biological phenomenon with implications in ecotoxicology and environmental safety. *Ecotoxicol. Environ. Saf.* **2018**, *148*, 1042–1053. [CrossRef]

80. Georgieva, K.; Doncheva, S.; Mihailova, G.; Petkova, S. Response of sun- and shade-adapted plants of *Haberlea rhodopensis* to desiccation. *Plant Growth Regul.* **2012**, *67*, 121–132. [CrossRef]
81. Jusovic, M.; Velitchkova, M.Y.; Misheva, S.P.; Börner, A.; Apostolova, E.L.; Dobrikova, A.G. Photosynthetic responses of a wheat mutant (*Rht-B1c*) with altered DELLA proteins to salt stress. *J. Plant Growth Regul.* **2018**, *37*, 645–656. [CrossRef]

 *materials*

Review

Fueling a Hot Debate on the Application of TiO$_2$ Nanoparticles in Sunscreen

Shweta Sharma [1,†], **Rohit K. Sharma** [2,†], **Kavita Gaur** [2,†], **José F. Cátala Torres** [2], **Sergio A. Loza-Rosas** [2], **Anamaris Torres** [3], **Manoj Saxena** [2], **Mara Julin** [4] and **Arthur D. Tinoco** [2,*]

[1] Department of Environmental Sciences, University of Puerto Rico Río Piedras, 17 AVE Universidad STE 1701, San Juan, PR 00925-2537, USA

[2] Department of Chemistry, University of Puerto Rico Río Piedras, 17 AVE Universidad STE 1701, San Juan, PR 00925-2537, USA

[3] Biochemistry & Pharmacology Department, San Juan Bautista School of Medicine, Caguas, PR 00726, USA

[4] Department of Chemistry, Syracuse University, Syracuse, NY 13244, USA

* Correspondence: atinoco9278@gmail.com; Tel.: +787-764-0000 (ext. 88566)

† These authors contributed equally to this work.

Received: 30 June 2019; Accepted: 17 July 2019; Published: 20 July 2019

Abstract: Titanium is one of the most abundant elements in the earth's crust and while there are many examples of its bioactive properties and use by living organisms, there are few studies that have probed its biochemical reactivity in physiological environments. In the cosmetic industry, TiO$_2$ nanoparticles are widely used. They are often incorporated in sunscreens as inorganic physical sun blockers, taking advantage of their semiconducting property, which facilitates absorbing ultraviolet (UV) radiation. Sunscreens are formulated to protect human skin from the redox activity of the TiO$_2$ nanoparticles (NPs) and are mass-marketed as safe for people and the environment. By closely examining the biological use of TiO$_2$ and the influence of biomolecules on its stability and solubility, we reassess the reactivity of the material in the presence and absence of UV energy. We also consider the alarming impact that TiO$_2$ NP seepage into bodies of water can cause to the environment and aquatic life, and the effect that it can have on human skin and health, in general, especially if it penetrates into the human body and the bloodstream.

Keywords: titanium dioxide; nanoparticles; solubility; toxicity; skin; safety

---

1. Introduction

Titanium is the ninth most abundant element in the earth's crust and is widely recognized for its strength, long-term endurance, and electronic properties, and for these reasons it is incorporated in many different materials [1]. The value of the metal has transcended to its successful use by humans for dental and orthopedic prosthetics [2,3] and in sunscreens as titanium dioxide (TiO$_2$) [4]. The metal, however, remains largely unappreciated for its biological importance despite many examples of its benefit to certain plants [5–7] and animals [8–10]. Even within the human body, there is strong evidence for a biological function—a structural templating role. Titanium features the property of osseointegration, a pioneering and serendipitous discovery made by Dr. Per-Ingvar Brånemark in the 1950s [11,12]. That is, the metal is able to integrate and be structurally accepted by bone without the requirement of soft tissue connection. For this reason, it is widely used in alloy form in different prosthetics. It essentially aids in the healing and regrowth of bones, and in many applications, substitutes for bones. In the context of titanium-containing prosthetics, osseointegration may be the result of the surface of the implant forming a layer of titanium oxide. This layer protects the implants from corrosion (an excellent feature for structural integrity) and favorably interacts with biomolecules of the body [13]. The surface becomes highly protein-covered due to strong protein affinity to the titanium oxide, a demonstration of excellent biocompatibility, and serves as a biomineralization template [14].

Human application of titanium in materials and skin products has long been driven by the belief that it is safe and is inert to biochemical reactivity. Recently though, increased levels and/or biotransformation of the metal within the human body has captured people's interest. The body's interaction with titanium-containing prosthetics can extend beyond a simple passive, biocompatible one. The metal from these materials demonstrates surprising reactivity in biological fluids, is able to be released and enter into the bloodstream in titanium (IV) (Ti(IV)) ion soluble form, and as TiO_2, is notoriously insoluble [15]. Older beliefs regarding titanium's inertness are likely due to overly simplistic stability studies of the metal (in pure or alloy form) performed in water at physiological pH values. Such solutions do not properly represent the diverse constituent of species in biological fluids and, thus, the contribution that biomolecules play in metal speciation in the body [16–18]. Biomolecules bind to titanium on implant surfaces or fragments dispersed due to wearing and can lead to its dissolution [19] and transportation throughout the body. That titanium leaches from implants and is present at significantly elevated levels in the blood of people with such implants [15] has led to growing concerns about the long-term stability of these products and their impediments to human health. Some of the potential problems reported are that the titanium can corrode and lead to implant breakage [20], can generate reactive oxygen species following release from implants [21], and can produce a type IV allergy toward the metal (rare) [22,23]. Severe health issues because of metal leaching from prosthetics have been reported particularly for cobalt, which has led to the formal medical term arthroprosthetic cobaltism [24–28]. Toxicological problems due to the leached soluble Ti(IV) ion form of the metal has been the subject of an extensive review by Piekoszewski et al. [29]. Another study determined the concentration of "leached" titanium in either soluble (using Ti(IV) tricitrate as an appropriate blood small-molecule model) or TiO_2 nanoparticle formulation that can lead to toxicity [19]. At concentrations ≥ 10 μg/mL, both formulations led to the significant antiproliferation of MC3T3 murine osteoblasts and human colorectal adenocarcinoma cell line HT29 [19]. Also at these levels (~200 μM), soluble Ti(IV) demonstrated cytotoxicity [19] most likely due to Ti(IV) binding and fragmentation of DNA [30]. Such concentrations largely exceed the concentrations of Ti typically found in the blood of people with Ti-containing implants (≤ 0.25 μM) [19], which suggests that toxicological concerns over leached elevated Ti levels may not be warranted except for possible localized high concentrations. Furthermore, we have recently proposed that citrate and the iron-transport protein serum transferrin may work in synergism in blood to regulate Ti(IV) ion uptake into cells and protect them from the cytotoxic properties of the metal [31–33].

An area where toxicological concerns regarding human application of Ti has not been as well explored is the use of TiO_2 in sunscreen. The function of sunscreen is to protect skin from harmful ultraviolet A (320–400 nm) and ultraviolet B (290–329 nm) radiation, which can cause mutations and metabolic effects [34,35]. UVB is particularly dangerous in long-term exposure because it is directly absorbed by DNA, giving rise to dimeric photoproducts between adjacent pyrimidine bases [36]. Active ingredients in sunscreen come in two forms, inorganic (mineral/physical blocker) and organic (chemical) filters. Inorganic filters like TiO_2 nanoparticles (NPs) display both light scattering (high refractive indices) and UV absorption properties [4]. In contrast, chemical filters in the form of organic compounds solely absorb UV radiation [37]. TiO_2 nanoparticles are widely used in sunscreen because of their excellent semiconducting properties, ease of processing, and the long-held belief that the material is biologically inert. Nonetheless, new considerations must be made regarding TiO_2 bioactivity especially in light of its seepage into bodies of water and the different routes by which it may enter the human body. This review will explore the potentially alarming impact that TiO_2 can have on aquatic life and on human health by evaluating its physicochemical and biochemical properties and identifying the molecular mechanisms that can affect its stability, solubility, and reactivity in living organisms (Figure 1).

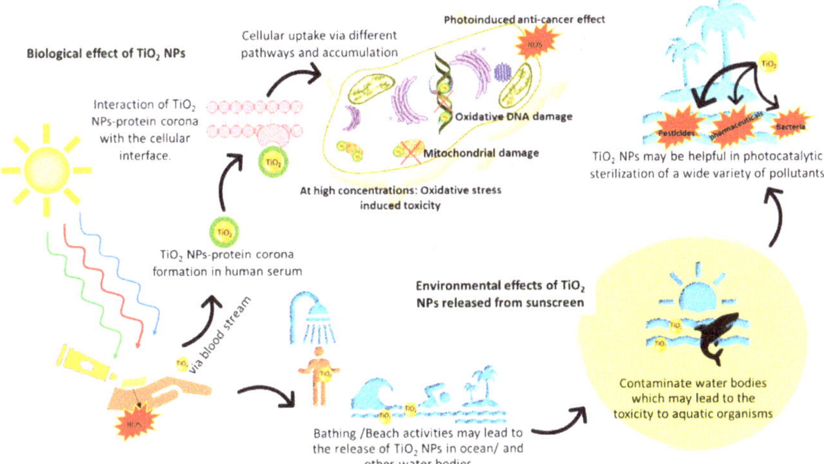

Figure 1. Environmental and biological effects of the application of TiO$_2$ nanoparticles (NPs) in sunscreen. The NPs can pollute water bodies and possibly hurt aquatic life, but also serve a beneficial photocatalytic sterilization function. In humans, the NPs may translocate into the body. There is evidence for proteins forming a protein corona around the NPs and influencing their cellular uptake. There are several UV and non-UV debilitating cellular effects caused by TiO$_2$ NPs. In both water bodies and humans, NP solubilization can occur, which produces Ti(IV) ions (not depicted) and effects similar to the NPs.

2. Sunscreen Exploits the Semiconducting Property of TiO$_2$

In sunscreen, TiO$_2$ can exist in conventional (amorphous) or nanoparticle forms. The conventional form creates a milky white appearance that, while effective for UV scattering, can be aesthetically unpleasing. The nanoparticle form appears transparent, retains its scattering ability, and has the added bonus of greater relative surface area allowing superior UV-absorbing capacity [38]. For these reasons, nanoparticles are more commonly used today in sunscreen formulations.

The UV protection provided by TiO$_2$ in sunscreen stems from its function as a semiconducting material. TiO$_2$ is an intrinsic N-type semiconductor due to oxygen vacancies in its lattice [39]. It is generally characterized by the band gap energy of ~3.2 eV [40]. In its anatase form, the band gap corresponds to a wavelength of 387 nm and in its rutile form, it is 405 nm. Light at or below these wavelengths can excite electrons from the valence band to the conduction band (Figure 2) [41]. In sunscreen, TiO$_2$ NPs absorb the UV-radiation from the sun by promoting electrons from its valence band to its conduction band [42]. This process results in photogenerated holes in its valence band. The photogenerated holes and excited electrons (Equation (1)) can either recombine or migrate to the particle surface and participate in different redox processes that lead to the formation of reactive oxygen species (ROS) (Equations (2)–(5)) [42]. Being a powerful oxidant, the valence band holes primarily target moisture present on the surface (Equation (2)), which produces hydroxyl radicals. The conduction band electrons are good reductants and for them, oxygen present at the surface acts as a primary electron acceptor producing superoxide and eventually hydrogen peroxide (Equation (3)) [43]. The conduction band electrons could also be rapidly trapped at Ti(IV) sites and then react with oxygen, yielding superoxide and hydrogen peroxide (Equations (4) and (5)) [43,44].

$$TiO_2 + h\upsilon \longrightarrow \left(e_{CB}^- + h_{VB}^+\right) TiO_2 \tag{1}$$

$$h_{VB}^+ + H_2O \longrightarrow H^+ + \bullet OH \tag{2}$$

$$e_{CB}^{-} + O_2 \longrightarrow O_2^{\bullet -} \longrightarrow\longrightarrow\longrightarrow H_2O_2 \tag{3}$$

$$e_{CB}^{-} + \text{Ti(IV)} \longrightarrow \text{Ti(III)} \tag{4}$$

$$\text{Ti(III)} + O_2 \longrightarrow \text{Ti(IV)} + O_2^{\bullet -} \longrightarrow\longrightarrow\longrightarrow H_2O_2 \tag{5}$$

If left uncontrolled, the ROS formed can target different substrates and be responsible for biological impairments. They can cause oxidative stress within cells including DNA damage, modification of proteins and sensitive thiols, and trigger redox activity of copper and iron ions [42,45–48]. Within sunscreen, radical scavengers and antioxidants are included or are coated on the surface of TiO_2 to suppress TiO_2-induced ROS generation and prevent harm to human skin [49]. Due to the strong redox reactivity of TiO_2, there has been great interest in its use as a photocatalyst for energy and biomedical applications, namely as anticancer [50] and antibacterial agents [51] (See Section 4).

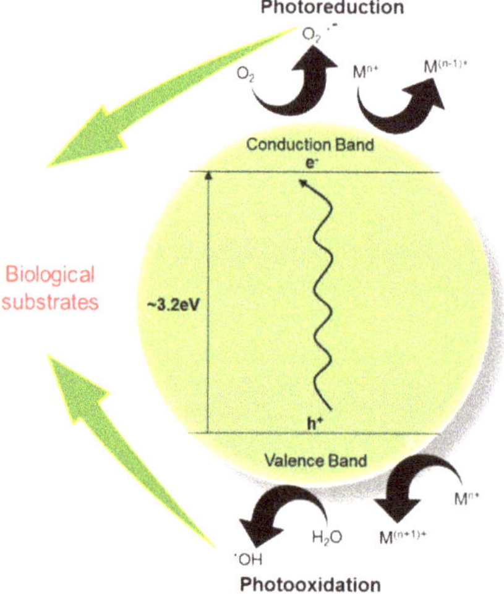

Figure 2. The semiconducting and photocatalytic properties of TiO_2 NPs.

3. The Biological Use and Solubility of TiO_2 and Its NPs

TiO_2 is structurally extremely stable especially in the anatase, brookite, and rutile crystalline forms [52], and exhibits very poor aqueous solubility [53]. These properties are key to the application of TiO_2 in sunscreen and other human cosmetics. Schmidt and Vogelsberger performed an extensive study to examine the aqueous solubility of TiO_2 in crystalline and amorphous hydrous forms under conditions that are physiologically relevant [54]. They found that crystalline forms are significantly less soluble under acidic pH and result in Ti(IV) concentrations of about 1 nanomolar in the pH 4 to 10 range. This finding suggests that TiO_2 NPs should be virtually insoluble in sunscreen particularly because of the water-resistant formulations of the sunscreen and, thus, not cause toxicity on account of dissolution-related phenomenon. Lack of solubility would retain the very high stability of these materials. It has long been thought that TiO_2 is biologically inert.

Several lines of evidence suggest that in a biological context, TiO_2 is active. In diatoms, an amorphous coating of it can be found on the SiO_2 crystalline lattice of the frustules. The diatoms are believed to take advantage of the photocatalytically activated antibacterial properties of TiO_2 to ward off predators [55]. TiO_2 NPs have demonstrated many benefits to plant growth, some of which appear to be species specific. They can increase seed germination rates and seedling growth, enhance root lengths, improve plant growth, and increase crop growth and yield [56]. They can also increase plant tolerance to abiotic and biotic stresses, including cold stress, heat stress, drought, and cadmium toxicity [56]. Additional benefits have been enhanced photosynthesis, increased chlorophyll content, and exploiting the photoactive antibacterial properties of TiO_2 to control bacterial and fungal pathogens in crop production [56]. A micro-X-ray absorption near edge structure (micro-XANES) study with a cucumber plant *Cucumis sativus (C. sativus)* showed a varied biodistribution of TiO_2 throughout the plant following treatment with a mixture of 19% rutile and 81% anatase [7]. In the xylem, the % of both rutile and anatase remained fairly consistent with the bulk material but within the phloem, TiO_2 was exclusively rutile. It was reasoned that the size of anatase restricted it from uptake beyond the roots. The TiO_2 improved the plant's growth possibly because of nitrogen activation [7] as a result of coordination to the metal.

Although there is no known function for Ti in bacteria, certain bacteria have the capacity to interact with the metal in TiO_2 form. The chemical proximity of Ti(IV) with iron(III) (Fe(III)) [33] could account for bacterial interaction with Ti as bacteria heavily depend on Fe for survival and have evolved numerous acquisition pathways for mobilizing and capturing the metal [57]. *Rhodococcus ruber (R. ruber)* GIN1, a Gram-positive species, strongly adheres to TiO_2 under a wide pH and temperature range [58]. A 52 kDa TiO_2-binding protein was isolated from the bacteria and identified to be a cell surface form of dihydrolipoamide dehydrogenase (rhDLDH) [59]. It adheres more effectively to the rutile than the anatase form [59]. A human homolog exists, which at pH 8.0, binds to TiO_2 and other metal oxides (ZnO, MgO, MnO, Al_2O_3, Fe_2O_3) [60]. The affinity is highest for TiO_2 and Fe_2O_3 [60]. The binding interaction between RhDLDH and hDLDH with metal oxides appears to be nonelectrostatic. A computational study was performed to identify the binding site for TiO_2 and this led to the determination of the involvement of a putative CHED motif (Cys, His, Glu, Asp) in both enzymes. CHED motifs are known to coordinately bind metal ions but how this motif coordinates to the Ti in TiO_2 is not yet known [60]. Very recently, hDLDH was coated onto the TiO_2 surface layer of a Ti-containing implant to enhance the osseointegration property of the implant [61]. The use of a coating of a human protein is expected to enhance the biocompatibility of the implant surface.

Other bacteria such as *Escherichia coli (E. coli)* and *Pseudomonas aeruginosa (P. aeruginosa)* have also been observed to interact with TiO_2 NPs with cell surface siderophores [62–64]. Siderophores are low molecular weight molecules that organisms (bacteria, yeast, algae, plants) synthesize and export to mobilize and/or sequester Fe(III) through different molecular pathways [57]. They consist of the functional groups catechols, hydroxamic acid, and α-hydroxy-carboxylic acids. The siderophore pyoverdine can enable *P. aeruginosa* to simultaneously bind to TiO_2 and iron oxide through direct coordination of the metal by its catechol moiety. It is speculated that the oxides provide a template for biofilm formation [62]. Siderophores can tightly bind Ti(IV) in ion form, producing coordination complexes comparable to their Fe(III) counterparts at physiologically relevant pH values (Figure 3). Enterobactin contains three catechol moieties that coordinate Ti(IV) in hexadentate fashion forming a 1:1 metal:ligand complex, the crystal structure of which has been reported [65]. Desferrioxamine B (DFOB), a trishydroxamic acid siderophore, also coordinates Ti(IV) in a hexadentate modality. Valentine et al. reported a density functional theory (DFT)-optimized structure for the 1:1 metal:ligand complex and explored its aqueous speciation over the pH 2 to 10 range [66]. Citric acid is considered a siderophore with α-hydroxy-carboxylic acid moieties [67]. There are many examples and structures of Ti(IV) citrate complexes, in which the ligand binds in bidentate fashion and can coordinatively saturate the metal [31,33,68–73]. It is important to note that citrate binds Fe(III) in a slightly different fashion by coordinating in tridentate mode using one of the carboxylic acid groups in the β position to provide the extra coordination site [67,74]. Valentine made the very important discovery that siderophores can dissolve TiO_2 by a coordination-induced mechanism [75]. The details of this study are pending but the work suggests that bacteria could mobilize Ti(IV) in ways comparable to its acquisition of Fe(III) and therefore transform the metal into a bioavailable species that may serve a function.

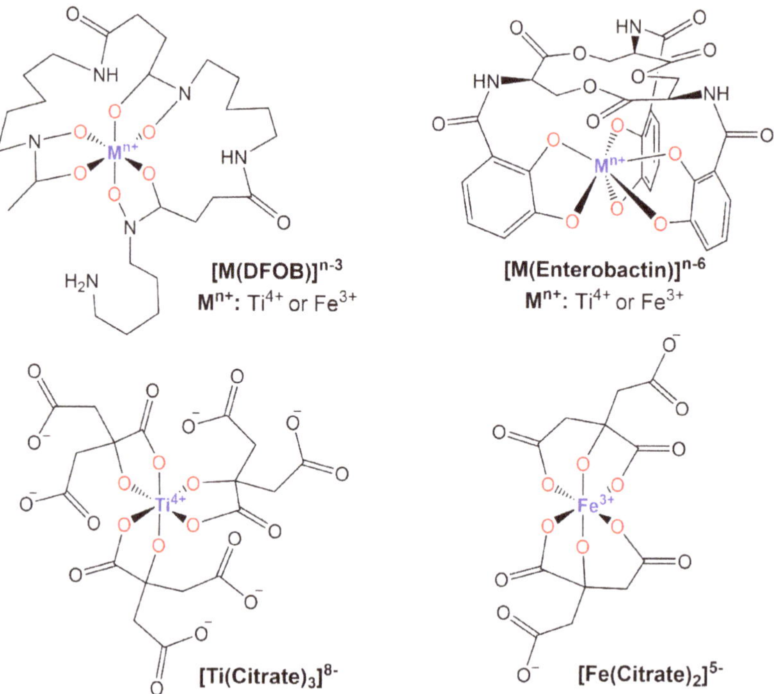

Figure 3. Coordination similarities of siderophore binding of Ti(IV) and Fe(III) at pH 7.4. The hexadentate siderophores, Desferrioxamine B (DFOB) and Enterobactin, form 1:1 metal:ligand complexes. Citrate coordinates Ti(IV) and Fe(III) in slightly different ways. Citrate serves as a tridentate ligand when coordinated to Fe(III) and as a bidentate ligand when coordinated to Ti(IV).

Biomolecular solubilization of TiO_2 suggests that the metal oxide might be far more soluble than previously considered in human products. Several skin care products, including sunscreen, contain hydroxyacids like salicylic acid and citric acid that can chelate Ti(IV) [31,33,68–73,76] and potentially lead to its dissolution. One study examined the solubility of TiO_2 in rutile and anatase form in an oil in water (o/w) weakly acidic emulsion that mimics cosmetic formulations. It was determined that after 1 to 2 days of 1 g TiO_2 suspended in 0.03% (w/w) citric acid mixed in the o/w emulsion, ~600 μM Ti(IV) was found present in solution. This is several orders of magnitude higher than the innate solubility of TiO_2. In human blood and synovial fluid (pH 7.4), citric acid is present as citrate in the concentration range of 100–200 μM. It is believed to contribute to the solubilization of some of the metal leached from Ti-containing implants by forming, at least transiently, the coordination complex Ti(IV) tricitrate. The complex is able to deliver Ti(IV) to the two metal binding sites of serum transferrin [31]. One citrate molecule can remain bound to the Ti(IV) at each site, enhancing the stabilization of the metal ion [31]. The interaction of citrate and sTf in regulating the blood speciation of Ti(IV) may account for the Ti blood levels of ≤0.25 μM in people with these implants, levels that are not expected to be toxic [19]. Citrate and sTf are hypothesized to engage in a synergistic molecular mechanism to decrease the cytotoxic properties of some anticancer Ti(IV) complexes by inducing their dissociation and scavenging the metal [31,77]. It is unknown whether Ti(IV) solubilization from skincare products, if it does occur in reality, could lead to elevated levels of the metal in the human body.

4. Applications of TiO$_2$ NPs Provide Further Insight into Its Functionality

TiO$_2$ NPs are one of the most manufactured nanomaterials worldwide with an estimated annual median of 3000 tons produced [78]. Of this total, 70%–80% of it is used in the cosmetic industries, which includes sunscreens [78]. There is a significant level of daily dietary intake of TiO$_2$ NPs in human beings because it is used as a whitening agent of certain foods [79]. Extensive production and growing applications are responsible for its exposure to the environment. In the United States, the daily consumption of TiO$_2$ has been estimated in the range of 0.2–2 mg/kg of body weight and per day [80]. During bathing activities, TiO$_2$ NPs enter the water bodies. A research group determined that the concentration of titanium ranges from 181 to 1233 µg/L in raw sewage obtained from 10 full-scale municipal centralized wastewater treatment plant municipalities in Arizona [81]. The treated water from these water plants flows into rivers and lakes where nanoparticles may cause an ecological risk. It has been found that the released TiO$_2$ NPs in the water bodies stay at the air–water interface for a short time and float on the water surface or hetero-aggregate with natural suspended particulate matter and sediment [82].

Despite its emerging status as a contaminant in water bodies, TiO$_2$ NPs is extensively studied as a material for several photocatalytic applications. Efforts are being directed at maximizing capturing the energy from sunlight. Sunlight's emission spectrum consists of only 4% UV light, whereas visible light constitutes a significantly larger percentage, approximately 40%. As documented in a recent review by Tan et al., the photocatalytic properties of TiO$_2$ NPs can be fine-tuned by structurally modifying it to decrease its bandgap in order to more effectively utilize visible light [83]. This can be achieved by introducing additional intrinsic defects, doping with a range of non-metal elements, shielding the particle through a suitable coating, by functionalizing the NPs, and testing different particle sizes [34,41,83–85].

A specific photocatalytic application of TiO$_2$ NPs being explored is sterilization because of its effectiveness in treating a wide variety of pollutants (e.g., pharmaceuticals, pesticides, antibiotics, endocrine-disrupting compounds), food, and bacteria. Carbamazepine is an antiepileptic pharmaceutical compound that is frequently found in water bodies and is believed to be a danger to aquatic life including bacteria, algae, invertebrates, and fishes, etc. [86]. It cannot be efficiently removed (<10%) by conventional wastewater treatment plants. Several studies have shown that it can be photodegraded using TiO$_2$-suspended NP photocatalysts [87–89]. Salicylic acid, another pollutant, can be degraded when subjected to UV irradiation in a photocatalytic reactor that uses TiO$_2$ NPs as a semiconductor [90].

The photocatalytic bactericidal effect of TiO$_2$ NPs has been an emerging field of investigation since the early 1990s [91–93]. In a study, researchers designed a photobioreactor to sterilize the selected foodborne pathogenic bacteria, *Salmonella choleraesuis (S. choleraesuis)*, *Vibrio parahaemolyticus (V. parahaemolyticus)*, and *Listeria monocytogenes (L. monocytogenes)* using various TiO$_2$ NP concentrations and ultraviolet (UV) illumination time [94]. The survival of all bacteria was decreased to ~20%–60% in the presence of UV-radiated TiO$_2$ NP (1.00 mg/mL) within 30 min. of illumination. Currently, research is focused on photoinactivation of various bacterial strains using doped TiO$_2$ NP photocatalysts, which include several different doping systems, for instance, nitrogen, silver, manganese, zinc oxide, sulfur, nickel, copper, and silicon [95–102]. The bactericidal mechanism is well characterized. The damage starts via bacterial cell membrane disruption caused by ROS, which results in the subsequent leakage of internal components from the damaged sites [103–105].

Investigations of TiO$_2$ NP use for their photocatalyzed anticancer properties has been another area of major interest. Zhang et al. have studied the photocatalytic killing effect of TiO$_2$ NPs on colon carcinoma cells and concluded that at concentrations lower than 200 µg/mL, they are effective protection against UVA irradiation alone, but above this concentration, there is a significant cytotoxic effect on these cells [106]. The mechanistic details underlying this cytotoxic behavior has been studied. Warner et al. observed that nucleic acids are the main target for photooxidative damage catalyzed by TiO$_2$ NPs. They observed the hydroxylation of guanine bases following calf thymus DNA reaction with TiO_2 NP while irradiated with UVA [107]. Jaeger et al. examined the putative pathway for TiO_2 NP-induced mitochondrial DNA damage in human HaCaT keratinocytes [108]. They found that ROS generation resulted in the mitochondrial common deletion of DNA base pairs in HaCaT cells. Lagopati et al. saw a significant induction of apoptosis in MDA-MB-468 cells when they irradiated the breast cancer epithelial cells using UV-A light (wavelength 350 nm) for 20 min in the presence of nanostructured TiO_2 sol-containing anatase NPs [109].

The utility of TiO_2 NPs as photosensitizers in targeted anticancer photodynamic therapy (PDT) is being explored. Zhang et al. compared the photosensitizer capacity for TiO_2 NPs with that of ZnO NPs and observed no differences in their anticancer potencies as both could generate ROS and lead to caspase-dependent apoptosis within the tumor cells [110]. Current research is more focused on the development of modified TiO_2 NPs to enhance their photocatalytic activity. Yang et al. synthesized Ce-doped TiO_2 nanocrystals by a modified sol-gel method for the treatment of deep-seated tumor [111]. These nanocrystals could serve as photosensitizers in PDT when activated by low-dose X-ray as they can generate intracellular ROS and lead to the apoptosis/necrosis of A549 cancer cells. The use of TiO_2 NPs as photosensitizers for photodynamic antibacterial therapy is also being investigated [112,113].

5. Elucidating the Impact that the Bioactivity of TiO_2 NPs from Sunscreen Use Could Have on the Aquatic Environment and Human Health

Industrial applications of photoexcited TiO_2 NPs demonstrate their potent redox reactivity and hint at the effect that they may have on the biological activity in prokaryotic and eukaryotic cells. TiO_2 NPs released into the environment could lead to the toxicity of aquatic organisms. Mueller and Nowack determined a predicted no effect concentration (PNEC) of <1 μg/L for TiO_2 exposure to aquatic organisms such as algae and daphnia [114]. Below this value, the TiO_2 content in water bodies is not expected to cause any toxicity. In 2010, the Environmental Protection Agency (EPA) issued a case study on nanoscale TiO_2 and its use in topical sunscreen [38]. It evaluated the stability of the material and its safety to the environment and people without making any definitive statements in support or against use of the material. The study revealed that at very high concentrations (in the mg/mL range), TiO_2 NPs is toxic to several algae, invertebrate organisms, and fish, a predictable result considering that the content far exceeds the PNEC value [38]. Ates et al. investigated the bioaccumulation and tissue distribution of TiO_2 NPs in goldfish (*Carassius auratus (C. auratus)*) [115]. In the study they found that a short period of exposure to 10 and 100 mg/L concentrations of TiO_2 NPs was not lethal, however physiological and behavioral changes were noticed at exposure to higher concentrations. The accumulation of TiO_2 NPs in the intestine was increased when the concentration of NPs was increased from 10 to 100 mg/L; conversely, the weight-wise growth of goldfish was decreased at higher concentrations. Mansfield et al. demonstrated the photo-induced toxicity of anatase TiO_2 NPs under natural sunlight to small planktonic crustacean *Daphnia magna (D. magna)* [116]. They determined the LC_{50} for NPs after 8 h of sunlight exposure. Under full intensity ambient natural sunlight, the LC_{50} was 139 ppb, under 50% natural sunlight, the LC_{50} was 778 ppb and >500 ppm under 10% natural sunlight. Kachenton et al. investigated the toxicological effects of TiO_2 NPs to the brine shrimp (*Artemia salina (A. salina)*), by determining 24 h LC_{50}, which was 1693.43 mg/L. Jovanovic et al. have conducted a series of studies probing the different types of detrimental effects that TiO_2 NPs can have on aquatic organisms. In one such study, they exposed Caribbean mountainous star coral (*Montastraea faveolata (M. faveolata)*) for 17 days in 0.1 mg/L and 10 mg/L TiO_2 NP suspensions. The coral exhibited symptoms of acute stress including expulsion of zooxanthella and a temporary increase in the expression of the heat-shock protein 70. In addition, bioaccumulation of the NPs was observed in the microflora of the coral [117]. In another study, Jovanovic examined the immunotoxicity of fish (*Fathead minnows; Pimephales promelas (P. promelas)*) induced by TiO_2 NPs [118]. Due to their antibacterial properties, the NPs were expected to serve as protective agents against predatory bacteria. However, they caused a reduction in the antibacterial activity of fish neutrophils (which function to eliminate bacteria by phagocytosis), histopathological effects, and an increase in mortality when challenged with two bacterial strains [118]. This suggests that fish with elevated levels of TiO$_2$ NPs would be liable to bacterial infection and increased mortality during disease outbreaks.

Many of the studies that report on the dangers of TiO$_2$ NPs focus on effects at high concentrations that may not reflect physical reality. The EPA reports that a number of environmental factors can contribute to the perceived effects of the NPs such as the UV index, the pH and chemical composition of a body of water, ambient temperatures, and ecological factors such as the storage and potential seepage of wastewater containing the NPs [38]. In addition, the size, crystallinity (whether anatase, rutile, or other forms), and surface coating of the NPs have a major influence [38]. Considering all of these factors is outside the scope of this work but they certainly impact the solid and solution state speciation [16–18] of the metal and its reactivity. Reports that focus on direct measurements of TiO$_2$ NPs in bodies of water provide a more realistic perspective on the safety of TiO$_2$ NPs. There is on average 46 mg of TiO$_2$ NP content present in per gram of sunscreen with an adult application of about 36 g, from which 25% of the total applied could wash off from the skin in the water during beach activities [119]. Sánchez et al. estimated the summer daily release of TiO$_2$ NP of approximately 4 kg at Palmira beach (Peguera, Majorca Island) and estimated an associated increase of net hydrogen peroxide production rate of H$_2$O$_2$ of 270 nM/day due to the redox activity of the material [119]. Venkatesan et al. used single-particle inductively coupled plasma mass spectrometry to measure Ti-containing particles in heavily frequented bathing areas in Arizona—the Salt River and five swimming pools [120]. Between 64 to 148 ng/L of TiO$_2$ NP were found in the pools whereas 260–659 ng/L were found in the Salt River, this number bordering very close to the PNEC limit. The concentration range for the Salt River is expected to be underestimated value because it is possible that larger-sized particles may have been filtered out in the sample preparation process and smaller-sized particles are undetectable by the instrument [120]. These particles are believed to originate from sunscreen products as TEM images compare favorably with Ti-containing NPs from commercially available sunscreen. Interestingly, the Ti content in the swimming pools was dominated (98.7%–99.8%) by dissolved Ti species [120]. Holbrook et al. also made a similar observation of dissolved Ti species in a swimming pool [121]. The source of this dissolved Ti and its speciation has not been characterized.

Whether TiO$_2$ NPs can dissolve in open water has not been established. There are a number of organisms that possess biomolecules with chelating moieties that have the capacity to bind Ti(IV) due to its hard Lewis acidic nature [33,122], such as siderophore-producing marine organisms [67], the dihydroxyphenylalanine (DOPA)-containing adhesive proteins of mussels [123–125], and the tunichromes of ascidians [126,127]. That said, whether chelation onto the metal in TiO$_2$ NPs can induce solubility needs to be examined. There are marine organisms like the brown algae *Fucus spiralis* (*F. spiralis*) (308 ppm) [128] and the ascidian *Eudistoma ritteri* (*E. ritteri*) (1512 ppm) [129] that can bioaccumulate Ti(IV) at several orders of magnitude greater than their local environment. This elevated concentration is presumably the product of biomolecular chelation although the reason for this binding has not been established. It is possible that the Ti(IV) is functionally useful to these living things. The profile for dissolved titanium in the open ocean suggests that the metal is biologically used. Its surface concentrations are quite low where there is an abundance of living organisms but are significantly higher at greater depths where life is less prevalent [130,131]. Were TiO$_2$ NPs to become solubilized in open waters, then perhaps the soluble Ti(IV) may not be too much of a concern if there are marine organisms that can scavenge and potentially utilize the metal unless the solubilized levels become too extreme or the speciation toxic.

A major issue for debate is the long-term effect of TiO$_2$ NPs on human skin and human health, in general, from sunscreen use. To address this matter, it is important to distinguish between potential effects from the NP form of these materials and any solubilized Ti(IV). Before doing so, let us consider routes of entry into the body. It is generally accepted that TiO$_2$ NPs and Ti(IV) ions can enter the human body primarily through inhalation (respiratory tract) and ingestion (gastrointestinal tract), the latter of which can lead to its circulation in blood [132] (Figure 4). It is much less clear how and if it actually penetrates the skin. Mammalian skin is structured in several layers: The stratum corneum (SC), epidermis, dermis, and the subcutaneous layer. SC is the rate limiting barrier against absorption/percutaneous penetration of topically applied substances [133]. The epidermis, the outermost layer of the skin, works as a barrier for the dermis, which contains connective tissue, sweat glands, hair follicles, and nerve endings. Some evidence suggests that TiO_2 NP may penetrate into or through human skin and can reach to the epidermis or dermis [132]. Most studies indicate that TiO_2 NP penetration is localized within the SC and hair follicles and much less penetration occurs at the epidermis or dermis. Sadrieh et al. showed that repeated application of 5% TiO_2 uncoated or coated particles of a 20–500 nm size range can penetrate the skin of mini pigs, leading to detectable levels of the particles in the dermis. It was unclear whether the presence of NPs in the dermal part of the skin resulted from viable skin penetration or from their presence in the hair follicles. The study of long-time (60 days) exposures of 4 and 60 nm TiO_2 NPs on hairless mice showed deeper penetration of TiO_2 [134]. The NPs were allocated in various tissues such as the lungs, spleen, and brain, indicative of potential crossing of the blood–brain barrier. Of the organs examined, the skin and liver exhibited the most severe pathological lesions, which are believed to be due to the oxidative stress caused by the NPs [134]. This work suggests that after repeated (long-term) application of sunscreen, TiO_2 NPs contained within may be able to translocate through human skin.

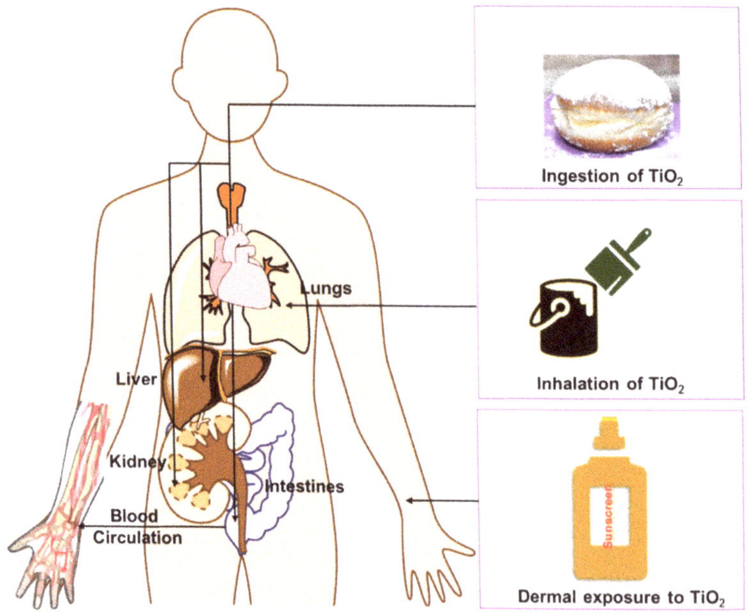

Figure 4. Different pathways by which TiO_2 nanoparticles can enter and distribute in the human body.

The interaction of the nanomaterials with macromolecules, for example, proteins is an interest for several points of view. The realization of this synergistic effect on nanotechnology and its subbranches (nanobiotechnology, protein nanotechnology, nanomaterial science, etc.) especially in improving the performance of proteins and enzymes for various applications, is evident from the numerous studies that have emerged in the last few years. A large emphasis is given to improving the performance of proteins when they are formulated in combination with other nanomaterials and, therefore, the interaction between proteins and nanomaterials are widely examined. However, despite the significant advancement in this area, comparatively less focus has been given to cellular uptake studies of nanomaterials facilitated by proteins in biological systems. Therefore, exact uptake pathways, mechanisms, and the final effect of nanomaterials inside the cell are poorly recognized. Nevertheless, some recent reports in this direction provide elucidate important insight, which is helpful not only for the smart utilization of the nanomaterial but also in testing biological responses towards nanoparticles and dose-dependent toxicity. The increasing use of TiO_2 NPs and, more recently, the discovery of the degradation of titanium implants resulting in the formation of said particles, has raised questions about its fate and effect in mammalian organisms. Several studies involving the injection and subsequent tracking of TiO_2 NPs in rats have discovered the deposition of these particles into the spleen, liver, and lung tissues. Although these studies linked the transportation of TiO_2 NPs to macrophages, the exact mechanism of uptake in these cells was not probed [135,136]. An important aspect to consider in the uptake of TiO_2 NPs is the formation of the protein corona (PC), a layer of proteins that rapidly covers nanoparticles when present in a protein-rich environment such as the human serum. The PC mediates the interactions between the NPs and cells, and so its composition plays an important role in the translocation of TiO_2 NPs [137].

Several blood and serum proteins have been studied so far to understand the effect of individual protein/s through cellular uptake routes such as phagocytosis, micropinocytosis, or endocytosis [137]. Serum/plasma proteins constitute a complex proteome system where different proteins will interact with nanoparticles at different physiological conditions (pH and ionic strength) and, therefore, injected nanoparticles might undergo modifications, which are not completely identified [137]. In a study by Tedja et al., TiO_2 NP uptake into the human lung cell lines A549 and H1299 was investigated with an emphasis towards PC function and composition [138]. By treating TiO_2 NP with fetal bovine serum (FBS), a protein-rich serum, and comparing it to the uptake of non-treated NPS, some insight was gained into the role of the PC. The size of the TiO_2 NPs suspended in FBS was smaller in comparison to those in PBS buffer alone. The reduction in the particles size is attributed to the coating of complex proteins-mix on the surface of the particles, which reduce aggregation by forming a steric layer. Similar results were observed by Allouni et al. by using anatase TiO_2 NPs adsorbed on three blood proteins; human serum albumin, γ-globulins, and fibrinogen for uptake studies using L929 mouse fibroblasts cells [139]. Although there was a larger initial uptake of the non-FBS-treated TiO_2 NPs, after a 24 h period, the FBS-treated particles showed a larger uptake into the cells, which was attributed to a second phase of particle uptake observed in the data [138]. Additionally, a difference in uptake was observed between the two cell lines, with the H1299 having a higher uptake than the A549 cells, highlighting the difference in biochemical composition of the cell membranes and consequently a difference in cellular uptake in cells from the same tissue of origin. Apart from looking at the TiO_2 NP uptake of these cell lines, its pathway was also probed. By subjecting the cells to cellular uptake inhibitory treatments, namely low temperature, adenosine triphosphate depletion, caveolae disruption by cholesterol sequestration, and hypertonic treatment, they were able to ascertain endocytosis as the main uptake mechanism in both cell lines. Furthermore, the data obtained suggested endocytosis in the A549 occurred via a clathrin-mediated pathway, while the H1299 uptake mechanism remained elusive. Another important aspect covered in this study was the potential role of the serum protein vitronectin, a serum glycoprotein associated with cellular adhesion to surfaces and the uptake of crocidolite asbestos in rabbit pleural mesothelial cells and A549 cells via the $\alpha_v\beta_5$ integrin receptors [140–142]. By using an antibody to remove vitronectin and measuring the uptake of TiO_2 NP into cells, the authors were able to pinpoint vitronectin as an important factor for TiO_2 NP absorption in A549 cells [138]. Although this effect was not observed in H1299 cells, this last result highlights the important role that PC composition plays in the interaction between TiO_2 NPs interaction with cells and their subsequent uptake.

Toll-like receptors (TLR3, TLR4, and TLR7) have also been studied in the uptake of TiO_2 NPs. These receptors are transmembrane proteins, which play an important role in cellular defense, ligand binding, and signaling pathways [143]. TLR4 and TLR7 were found to be able to transport the NPs [143]. The involvement of TLR4 and other various receptors and proteins in uptake pathways was also investigated in the sea urchin *Paracentrotus lividus (P. lividus)* immune cell [144]. High expression of the TLR gene and in the levels of related proteins in immune cells was observed when TiO_2 NPs encounter the immune cells. The size of the NPs is important for such uptake studies as the NP size generally falls within the range of bacteria and fungi, which is perhaps related to identifying the foreign material [144,145].

Some studies have moved toward PC characterization, and have identified serum and plasma proteins such as immunoglobulin, human serum transferrin (hTf), and human serum albumin (HSA) in the TiO$_2$ PC [146–148]. Due to its metal-binding ability and abundance in human serum, HSA and its interaction with TiO$_2$ NPS has been extensively studied. It has been shown that in different media, the presence of the HSA model protein, bovine serum albumin (BSA), FBS, or a mixture of serum proteins (BSA, γ-globulin, and apo-hTf) improved the dispersion of TiO$_2$ NPS. This study, which used high-throughput dynamic light scattering (HT-DLS) to determine nanoparticle size distribution and state of agglomeration-dispersion, was able to observe the effect of the PC on the hydrophobic and electrostatic interactions, which govern TiO$_2$ aggregation or dispersion. In almost every case, BSA aided in TiO$_2$ dispersion but most noteworthy is the synergistic effect observed in the FBS and the mixture of serum proteins, which had a higher stabilizing effect on all culture media, highlighting the important role of the diversity of the PC in stabilization of TiO$_2$ NPs [149]. Due to the nature of the interactions governing PC formation, many factors have to be taken into account. Environmental factors such as pH and salt content have been shown to have an effect in the binding of HSA and possibly other serum proteins, mainly due the protonation/deprotonation of the TiO$_2$ NPs surface and the change in protein charge induced by the pH [149,150]. Additionally, the morphology of the particles has been shown to be an important factor in protein–NPs interaction. Zaquot et al. studied the binding of serum proteins to the anatase, rutile, and polymorph forms of TiO$_2$ and found that the polymorph form had a greater adsorption of serum proteins, although the exact mechanism of these interactions could not be determined [151]. The complex nature of the PC and its interaction with TiO$_2$ NPs makes it difficult to study in vivo, although they are certain to be involved in its transport through the blood stream and translocation to systemic organs [135,136]. Despite their clear limitations, in vitro studies have proven to be useful in observing these interactions and help us work towards an understanding of TiO$_2$ NP–protein interactions and transport into the cells and through the human body.

Whether or not TiO$_2$ NPs from sunscreen use are able to penetrate the skin, it is important to consider the different effects that they might have on human cells. As already established, by carefully regulating their redox activity, TiO$_2$ NPs can be designed as potentially excellent anticancer agents. However, the absence of control over this activity could lead to many issues to cells and tissue. The excessive generation of ROS and reactive nitrogen species (RNS) could lead to inflammation, fibrosis, and pulmonary damage [152]. At the cellular level, oxidative stress can occur, resulting in chemical and structural modifications of macromolecules and interference with signal transduction pathways and changes to transcription factors. In extreme cases, oxidative DNA damage occurs that results in cytotoxicity or in mutations that can cause cancer. Jin et al. observed that following treatment of mice fibroblast cells (L929) with colloidal TiO$_2$ NPs (3–600 µg/mL), the cells became round, shrank, and lost their ability to adhere to one another and to proliferate. They exhibited severe DNA damage from oxidative stress and were either in the late stages of apoptosis or necrotic. The TiO$_2$ NPs appeared to alter the release and structural composition of lysosomes, which in turn, led to increased lysosomal permeability and to damage and destruction of organelles, to changes in mitochondria, and to triggering of apoptosis [153]. Concerns regarding the cytotoxicity of TiO$_2$ NPs may be more directed at human skin and layers beneath the surface to the extent at which UV light can penetrate although the NPs can certainly produce oxidative stress at high concentrations, without UV activation. As for the cancer-causing ability of TiO$_2$ NPs, the International Agency for Research on Cancer, which is part of the World Health Organization, classified it as carcinogenic to animals [48]. The evidence from epidemiological studies is extremely poor for classifying the NPs as carcinogenic to humans. However, it is labeled as potentially carcinogenic to humans and this potential appears to be targeted to lungs [48]. TiO$_2$ is found at its highest levels in the lungs of the human body (3.7 µg/g of body) due to the ubiquitous nature of TiO$_2$ particles in the air [29,154]. A single multi-country study of TiO$_2$ NP production workers found that the workers had higher TiO$_2$ levels and a slightly higher risk for lung cancer than the general population and exhibited a clear dose-response to inhalation exposure [48]. For this reason, the use of spray-on sunscreen that contains TiO_2 NPs is not recommended because it can lead to increased inhalation of the particles [38,48]. Throughout the world, the safety classification of TiO_2 NPs is greatly debated. Very recently (October 2019), France was the first nation to publish a law suspending the use of food additive TiO_2 (E171) due to its numerous health risks. Several non-governmental organizations, including the European Environmental Bureau (EEB), ClientEarth, and the Center for International Environmental Law (Ciel) are pushing for similar laws within the European Union to classify TiO_2 NPs in all of its forms as a category 2 carcinogen. In contrast, the U.S. Food and Drug Administration (FDA) recently evaluated the physicochemical properties, reactivity, and skin contact mobility of TiO_2 NPs and deemed their use in over-the-counter sunscreens as generally recognized as safe and effective (GRASE) [155].

There are additional safety factors that should be evaluated. In the absence of UV activation, TiO_2 NPs can still have a debilitating effect on human cells. UV activation enables the NPs to exhibit prophylactic activity against bacteria as previously described but surprisingly, sans the UV energy, they double the risk of human cells for bacterial infection similar to what has been reported for fish [118]. Mironava et al. examined the non-UV effect of TiO_2 NP treatment of human cervix adenocarcinoma (HeLa) cells at levels that do not induce ROS and in the presence of *Staphylococcus aureus (S. aureus)* [156]. This gram-positive bacteria can be found on human skin. While there was no observed change in cell size, the HeLa cells were more porous when treated with anatase and rutile than normal and showed loss of membrane integrity. Both anatase and rutile caused a similarly elevated level of bacteria in the HeLa cells compared to control (Figure 5) but, unexpectedly, not in macrophages. This suggests that they induce a compromised immune response, comparable to findings with fish [118]. The increased porosity of human cells by TiO_2 NPs leads to the release of lactate dehydrogenase, which Mironava et al. believe produces a favorable extracellular environment that bacteria may be attracted to [156]. It does not appear as though the higher attraction for these cells is because the bacteria are drawn to the Ti itself, but this factor needs to be examined in light of evidence that bacterial biomolecules can interact with Ti(IV) (See Section 3). In a related work, Wang et al. found that TiO_2 NP dietary exposure induced an abnormal state of macrophages characterized by excessive inflammation and suppressed innate immune function [157]. The macrophages exhibited decreased chemotactic, phagocytic, and antibacterial activity, which make people more susceptible to infections. Piedimonte et al. have demonstrated the enhanced susceptibility to respiratory syncytial virus infections in human bronchial epithelial cells exposed to TiO_2 NPs [158].

TiO_2 NPs may also influence bacteria in humans in another significant manner. Wu and Xing et al. investigated the impact of oral consumption of anatase and rutile NPs found as additives in sweets, on mice gut microbiota [159]. The treatment did not affect the microbiota diversity but shifted the amounts of each strain in a time-dependent manner, which could, to an extent, actually be due to different bacteria having a propensity for the material. The impact of rutile NPs was more pronounced than that of anatase NPs [159]. TiO_2 NPs applied to human skin could effect skin microbiota (and even gut microbiota from ingestion) in an analogous manner. The symbiotic relationship of the human microbiome plays an extremely important function in helping to regulate the immune system [160,161]. While the cutaneous innate and adaptive immune response modulates the skin microbiota, the skin microbiota informs the immune system of the external environment and foreign bodies. Any change due to nanoparticle interaction with skin microbiota can disrupt the innate and adaptive immune response. Billions of T cells are found in the skin that are responsible for responding to pathogenic micro-organism. Animal studies have shown that TiO_2 NPs (<100 nm) can translocate to the lymph nodes by lymphatic vessels and can activate dendritic cells [162], messenger cells between the innate and adaptive immune response These studies have also shown NP accumulation at hair follicles [162]. Hair follicles are now believed to help regulate the trafficking of immune cells [163,164]. Not much is understood about how TiO_2 NPs affect the function of immune cells but it has been observed that the NPs can bind antigens and increase their persistance, possibly leading to an increased antigenicity [165].

This behavior could be the source of the small number of reported allergies toward Ti-containing implants [22,23]. A recent study evaluating the biological fate of TiO$_2$ NPs in pigment used in tattooed human skin by synchrotron X-ray fluorescence (XRF) revealed that they translocate to lymph nodes (Figure 6) [166].

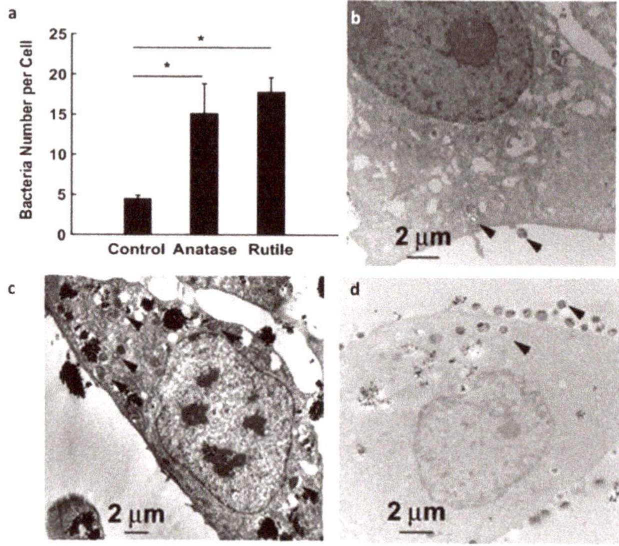

Figure 5. The effect of TiO$_2$ NPs on infectious bacteria *S. aureus* in HeLa cells. The number of bacteria *S. aureus* in control vs. anatase and rutile exposed HeLa cells (**a**). TEM cross-sections of HeLa control cells (**b**), cells exposed to 0.1 mg/mL anatase (**c**), and 0.1 mg/mL rutile TiO2 (**d**) followed by exposure to *S. aureus* bacteria for 90 min. Arrows point towards bacteria. * Means $p < 0.05$. This figure was modified from *Journal of Nanobiotechnology*, 14:34, Copyright 2016, Springer Nature. This work was published under a CC BY 4.0 license (http://creativecommons.org/licenses/by/4.0/) [156].

The possibility for a portion of TiO$_2$ NPs to become solubilized exists because of the presence of hydroxyacids such as citrate, as previously discussed, that can induce solubilization by chelation [31,33,68–73,76]. In a related study, it has been shown that the Ti(IV) tricitrate complex can be photoreduced by UV, producing a Ti(III) species as confirmed by electron paramagnetic resonance [167]. The structure of this species has not been fully characterized, although Ti(III) citrate species are notoriously excellent reducing agents [168]. An anaerobic environment was used to generate the Ti(III) product in addition to several hours of UV irradiation but it is not known whether it could form under aerobic conditions. If Ti ions are generated within sunscreen, then it is likely to be of the Ti(IV) form. At present, we can only speculate on the movement of Ti ions into skin cells and their translocation into the body. The Fe(III)-binding transferrin family of proteins have been attributed to the binding and transport of Ti(IV) within the human body [31–33,73,169–173]. One other member of this family that may play a role is melanotransferrin (MTf). MTf exists mainly in a glycosylphosphatidylinositol-anchored membrane form predominantly in the epidermis of the skin, although a secreted form does exist [174–176]. Although the innate functions of MTf are not yet clear, it does appear to play a role in Fe(III) cellular uptake [176] and presumably should be able to do the same for Ti(IV) in skin cells. If solubilized Ti(IV) ions translocate with TiO$_2$ NPs into the body and eventually the bloodstream, then citrate molecules and serum transferrin would be expected to capture this pool of Ti(IV) and regulate its blood speciation. Soluble Ti(IV) ions in human cells and tissue can demonstrate detrimental effects similar to TiO$_2$ NPs and several that are distinct. Ti ions have been reported to induce substantial damage in macrophages by interrupting the cell division, oxidative stress, and other inflammatory reactions [177]. They can cause DNA structural modifications and result in DNA fragmentation [30] possibly by phosphate hydrolysis [170,178]. Piekoszewski et al. reviewed several of the potential problems that soluble Ti(IV) can cause in the body [29]. It is important to remember that several of these issues may be overcome by the synergistic regulation of Ti(IV) by citrate and sTf.

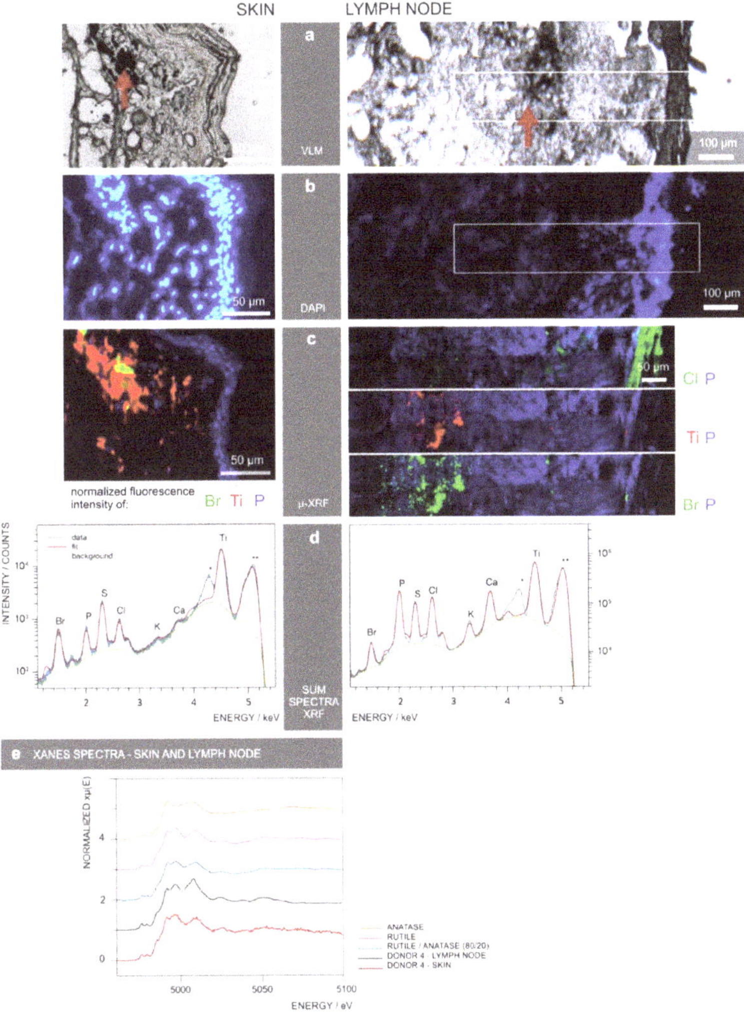

Figure 6. Synchrotron X-ray fluorescence was used to identify and locate tattoo particle elements in skin and lymph nodes of a donor. (**a**) Visible light microscopy (VLM) images of the area were mapped by µ-XRF and tattoo pigments were indicated by a red arrow. (**b**) 4′,6-diamidino-2-phenylindole (DAPI) staining of the tissues showing the cell nuclei. (**c**) µ-XRF maps of P, Ti, Cl, and/or Br. For the lymph node, these areas are marked in (**a**,**b**). (**d**) Average µ-XRF spectra over the full area displayed in (**c**) * diffraction peak from sample support; ** scatter peak of the incoming beam. (**e**) Ti K-edge micro-X-ray absorption near edge structure (µ-XANES) spectra of skin and lymph node compared to the spectra of rutile, anatase, and an 80/20 rutile/anatase mixture calculation. This figure was obtained from *Scientific Reports*, 11395, Copyright 2017, Springer Nature. This work was published under a CC BY 4.0 license (http://creativecommons.org/licenses/by/4.0/) [166].

6. Conclusions

The diverse use of TiO_2 NPs is increasing every day by a variety of industries around the world, especially in the food and cosmetic fields. The tremendous use of these materials poses health hazards but in our opinion, TiO_2 NPs should be used strategically with great consideration for their formulation in sunscreens to avoid a detrimental effect on a wide range of living organisms and the environment at large. While these particles display a variety of bioactive properties that can be fine-tuned for beneficial human use and to clean up the environment, if not designed properly, they can undergo uncontrolled release and even solubilization that can lead to unpredictable speciation and ultimately devastating effects.

Author Contributions: S.S., R.K.S., K.G., J.F.C.T., S.A.L.-R., A.T., M.S., M.J., and A.D.T. participated in the conceptualization, investigation, and writing of the manuscript.

Funding: We are quite grateful to the different sources of funding that supported this work. A.D.T. and K.G. were supported by the NIH 5SC1CA190504 grant (which also funded S.S., R.K.S., M.S. and S.A.L.-R.), the UPR RP FIPI and PBDT Grants from the office of the DEGI. J.F.C.T. was supported by the NIH RISE 5R25GM061151-17 grant. M.J. and A.T. was funded by the NSF REU 1560278 and 1757365 grants, respectively.

Conflicts of Interest: The authors declare no conflict of interest. The funders had no role in the design of the study; in the collection, analyses, or interpretation of data; in the writing of the manuscript, and in the decision to publish the results.

References

1. Oshida, Y. *Bioscience and Bioengineering of Titanium Materials*; Elsevier Ltd.: Amsterdam, The Netherlands, 2007.
2. Ratner, B.; Hoffman, A.; Schoen, F.; Lemons, J. *Biomaterials Science: An Introduction to Materials in Medicine*, 3rd ed.; Elsevier Inc.: Amsterdam, The Netherlands, 2013.
3. Nair, M.; Elizabeth, E. Applications of Titania Nanotubes in Bone Biology. *J. Nanosci. Nanotechnol.* **2015**, *15*, 939–955. [CrossRef] [PubMed]
4. Egerton, A.; Tooley, I.R. UV absorption and scattering properties of inorganic-based sunscreens. *Int. J. Cosmet. Sci.* **2011**, *34*, 117–122. [CrossRef] [PubMed]
5. Carvajal, M.; Alcaraz, C.F. Why titanium is a beneficial element for plants. *J. Plant Nutr.* **1998**, *21*, 655–664. [CrossRef]
6. Hruby, M.; Cigler, P.; Kuzel, S. Contribution to understanding the mechanism of titanium action in plant. *J. Plant Nutr.* **2002**, *25*, 577–598. [CrossRef]
7. Servin, A.D.; Castillo-Michel, H.; Hernandez-Viezcas, J.A.; Diaz, B.C.; Peralta-Videa, J.R.; Gardea-Torresdey, J.L. Synchrotron Micro-XRE and Micro-XANES Confirmation of the Uptake and Translocation of TiO_2 Nanoparticles in Cucumber (Cucumis sativus) Plants. *Environ. Sci. Technol.* **2012**, *46*, 7637–7643. [CrossRef]
8. Istvan, P.; Feher, M.; Bokori, J.; Nagy, B. Physilogically beneficial effects of titanium. *Water Air Soil Pollut.* **1991**, *57–58*, 675–680. [CrossRef]
9. Yaghoubi, S.; Schwietert, C.W.; McCue, J.P. Biological roles of titanium. *Biol. Trace Elem. Res.* **2000**, *78*, 205–217. [CrossRef]
10. Schwietert, C.W.; Yaghoubi, S.; Gerber, N.C.; McSharry, J.J.; McCue, J.P. Dietary titanium and infant growth. *Biol. Trace Elem. Res.* **2001**, *83*, 149–167. [CrossRef]
11. Oosthuizen, S.J. Titanium: The innovators' metal-Historical case studies tracing titanium process and product innovation. *J. S. Afr. Inst. Min. Metall.* **2011**, *111*, 781–786.
12. Sansone, V.; Pagani, D.; Melato, M. The effects on bone cells of metal ions released from orthopaedic implants. A review. *Clin. Cases Miner. Bone Metab.* **2013**, *10*, 34–40. [CrossRef]
13. Jung, C. About Oxygen, Cytochrome P450 and Titanium: Learning from Ron Estabrook. *Drug Metab. Rev.* **2007**, *39*, 501–513. [CrossRef] [PubMed]
14. Kim, J.-S.; Kang, S.-M.; Seo, K.-W.; Nahm, K.-Y.; Chung, K.-R.; Kim, S.-H.; Ahn, J.-P. Nanoscale bonding between human bone and titanium surfaces: Osseohybridization. *BioMed. Res. Int.* **2015**. [CrossRef] [PubMed]

15. Nuevo-Ordonez, Y.; Montes-Bayon, M.; Blanco-Gonzalez, E.; Paz-Aparicio, J.; Raimundez, J.D.; Tejerina, J.M.; Pena, M.A.; Sanz-Medel, A. Titanium release in serum of patients with different bone fixation implants and its interaction with serum biomolecules at physiological levels. *Anal. Bioanal. Chem.* **2011**, *401*, 2747–2754. [CrossRef] [PubMed]

16. Crans, D.C.; Woll, K.A.; Prusinskas, K.; Johnson, M.D.; Norkus, E. Metal Speciation in Health and Medicine Represented by Iron and Vanadium. *Inorg. Chem.* **2013**, *52*, 12262–12275. [CrossRef] [PubMed]

17. Doucette, K.A.; Hassell, K.N.; Crans, D.C. Selective speciation improves efficacy and lowers toxicity of platinum anticancer and vanadium antidiabetic drugs. *J. Inorg. Biochem.* **2016**, *165*, 56–70. [CrossRef] [PubMed]

18. Levina, A.; Crans, D.C.; Lay, P.A. Speciation of metal drugs, supplements and toxins in media and bodily fluids controls in vitro activities. *Coord. Chem. Rev.* **2017**, *352*, 473–498. [CrossRef]

19. Soto-Alvaredo, J.; Blanco, E.; Bettmer, J.; Hevia, D.; Sainz, R.M.; Chaves, C.L.; Sanchez, C.; Llopis, J.; Sanz-Medel, A.; Montes-Bayon, M. Evaluation of the biological effect of Ti generated debris from metal implants: Ions and nanoparticles. *Metallomics* **2014**, *6*, 1702–1708. [CrossRef] [PubMed]

20. Tagger Green, N.; Machtei, E.; Horwitz, J.; Peled, M. Fracture of dental implants: Literature review and report of a case. *Implant Dent.* **2002**, *11*, 137–143. [CrossRef]

21. Olmedo, D.G.; Tasat, D.R.; Guglielmotti, M.B.; Cabrini, R.L. Biodistribution of titanium dioxide from biologic compartments. *J. Mater. Sci. Mater. Med.* **2008**, *19*, 3049–3056. [CrossRef]

22. Thomas, P.; Bandl, W.D.; Maier, S.; Summer, B.; Przybilla, B. Hypersensitivity to titanium osteosynthesis with impaired fracture healing, eczema, and T-cell hyperresponsiveness in vitro: Case report and review of the literature. *Contact Dermat.* **2006**, *55*, 199–202. [CrossRef]

23. Goutam, M.; Giriyapura, C.; Mishra, S.K.; Gupta, S. Titanium allergy: A literature review. *Indian J. Dermatol.* **2014**, *59*, 630. [CrossRef] [PubMed]

24. Tower, S.S. Arthroprosthetic cobaltism: Neurological and cardiac manifestations in two patients with metal-on-metal arthroplasty: A case report. *J. Bone Jt. Surg. Am.* **2010**, *92*, 2847–2851. [CrossRef] [PubMed]

25. Jacobs, J.J. Commentary on an article by Stephen S. Tower, MD: "Arthroprosthetic cobaltism: Neurological and cardiac manifestations in two patients with metal-on-metal arthroplasty. A case report". *J. Bone Jt. Surg.* **2010**, *92*, e35. [CrossRef] [PubMed]

26. Sotos, J.G.; Tower, S.S. Systemic disease after hip replacement: Aeromedical implications of arthroprosthetic cobaltism. *Aviat. Space Environ. Med.* **2013**, *84*, 242–245. [CrossRef] [PubMed]

27. Bradberry, S.M.; Wilkinson, J.M.; Ferner, R.E. Systemic toxicity related to metal hip prostheses. *Clin. Toxicol.* **2014**, *52*, 837–847. [CrossRef] [PubMed]

28. Dirk, K. *The Bleeding Edge*; Netflix: Los Gatos, CA, USA, 2018.

29. Golasik, M.; Herman, M.; Piekoszewski, W. Toxicological aspects of soluble titanium—A review of in vitro and in vivo studies. *Metallomics* **2016**, *8*, 1227–1242. [CrossRef]

30. Christodoulou, C.V.; Eliopoulos, A.G.; Young, L.S.; Hodgkins, L.; Ferry, D.R.; Kerr, D.J. Anti proliferative activity and mechanism of action of titanocene dichloride. *Br. J. Cancer* **1998**, *77*, 2088–2097. [CrossRef]

31. Tinoco, A.D.; Saxena, M.; Sharma, S.; Noinaj, N.; Delgado, Y.; Gonzalez, E.P.Q.; Conklin, S.E.; Zambrana, N.; Loza-Rosas, S.A.; Parks, T.B. Unusual synergism of transferrin and citrate in the regulation of Ti(IV) speciation, transport, and toxicity. *J. Am. Chem. Soc.* **2016**, *138*, 5659–5665. [CrossRef]

32. Loza-Rosas, S.A.; Saxena, M.; Delgado, Y.; Gaur, K.; Pandrala, M.; Tinoco, A.D. A ubiquitous metal, difficult to track: Towards an understanding of the regulation of titanium(iv) in humans. *Metallomics* **2017**, *9*, 346–356. [CrossRef]

33. Saxena, M.; Loza Rosas, S.; Gaur, K.; Sharma, S.; Perez Otero, S.C.; Tinoco, A.D. Exploring titanium(IV) chemical proximity to iron(III) to elucidate a function for Ti(IV) in the human body. *Coord. Chem. Rev.* **2018**, *363*, 109–125. [CrossRef]

34. Smijs, T.G.; Pavel, S. Titanium dioxide and zinc oxide nanoparticles in sunscreens: Focus on their safety and effectiveness. *Nanotechnol. Sci. Appl.* **2011**, *4*, 95–112. [CrossRef] [PubMed]

35. Norval, M.; Lucas, R.M.; Cullen, A.P.; de Gruijl, F.R.; Longstreth, J.; Takizawa, Y.; van der Leun, J.C. The human health effects of ozone depletion and interactions with climate change. *Photochem. Photobiol. Sci.* **2011**, *2*, 199–225. [CrossRef] [PubMed]

36. Antoniou, C.; Kosmadaki, M.; Stratigos, A.J.; Katsambas, A.D. Sunscreens—What's important to know. *JEADV* **2008**, *22*, 1110–1119. [PubMed]

37. Dransfield, G.P. Inorganic Sunscreens. *Radiat. Prot. Dosim.* **2000**, *91*, 271–273. [CrossRef]

38. Davis, J.; Wang, A.; Shtakin, J. *Nanomaterial Case Studies: Nanoscale Titanium Dioxide in Water Treatment and in Topical Sunscreen*; US EPA: Research Triangle Park, NC, USA, 2010.

39. Morgan, B.J.; Watson, G.W. Intrinsic n-type Defect Formation in TiO_2: A Comparison of Rutile and Anatase from GGA+U Calculations. *J. Phys. Chem. C* **2010**, *114*, 2321–2328. [CrossRef]

40. Marschall, R. Semiconductor composites: Strategies for enhancing charge carrier separation to improve photocatalytic activity. *Adv. Funct. Mater.* **2014**, *24*, 2421–2440. [CrossRef]

41. Cantrell, A.; McGarvey, D.J.; Truscott, T.G. *Comprehensive Series in Photosciences*; Elsevier: Amsterdam, The Netherlands, 2001; Volume 3, pp. 495–519.

42. Tanvir, S.; Pulvin, S.; Anderson, W. Toxicity associated with the photo catalytic and photo stable forms of titanium dioxide nanoparticles used in sunscreen. *MOJ Toxicol.* **2015**, *1*, 00011.

43. Hoffmann, M.R.; Martin, S.T.; Choi, W.; Bahnemann, D.W. Environmental applications of semiconductor photocatalysis. *Chem. Rev.* **1995**, *95*, 69–96. [CrossRef]

44. Dodd, N.J.; Jha, A.N. Photoexcitation of aqueous suspensions of titanium dioxide nanoparticles: An electron spin resonance spin trapping study of potentially oxidative reactions. *Photochem. Photobiol.* **2011**, *87*, 632–640. [CrossRef]

45. Dunford, R.; Salinaro, A.; Cai, L.; Serpone, N.; Horikoshi, S.; Hidaka, H.; Knowland, J. Chemical oxidation and DNA damage catalysed by inorganic sunscreen ingredients. *FEBS Lett.* **1997**, *418*, 87–90. [CrossRef]

46. Lewicka, Z.A.; William, W.Y.; Oliva, B.L.; Contreras, E.Q.; Colvin, V.L. Photochemical behavior of nanoscale TiO_2 and ZnO sunscreen ingredients. *J. Photochem. Photobiol. A Chem.* **2013**, *263*, 24–33. [CrossRef]

47. Armand, L.; Tarantini, A.; Beal, D.; Biola-Clier, M.; Bobyk, L.; Sorieul, S.; Pernet-Gallay, K.; Marie-Desvergne, C.; Lynch, I.; Herlin-Boime, N. Long-term exposure of A549 cells to titanium dioxide nanoparticles induces DNA damage and sensitizes cells towards genotoxic agents. *Nanotoxicology* **2016**, *10*, 913–923. [CrossRef] [PubMed]

48. IARC. Carbon Black, Titanium Dioxide, and Talc. In *IARC Monographs on the Evaluation of Carcinogenic Risks to Humans*; International Agency for Research on Cancer: Lyon, France, 2010; Volume 93.

49. Brezová, V.; Gabčová, S.; Dvoranová, D.; Staško, A. Reactive oxygen species produced upon photoexcitation of sunscreens containing titanium dioxide (an EPR study). *J. Photochem. Photobiol. B-Biol.* **2005**, *79*, 121–134.

50. Fujiwara, R.; Luo, Y.; Sasaki, T.; Fujii, K.; Ohmori, H.; Kuniyasu, H. Cancer Therapeutic Effects of Titanium Dioxide Nanoparticles Are Associated with Oxidative Stress and Cytokine Induction. *Pathobiology* **2015**, *82*, 243–251. [CrossRef] [PubMed]

51. Kubacka, A.; Diez, M.S.; Rojo, D.; Bargiela, R.; Ciordia, S.; Zapico, I.; Albar, J.P.; Barbas, C.; Martins dos Santos, V.A.P.; Fernández-García, M.; et al. Understanding the antimicrobial mechanism of TiO_2-based nanocomposite films in a pathogenic bacterium. *Sci. Rep.* **2014**, *4*, 4134. [CrossRef] [PubMed]

52. Schmets, J.; Van Muylder, J.; Pourbaix, M. *Atlas of Electrochemical Equilibria in Aqueous Solutions*; Pourbaix, M., Ed.; Pergamon Press: Oxford, UK, 1996.

53. Knauss, K.G.; Dibley, M.J.; Bourcier, W.L.; Shaw, H.F. Ti(IV) hydrolysis constants derived from rutile solubility measurements made from 100 to 300 degrees C. *Appl. Geochem.* **2001**, *16*, 1115–1128. [CrossRef]

54. Schmidt, J.; Vogelsberger, W. Aqueous Long-Term Solubility of Titania Nanoparticles and Titanium(IV) Hydrolysis in a Sodium Chloride System Studied by Adsorptive Stripping Voltammetry. *J. Solution Chem.* **2009**, *38*, 1267–1282. [CrossRef]

55. Lang, Y.; del Monte, F.; Rodriguez, B.J.; Dockery, P.; Finn, D.P.; Pandit, A. Integration of TiO2 into the diatom Thalassiosira weissflogii during frustule synthesis. *Sci. Rep.* **2013**, *3*, 3205. [CrossRef]

56. Lyu, S.; Wei, X.; Chen, J.; Wang, C.; Wang, X.; Pan, D. Titanium as a Beneficial Element for Crop Production. *Front. Plant Sci.* **2017**, *8*, 597. [CrossRef]

57. Bertini, I.; Gray, H.B.; Stiefel, E.I.; Valentine, J.S. *Biological Inorganic Chemistry: Structure and Reactivity*; University Science Books: Sausalit, CA, USA, 2007.

58. Shabtai, Y.; Fleminger, G. Adsorption of Rhodococcus Strain GIN-1 (NCIMB 40340) on Titanium Dioxide and Coal Fly Ash Particles. *Appl. Environ. Microbiol.* **1994**, *60*, 3079–3088.

59. Siegmann, A.; Komarska, A.; Betzalel, Y.; Brudo, I.; Jindou, S.; Mor, G.; Fleminger, G. The titanium binding protein of Rhodococcus ruber GIN1 (NCIMB 40340) is a cell-surface homolog of the cytosolic enzyme dihydrolipoamide dehydrogenase. *J. Mol. Recognit.* **2009**, *22*, 138–145. [CrossRef]

60. Dayan, A.; Babin, G.; Ganoth, A.; Kayouf, N.S.; Nitoker Eliaz, N.; Mukkala, S.; Tsfadia, Y.; Fleminger, G. The involvement of coordinative interactions in the binding of dihydrolipoamide dehydrogenase to titanium dioxide-Localization of a putative binding site. *J. Mol. Recognit.* **2017**, *30*. [CrossRef]

61. Dayan, A.; Lamed, R.; Benayahu, D.; Fleminger, G. RGD-modified dihydrolipoamide dehydrogenase as a molecular bridge for enhancing the adhesion of bone forming cells to titanium dioxide implant surfaces. *J. Biomed. Mater. Res. Part A* **2019**, *107*, 545–551. [CrossRef]

62. McWhirter, M.J.; Bremer, P.J.; Lamont, I.L.; McQuillan, A.J. Siderophore-Mediated Covalent Bonding to Metal (Oxide) Surfaces during Biofilm Initiation by Pseudomonas aeruginosa Bacteria. *Langmuir* **2003**, *19*, 3575–3577. [CrossRef]

63. Petrone, L. Molecular surface chemistry in marine bioadhesion. *Adv. Colloid Interface Sci.* **2013**, *195–196*, 1–18. [CrossRef]

64. Horst, A.M.; Neal, A.C.; Mielke, R.E.; Sislian, P.R.; Suh, W.H.; Madler, L.; Stucky, G.D.; Holden, P.A. Dispersion of TiO$_2$ nanoparticle agglomerates by Pseudomonas aeruginosa. *Appl. Environ. Microbiol.* **2010**, *76*, 7292–7298. [CrossRef]

65. Baramov, T.; Keijzer, K.; Irran, E.; Mosker, E.; Baik, M.H.; Sussmuth, R. Synthesis and Structural Characterization of Hexacoordinate Silicon, Germanium, and Titanium Complexes of the E-coli Siderophore Enterobactin. *Chem. Eur. J.* **2013**, *19*, 10536–10542. [CrossRef]

66. Jones, K.E.; Batchler, K.L.; Zalouk, C.; Valentine, A.M. Ti(IV) and the Siderophore Desferrioxamine B: A Tight Complex Has Biological and Environmental Implications. *Inorg. Chem.* **2017**, *56*, 1264–1272. [CrossRef]

67. Butler, A.; Theisen, R.M. Iron(III)-siderophore coordination chemistry: Reactivity of marine siderophores. *Coord. Chem. Rev.* **2010**, *254*, 288–296. [CrossRef]

68. Dakanali, M.; Kefalas, E.T.; Raptopoulou, C.P.; Terzis, A.; Voyiatzis, G.; Kyrikou, I.; Mavromoustakos, T.; Salifoglou, A. A new dinuclear Ti(IV)-peroxo-citrate complex from aqueous solutions. Synthetic, structural, and spectroscopic studies in relevance to aqueous titanium(IV)-peroxo-citrate speciation. *Inorg. Chem.* **2003**, *42*, 4632–4639. [CrossRef]

69. Deng, Y.F.; Zhou, Z.H.; Wan, H.L. pH-dependent isolations and spectroscopic, structural, and thermal studies of titanium citrate complexes. *Inorg. Chem.* **2004**, *43*, 6266–6273. [CrossRef]

70. Collins, J.M.; Uppal, R.; Incarvito, C.D.; Valentine, A.M. Titanium(IV) Citrate Speciation and Structure under Environmentally and Biologically Relevant Conditions. *Inorg. Chem.* **2005**, *44*, 3431–3440. [CrossRef]

71. Panagiotidis, P.; Kefalas, E.T.; Raptopoulou, C.P.; Terzis, A.; Mavromoustakos, T.; Salifoglou, A. Delving into the complex picture of Ti(IV)-citrate speciation in aqueous media: Synthetic, structural, and electrochemical considerations in mononuclear Ti(IV) complexes containing variably deprotonated citrate ligands. *Inorg. Chim. Acta* **2008**, *361*, 2210–2224. [CrossRef]

72. Uppal, R.; Incarvito, C.D.; Lakshmi, K.V.; Valentine, A.M. Aqueous spectroscopy and redox properties of carboxylate-bound titanium. *Inorg. Chem.* **2006**, *45*, 1795–1804. [CrossRef]

73. Tinoco, A.D.; Eames, E.V.; Valentine, A.M. Reconsideration of serum Ti(IV) transport: Albumin and transferrin trafficking of Ti(IV) and its complexes. *J. Am. Chem. Soc.* **2008**, *130*, 2262–2270. [CrossRef]

74. Silva, A.M.N.; Kong, X.; Parkin, M.C.; Cammack, R.; Hider, R.C. Iron(III) citrate speciation in aqueous solution. *Dalton Trans.* **2009**, 8616–8625. [CrossRef]

75. Valentine, A.M. Siderophore-promoted release of titanium(IV) from metal oxide materials. *Abstr. Pap. Am. Chem. Soc.* **2017**, *253*, 1.

76. Dey, R.; Mukharjee, K.; Langer, V.; Roychowdhury, P. A titanium salicylate, Na$_4$[Ti(C$_7$H$_4$O$_3$)$_3$]$_2$ 11H$_2$O. *Acta Crystallogr. Sect. E Struct Rep. Online* **2005**, *61*, M1495–M1497. [CrossRef]

77. Loza-Rosas, S.A.; Vazquez, A.M.; Rivero, K.I.; Negrón, L.J.; Delgado, Y.; Benjamin-Rivera, J.A.; Vazquez-Maldonado, A.L.; Parks, T.B.; Munet-Colon, C.; Tinoco, A.D. Expanding the therapeutic potential of the iron chelator deferasirox in the development of aqueous stable Ti(IV) anticancer complexes. *Inorg. Chem.* **2017**, *56*, 7788–7802. [CrossRef]

78. Piccinno, F.; Gottschalk, F.; Seeger, S.; Nowack, B. Industrial production quantities and uses of ten engineered nanomaterials in Europe and the world. *J. Nanopart. Res.* **2012**, *14*, 1109. [CrossRef]

79. Bettini, S.; Boutet-Robinet, E.; Cartier, C.; Coméra, C.; Gaultier, E.; Dupuy, J.; Naud, N.; Taché, S.; Grysan, P.; Reguer, S. Food-grade TiO$_2$ impairs intestinal and systemic immune homeostasis, initiates preneoplastic lesions and promotes aberrant crypt development in the rat colon. *Sci. Rep.* **2017**, *7*, 40373. [CrossRef]

80. Weir, A.; Westerhoff, P.; Fabricius, L.; Hristovski, K.; von Goetz, N. Titanium Dioxide Nanoparticles in Food and Personal Care Products. *Environ. Sci. Technol.* **2012**, *46*, 2242–2250. [CrossRef]

81. Westerhoff, P.; Song, G.; Hristovski, K.; Kiser, M.A. Occurrence and removal of titanium at full scale wastewater treatment plants: Implications for TiO_2 nanomaterials. *J. Environ. Monit.* **2011**, *13*, 1195–1203. [CrossRef]

82. Gondikas, A.; von der Kammer, F.; Kaegi, R.; Borovinskaya, O.; Neubauer, E.; Navratilova, J.; Praetorius, A.; Cornelis, G.; Hofmann, T. Where is the nano? Analytical approaches for the detection and quantification of TiO 2 engineered nanoparticles in surface waters. *Environ. Sci. Nano.* **2018**, *5*, 313–326. [CrossRef]

83. Kang, X.; Liu, S.; Dai, Z.; He, Y.; Song, X.; Tan, Z. Titanium Dioxide: From Engineering to Applications. *Catalysts* **2019**, *9*, 191. [CrossRef]

84. Kumar, S.; Verma, N.K.; Singla, M.L. Study on reflectivity and photostability of Al-doped TiO_2 nanoparticles and their reflectors. *J. Mater. Res.* **2013**, *28*, 521–528. [CrossRef]

85. Schug, H.; Isaacson, C.W.; Sigg, L.; Ammann, A.A.; Schirmer, K. Effect of TiO_2 nanoparticles and UV radiation on extracellular enzyme activity of intact heterotrophic biofilms. *Environ. Sci. Technol.* **2014**, *48*, 11620–11628. [CrossRef]

86. Fagan, R.; McCormack, D.E.; Dionysiou, D.D.; Pillai, S.C. A review of solar and visible light active TiO_2 photocatalysis for treating bacteria, cyanotoxins and contaminants of emerging concern. *Mater. Sci. Semicond. Process.* **2016**, *42*, 2–14. [CrossRef]

87. Carabin, A.; Drogui, P.; Robert, D. Photo-degradation of carbamazepine using TiO_2 suspended photocatalysts. *J. Taiwan Ins. Chem. Eng.* **2015**, *54*, 109–117. [CrossRef]

88. Achilleos, A.; Hapeshi, E.; Xekoukoulotakis, N.; Mantzavinos, D.; Fatta-Kassinos, D. UV-A and solar photodegradation of ibuprofen and carbamazepine catalyzed by TiO_2. *Sep. Sci. Technol.* **2010**, *45*, 1564–1570. [CrossRef]

89. Luster, E.; Avisar, D.; Horovitz, I.; Lozzi, L.; Baker, M.; Grilli, R.; Mamane, H. N-doped TiO_2-coated ceramic membrane for carbamazepine degradation in different water qualities. *Nanomaterials* **2017**, *7*, 206. [CrossRef]

90. Djouder, R.; Laoufi, A.; Bentahar, F. Photodegradation of salicylic acid in aqueous phase by TiO_2/UV System. *Revue des Energies Renouvelables* **2012**, *15*, 179–185.

91. Saito, T.; Iwase, T.; Horie, J.; Morioka, T. Mode of photocatalytic bactericidal action of powdered semiconductor TiO2 on mutans streptococci. *J. Photochem. Photobiol. B Biol.* **1992**, *14*, 369–379. [CrossRef]

92. Gogniat, G.; Thyssen, M.; Denis, M.; Pulgarin, C.; Dukan, S. The bactericidal effect of TiO2 photocatalysis involves adsorption onto catalyst and the loss of membrane integrity. *FEMS Microbiol. Lett.* **2006**, *258*, 18–24. [CrossRef]

93. Maness, P.-C.; Smolinski, S.; Blake, D.M.; Huang, Z.; Wolfrum, E.J.; Jacoby, W.A. Bactericidal activity of photocatalytic TiO_2 reaction: Toward an understanding of its killing mechanism. *Appl. Environ. Microbiol.* **1999**, *65*, 4094–4098.

94. Kim, B.; Kim, D.; Cho, D.; Cho, S. Bactericidal effect of TiO_2 photocatalyst on selected food-borne pathogenic bacteria. *Chemosphere* **2003**, *52*, 277–281. [CrossRef]

95. Zane, A.; Zuo, R.; Villamena, F.A.; Rockenbauer, A.; Foushee, A.M.D.; Flores, K.; Dutta, P.K.; Nagy, A. Biocompatibility and antibacterial activity of nitrogen-doped titanium dioxide nanoparticles for use in dental resin formulations. *Int. J. Nanomed.* **2016**, *11*, 6459. [CrossRef]

96. Florez, F.L.E.; Hiers, R.D.; Larson, P.; Johnson, M.; O'Rear, E.; Rondinone, A.J.; Khajotia, S.S. Antibacterial dental adhesive resins containing nitrogen-doped titanium dioxide nanoparticles. *Mater. Sci. Eng. C.* **2018**, *93*, 931–943. [CrossRef]

97. Chambers, C.; Stewart, S.; Su, B.; Jenkinson, H.; Sandy, J.; Ireland, A. Silver doped titanium dioxide nanoparticles as antimicrobial additives to dental polymers. *Dent. Mater.* **2017**, *33*, e115–e123. [CrossRef]

98. Lakshmi, K.D.; Rao, T.S.; Padmaja, J.S.; Raju, I.M.; Kumar, M.R. Structure, photocatalytic and antibacterial activity study of Meso porous Ni and S co-doped TiO_2 nano material under visible light irradiation. *Chin. J. Chem. Eng.* **2018**, *10*, 494–504. [CrossRef]

99. Zahid, M.; Papadopoulou, E.L.; Suarato, G.; Binas, V.D.; Kiriakidis, G.; Gounaki, I.; Moira, O.; Venieri, D.; Bayer, I.S.; Athanassiou, A. Fabrication of visible light-induced antibacterial and self-cleaning cotton fabrics using manganese doped TiO_2 nanoparticles. *ACS Appl. Bio Mater.* **2018**, *1*, 1154–1164. [CrossRef]

100. Kaviyarasu, K.; Geetha, N.; Kanimozhi, K.; Magdalane, C.M.; Sivaranjani, S.; Ayeshamariam, A.; Kennedy, J.; Maaza, M. In vitro cytotoxicity effect and antibacterial performance of human lung epithelial cells A549 activity of zinc oxide doped TiO_2 nanocrystals: Investigation of bio-medical application by chemical method. *Mater. Sci. Eng. C* **2017**, *74*, 325–333. [CrossRef]

101. Mathew, S.; Ganguly, P.; Rhatigan, S.; Kumaravel, V.; Byrne, C.; Hinder, S.; Bartlett, J.; Nolan, M.; Pillai, S. Cu-doped TiO2: Visible light assisted photocatalytic antimicrobial activity. *Appl. Sci.* **2018**, *8*, 2067. [CrossRef]

102. Guo, J.; Li, S.; Duan, L.; Guo, P.; Li, X.; Cui, Q.; Wang, H.; Jiang, Q. Preparation of Si doped molecularly imprinted TiO_2 photocatalyst and its degradation to antibiotic wastewater. *Integr. Ferroelectr.* **2016**, *168*, 170–182. [CrossRef]

103. Leyland, N.S.; Podporska-Carroll, J.; Browne, J.; Hinder, S.J.; Quilty, B.; Pillai, S.C. Highly efficient F, Cu doped TiO_2 anti-bacterial visible light active photocatalytic coatings to combat hospital-acquired infections. *Sci. Rep.* **2016**, *6*, 24770. [CrossRef]

104. Yadav, H.M.; Kim, J.-S.; Pawar, S.H. Developments in photocatalytic antibacterial activity of nano TiO_2: A review. *Korean J. Chem. Eng.* **2016**, *33*, 1989–1998. [CrossRef]

105. Yadav, H.M.; Otari, S.V.; Bohara, R.A.; Mali, S.S.; Pawar, S.H.; Delekar, S.D. Synthesis and visible light photocatalytic antibacterial activity of nickel-doped TiO_2 nanoparticles against Gram-positive and Gram-negative bacteria. *J. Photochem. Photobiol. A Chem.* **2014**, *294*, 130–136. [CrossRef]

106. Zhang, A.-P.; Sun, Y.-P. Photocatalytic killing effect of TiO2 nanoparticles on Ls-174-t human colon carcinoma cells. *World J. Gastroenterol.* **2004**, *10*, 3191. [CrossRef]

107. Wamer, W.G.; Yin, J.-J.; Wei, R.R. Oxidative damage to nucleic acids photosensitized by titanium dioxide. *Free Radic. Biol. Med.* **1997**, *23*, 851–858. [CrossRef]

108. Jaeger, A.; Weiss, D.G.; Jonas, L.; Kriehuber, R. Oxidative stress-induced cytotoxic and genotoxic effects of nano-sized titanium dioxide particles in human HaCaT keratinocytes. *Toxicology* **2012**, *296*, 27–36. [CrossRef]

109. Lagopati, N.; Kitsiou, P.; Kontos, A.; Venieratos, P.; Kotsopoulou, E.; Kontos, A.; Dionysiou, D.; Pispas, S.; Tsilibary, E.; Falaras, P. Photo-induced treatment of breast epithelial cancer cells using nanostructured titanium dioxide solution. *J. Photochem. Photobiol. A Chem.* **2010**, *214*, 215–223. [CrossRef]

110. Zhang, H.; Shan, Y.; Dong, L. A comparison of TiO_2 and ZnO nanoparticles as photosensitizers in photodynamic therapy for cancer. *J. Biomed. Nanotechnol.* **2014**, *10*, 1450–1457. [CrossRef]

111. Yang, C.-C.; Sun, Y.-J.; Chung, P.-H.; Chen, W.-Y.; Swieszkowski, W.; Tian, W.; Lin, F.-H. Development of Ce-doped TiO_2 activated by X-ray irradiation for alternative cancer treatment. *Ceram. Int.* **2017**, *43*, 12675–12683. [CrossRef]

112. Xie, J.; Pan, X.; Wang, M.; Yao, L.; Liang, X.; Ma, J.; Fei, Y.; Wang, P.-N.; Mi, L. Targeting and Photodynamic Killing of Cancer Cell by Nitrogen-Doped Titanium Dioxide Coupled with Folic Acid. *Nanomaterials* **2016**, *6*, 113. [CrossRef]

113. Glass, S.; Trinklein, B.; Abel, B.; Schulze, A. TiO2 as Photosensitizer and Photoinitiator for Synthesis of Photoactive TiO_2-PEGDA Hydrogel Without Organic Photoinitiator. *Front. Chem.* **2018**, *6*, 340. [CrossRef]

114. Mueller, N.C.; Nowack, B. Exposure Modeling of Engineered Nanoparticles in the Environment. *Environ. Sci. Technol.* **2008**, *42*, 4447–4453. [CrossRef]

115. Ates, M.; Demir, V.; Adiguzel, R.; Arslan, Z. Bioaccumulation, subacute toxicity, and tissue distribution of engineered titanium dioxide nanoparticles in goldfish (Carassius auratus). *J. Nanomater.* **2013**, *2013*, 9. [CrossRef]

116. Mansfield, C.; Alloy, M.; Hamilton, J.; Verbeck, G.; Newton, K.; Klaine, S.; Roberts, A. Photo-induced toxicity of titanium dioxide nanoparticles to Daphnia magna under natural sunlight. *Chemosphere* **2015**, *120*, 206–210. [CrossRef]

117. Jovanovic, B.; Guzman, H.M. Effects of titanium dioxide (TiO_2) nanoparticles on caribbean reef-building coral (Montastraea faveolata). *Environ. Toxicol. Chem.* **2014**, *33*, 1346–1353. [CrossRef]

118. Jovanović, B.; Whitley, E.M.; Kimura, K.; Crumpton, A.; Palić, D. Titanium dioxide nanoparticles enhance mortality of fish exposed to bacterial pathogens. *Environ. Pollut.* **2015**, *203*, 153–164. [CrossRef]

119. Sánchez-Quiles, D.; Tovar-Sánchez, A. Sunscreens as a source of hydrogen peroxide production in coastal waters. *Environ. Sci. Technol.* **2014**, *48*, 9037–9042. [CrossRef]

120. Venkatesan, A.K.; Reed, R.B.; Lee, S.; Bi, X.; Hanigan, D.; Yang, Y.; Ranville, J.F.; Herckes, P.; Westerhoff, P. Detection and Sizing of Ti-Containing Particles in Recreational Waters Using Single Particle ICP-MS. *Bull. Environ. Contam. Toxicol.* **2018**, *100*, 120–126. [CrossRef]

121. David Holbrook, R.; Motabar, D.; Quinones, O.; Stanford, B.; Vanderford, B.; Moss, D. Titanium distribution in swimming pool water is dominated by dissolved species. *Environ. Pollut.* **2013**, *181*, 68–74. [CrossRef]

122. Buettner, K.M.; Valentine, A.M. Bioinorganic chemistry of titanium. *Chem. Rev.* **2011**, *112*, 1863–1881. [CrossRef]

123. Silverman, H.G.; Roberto, F.F. Understanding marine mussel adhesion. *Mar. Biotechnol. (N. Y.)* **2007**, *9*, 661–681. [CrossRef]

124. Sever, M.J.; Wilker, J.J. Absorption spectroscopy and binding constants for first-row transition metal complexes of a DOPA-containing peptide. *Dalton Trans.* **2005**, *6*, 813–822. [CrossRef]

125. Guo, M.; Harvey, I.; Campopiano, D.J.; Sadler, P.J. Short Oxo–Titanium (iv) Bond in Bacterial Transferrin: A Protein Target for Metalloantibiotics. *Angew. Chem. Int. Edit.* **2006**, *45*, 2758–2761. [CrossRef]

126. Sugumaran, M.; Robinson, W.E. Structure, biosynthesis and possible function of tunichromes and related compounds. *Comp. Biochem. Physiol. B Biochem. Mol. Biol.* **2012**, *163*, 1–25. [CrossRef]

127. Uppal, R.; Israel, H.P.; Incarvito, C.D.; Valentine, A.M. Titanium(IV) Complexes with N,N′-Dialkyl-2,3-dihydroxyterephthalamides and 1-Hydroxy-2(1H)-pyridinone as Siderophore and Tunichrome Analogues. *Inorg. Chem.* **2009**, *48*, 10769–10779. [CrossRef]

128. Black, W.A.P.; Mitchell, R.L. Trace elements in the common brown algae and in sea water. *J. Mar. Biol. Assoc. UK* **1952**, *30*, 575–584. [CrossRef]

129. Levine, E.P. Occurrence of titanium, vanadium, chromium, and sulfuric acid in ascidian *Eudistoma ritteri*. *Science* **1961**, *133*, 1352–1353. [CrossRef]

130. Orians, K.J.; Boyle, E.A.; Bruland, K.W. Dissolved titanium in the open ocean. *Nature.* **1990**, *348*, 322–325. [CrossRef]

131. Butler, A. Acquisition and utilization of transition metal ions by marine organisms. *Science* **1998**, *281*, 207–210. [CrossRef]

132. Oberdorster, G.; Oberdorster, E.; Oberdorster, J. Nanotoxicology: An emerging discipline evolving from studies of ultrafine particles. *Environ. Health Perspect.* **2005**, *113*, 823–839. [CrossRef]

133. Nohynek, G.J.; Dufour, E.K.; Roberts, M.S. Nanotechnology, cosmetics and the skin: Is there a health risk? *Skin Pharmacol. Physiol.* **2008**, *21*, 136–149. [CrossRef]

134. Wu, J.; Liu, W.; Xue, C.; Zhou, S.; Lan, F.; Bi, L.; Xu, H.; Yang, X.; Zeng, F.D. Toxicity and penetration of TiO2 nanoparticles in hairless mice and porcine skin after subchronic dermal exposure. *Toxicol. Lett.* **2009**, *191*, 1–8. [CrossRef]

135. Jani, P.U.; McCarthy, D.E.; Florence, A.T. Titanium dioxide (rutile) particle uptake from the rat GI tract and translocation to systemic organs after oral administration. *Int. J. Pharm.* **1994**, *105*, 157–168. [CrossRef]

136. Olmedo, D.G.; Tasat, D.; Guglielmotti, M.B.; Cabrini, R.L. Titanium transport through the blood stream. An experimental study on rats. *J. Mater. Sci. Mater. Med.* **2003**, *14*, 1099–1103. [CrossRef]

137. Saptarshi, S.R.; Duschl, A.; Lopata, A.L. Interaction of nanoparticles with proteins: Relation to bio-reactivity of the nanoparticle. *J. Nanobiotechnol.* **2013**, *11*, 26. [CrossRef]

138. Tedja, R.; Lim, M.; Amal, R.; Marquis, C. Effects of Serum Adsorption on Cellular Uptake Profile and Consequent Impact of Titanium Dioxide Nanoparticles on Human Lung Cell Lines. *ACS Nano* **2012**, *6*, 4083–4093. [CrossRef]

139. Allouni, Z.E.; Gjerdet, N.R.; Cimpan, M.R.; Høl, P.J. The effect of blood protein adsorption on cellular uptake of anatase TiO$_2$ nanoparticles. *Int. J. Nanomed.* **2015**, *10*, 687.

140. Preissner, K.T. Structure and biological role of vitronectin. *Annu. Rev. Cell Biol.* **1991**, *7*, 275–310. [CrossRef]

141. Boylan, A.M.; Sanan, D.A.; Sheppard, D.; Broaddus, V.C. Vitronectin enhances internalization of crocidolite asbestos by rabbit pleural mesothelial cells via the integrin alpha v beta 5. *J. Clin. Investig.* **1995**, *96*, 1987–2001. [CrossRef]

142. Pande, P.; Mosleh, T.A.; Aust, A.E. Role of $\alpha v \beta 5$ integrin receptor in endocytosis of crocidolite and its effect on intracellular glutathione levels in human lung epithelial (A549) cells. *Toxicol. Appl. Pharmacol.* **2006**, *210*, 70–77. [CrossRef]

143. Chen, P.; Kanehira, K.; Taniguchi, A. Role of toll-like receptors 3, 4 and 7 in cellular uptake and response to titanium dioxide nanoparticles. *Sci. Technol. Adv. Mater.* **2013**, *14*, 015008. [CrossRef]

144. Pinsino, A.; Russo, R.; Bonaventura, R.; Brunelli, A.; Marcomini, A.; Matranga, V. Titanium dioxide nanoparticles stimulate sea urchin immune cell phagocytic activity involving TLR/p38 MAPK-mediated signalling pathway. *Sci. Rep.* **2015**, *5*, 14492. [CrossRef]

145. Sweet, M.J.; Chessher, A.; Singleton, I. *Advances in Applied Microbiology*; Elsevier: Amsterdam, The Netherlands, 2012; Volume 80, pp. 113–142.

146. Sund, J.; Alenius, H.; Vippola, M.; Savolainen, K.; Puustinen, A. Proteomic characterization of engineered nanomaterial–protein interactions in relation to surface reactivity. *ACS Nano* **2011**, *5*, 4300–4309. [CrossRef]

147. Deng, Z.J.; Mortimer, G.; Schiller, T.; Musumeci, A.; Martin, D.; Minchin, R.F. Differential plasma protein binding to metal oxide nanoparticles. *Nanotechnology* **2009**, *20*, 455101. [CrossRef]

148. Maurer, M.M.; Donohoe, G.C.; Maleki, H.; Yi, J.; McBride, C.; Nurkiewicz, T.R.; Valentine, S.J. Comparative plasma proteomic studies of pulmonary TiO_2 nanoparticle exposure in rats using liquid chromatography tandem mass spectrometry. *J. Proteom.* **2016**, *130*, 85–93. [CrossRef]

149. Ji, Z.; Jin, X.; George, S.; Xia, T.; Meng, H.; Wang, X.; Suarez, E.; Zhang, H.; Hoek, E.M.; Godwin, H. Dispersion and stability optimization of TiO_2 nanoparticles in cell culture media. *Environ. Sci. Technol.* **2010**, *44*, 7309–7314. [CrossRef]

150. Givens, B.E.; Xu, Z.; Fiegel, J.; Grassian, V.H. Bovine serum albumin adsorption on SiO_2 and TiO_2 nanoparticle surfaces at circumneutral and acidic pH: A tale of two nano-bio surface interactions. *J. Colloid Interface Sci.* **2017**, *493*, 334–341. [CrossRef]

151. Zaqout, M.S.; Sumizawa, T.; Igisu, H.; Higashi, T.; Myojo, T. Binding of human serum proteins to titanium dioxide particles in vitro. *J. Occup. Health* **2011**, *53*, 75. [CrossRef]

152. Afaq, F.; Abidi, P.; Matin, R.; Rahman, Q. Cytotoxicity, pro-oxidant effects and antioxidant depletion in rat lung alveolar macrophages exposed to ultrafine titanium dioxide. *J. Appl. Toxicol.* **1998**, *18*, 307–312. [CrossRef]

153. Jin, C.Y.; Zhu, B.S.; Wang, X.F.; Lu, Q.H. Cytotoxicity of titanium dioxide nanoparticles in mouse fibroblast cells. *Chem. Res. Toxicol.* **2008**, *21*, 1871–1877. [CrossRef]

154. Shi, H.; Magaye, R.; Castranova, V.; Zhao, J. Titanium dioxide nanoparticles: A review of current toxicological data. *Part. Fibre Toxicol.* **2013**, *10*, 1–33. [CrossRef]

155. Food and Drug Administration. *Sunscreen Drug Products for Over-the-Counter Human Use*; Food and Drug Administration: Silver Spring, MD, USA, 2019; pp. 6204–6275.

156. Xu, Y.; Wei, M.-T.; Ou-Yang, H.D.; Walker, S.G.; Wang, H.Z.; Gordon, C.R.; Guterman, S.; Zawacki, E.; Applebaum, E.; Brink, P.R. Exposure to TiO_2 nanoparticles increases Staphylococcus aureus infection of HeLa cells. *J. Nanobiotechnol.* **2016**, *14*, 34. [CrossRef]

157. Huang, C.; Sun, M.; Yang, Y.; Wang, F.; Ma, X.; Li, J.; Wang, Y.; Ding, Q.; Ying, H.; Song, H. Titanium dioxide nanoparticles prime a specific activation state of macrophages. *Nanotoxicology* **2017**, *11*, 737–750. [CrossRef]

158. Chakraborty, S.; Castranova, V.; Perez, M.K.; Piedimonte, G. Nanoparticles increase human bronchial epithelial cell susceptibility to respiratory syncytial virus infection via nerve growth factor-induced autophagy. *Physiol. Rep.* **2017**, *5*, e13344. [CrossRef]

159. Li, J.; Yang, S.; Lei, R.; Gu, W.; Qin, Y.; Ma, S.; Chen, K.; Chang, Y.; Bai, X.; Xia, S. Oral administration of rutile and anatase TiO_2 nanoparticles shifts mouse gut microbiota structure. *Nanoscale* **2018**, *10*, 7736–7745. [CrossRef]

160. Belkaid, Y.; Hand, T.W. Role of the microbiota in immunity and inflammation. *Cell.* **2014**, *157*, 121–141. [CrossRef]

161. Grice, E.A.; Segre, J.A. The skin microbiome. *Nat. Rev. Microbiol.* **2011**, *9*, 244–253. [CrossRef]

162. Yoshioka, Y.; Kuroda, E.; Hirai, T.; Tsutsumi, Y.; Ishii, K.J. Allergic Responses Induced by the Immunomodulatory Effects of Nanomaterials upon Skin Exposure. *Front. Immunol.* **2017**, *8*, 169. [CrossRef]

163. Adachi, T.; Kobayashi, T.; Sugihara, E.; Yamada, T.; Ikuta, K.; Pittaluga, S.; Saya, H.; Amagai, M.; Nagao, K. Hair follicle-derived IL-7 and IL-15 mediate skin-resident memory T cell homeostasis and lymphoma. *Nat. Med.* **2015**, *21*, 1272–1279. [CrossRef]

164. Nagao, K.; Kobayashi, T.; Moro, K.; Ohyama, M.; Adachi, T.; Kitashima, D.Y.; Ueha, S.; Horiuchi, K.; Tanizaki, H.; Kabashima, K.; et al. Stress-induced production of chemokines by hair follicles regulates the trafficking of dendritic cells in skin. *Nat. Immunol.* **2012**, *13*, 744–752. [CrossRef]

165. Larsen, S.T.; Roursgaard, M.; Jensen, K.A.; Nielsen, G.D. Nano titanium dioxide particles promote allergic sensitization and lung inflammation in mice. *Basic Clin. Pharmacol. Toxicol.* **2010**, *106*, 114–117. [CrossRef]

166. Schreiver, I.; Hesse, B.; Seim, C.; Castillo-Michel, H.; Villanova, J.; Laux, P.; Dreiack, N.; Penning, R.; Tucoulou, R.; Cotte, M.; et al. Synchrotron-based ν-XRF mapping and μ-FTIR microscopy enable to look into the fate and effects of tattoo pigments in human skin. *Sci. Rep.* **2017**, *7*, 11395. [CrossRef]

167. Paradies, J.; Crudass, J.; MacKay, F.; Yellowlees, L.J.; Montgomery, J.; Parsons, S.; Oswald, I.; Robertson, N.; Sadler, P.J. Photogeneration of titanium(III) from titanium(IV) citrate in aqueous solution. *J. Inorg. Biochem.* **2006**, *100*, 1260–1264. [CrossRef]

168. Zehnder, A.; Wuhrmann, K. Titanium (III) citrate as a nontoxic oxidation-reduction buffering system for the culture of obligate anaerobes. *Science* **1976**, *194*, 1165–1166. [CrossRef]

169. Sun, H.Z.; Li, H.Y.; Weir, R.A.; Sadler, P.J. The first specific Ti-IV-protein complex: Potential relevance to anticancer activity of titanocenes. *Angew. Chem. Int. Edit.* **1998**, *37*, 1577–1579. [CrossRef]

170. Guo, M.L.; Sun, H.Z.; McArdle, H.J.; Gambling, L.; Sadler, P.J. Ti(IV) uptake and release by human serum transferrin and recognition of Ti(IV) transferrin by cancer cells: Understanding the mechanism of action of the anticancer drug titanocene dichloride. *Biochemistry* **2000**, *39*, 10023–10033. [CrossRef]

171. Messori, L.; Orioli, P.; Banholzer, V.; Pais, I.; Zatta, P. Formation of titanium(IV) transferrin by reaction of human serum apotransferrin with titanium complexes. *FEBS Lett.* **1999**, *442*, 157–161. [CrossRef]

172. Moshtaghie, A.A.; Ani, M.; Arabi, M.H. Spectroscopic Studies on Titanium Ion Binding to the Apolactoferrin. *Iran. Biomed. J.* **2006**, *10*, 93–98.

173. Tinoco, A.D.; Incarvito, C.D.; Valentine, A.M. Calorimetric, Spectroscopic, and Model Studies Provide Insight into the Transport of Ti(IV) by Human Serum Transferrin. *J. Am. Chem. Soc.* **2007**, *129*, 3444–3454. [CrossRef]

174. Sekyere, E.; Richardson, D.R. The membrane-bound transferrin homologue melanotransferrin: Roles other than iron transport? *FEBS Lett.* **2000**, *483*, 11–16. [CrossRef]

175. Sekyere, E.O.; Dunn, L.L.; Richardson, D.R. Examination of the distribution of the transferrin homologue, melanotransferrin (tumour antigen p97), in mouse and human. *Biochim. Biophys. Acta* **2005**, *1722*, 131–142. [CrossRef]

176. Suryo Rahmanto, Y.; Bal, S.; Loh, K.H.; Yu, Y.; Richardson, D.R. Melanotransferrin: Search for a function. *Biochim. Biophys. Acta* **2012**, *1820*, 237–243. [CrossRef]

177. Taira, M.; Sasaki, K.; Saitoh, S.; Nezu, T.; Sasaki, M.; Kimura, S.; Terasaki, K.; Sera, K.; Narushima, T.; Araki, Y. Accumulation of Element Ti in Macrophage-like RAW264 Cells Cultured in Medium with 1 ppm Ti and Effects on Cell Viability, SOD Production and TNF-α Secretion. *Dent. Mater. J.* **2006**, *25*, 726–732. [CrossRef]

178. Guo, M.L.; Guo, Z.J.; Sadler, P.J. Titanium(IV) targets phosphoesters on nucleotides: Implications for the mechanism of action of the anticancer drug titanocene dichloride. *J. Biol. Inorg. Chem.* **2001**, *6*, 698–707. [CrossRef]

MDPI
St. Alban-Anlage 66
4052 Basel
Switzerland
Tel. +41 61 683 77 34
Fax +41 61 302 89 18
www.mdpi.com

Materials Editorial Office
E-mail: materials@mdpi.com
www.mdpi.com/journal/materials